THE
ECONOMICS
OF
NUCLEAR
ENERGY

THE ECONOMICS OF NUCLEAR ENERGY

Edited by
LEONARD G. BROOKES
Economic Consultant

and

HOMA MOTAMEN
*Imperial College
London University*

LONDON   NEW YORK
CHAPMAN AND HALL

First published 1984 by
Chapman and Hall Ltd
11 New Fetter Lane, London EC4P 4EE
Published in the USA by
Chapman and Hall
733 Third Avenue, New York NY10017

© 1984 Chapman and Hall

Photoset by Enset Ltd
Midsomer Norton, Bath, Avon
Printed in Great Britain by
J.W. Arrowsmith Ltd, Bristol

ISBN 0 412 24350 4

All rights reserved. No part of this book may be reprinted, or reproduced or utilized in any form or by any electronic, mechanical or other means, now known or hereafter invented, including photocopying and recording, or in any information storage and retrieval system, without permission in writing from the publisher.

ERRATA

Page 12, line 33: *for* factors *read* reactors
Page 406, line 19: *after* possible *insert* except in a limited way
line 21: *for* reprocessing *read* recycling and consuming most of the
Page 407, line 7: *for* could more than compensate *read* could be more than compensated by

The Economics of Nuclear Energy Edited by Leonard G. Brookes and Homa Motamen   Chapman and Hall   1984

British Library Cataloguing in Publication Data

The Economics of nuclear energy
 1. Atomic power—Economic aspects
 I. Brookes, Leonard G.   II. Motamen, Homa
 338.4'7'62148   HD9698

ISBN 0-412-24350-4

Library of Congress Cataloging in Publication Data

Main entry under title:

Economics of nuclear energy.

   Bibliography: p.
   Includes index.
   1. Atomic power industry. 2. Atomic power-plants—Economic aspects. 3. Electric utilities. 4. Energy policy. I. Brookes, Leonard G. II. Motamen, Homa.
   HD9698.A2E24   1984   33.79'24   83-14281
   ISBN 0-412-24350-4

CONTENTS

Preface   *page* vii
Acknowledgements   viii
Contributors   ix

Introduction   1
L.G. Brookes and H. Motamen

THE ROLE OF NUCLEAR POWER IN ENERGY SYSTEMS   29

1. The role of centralized energy in national energy systems   30
   C. Starr and O. Yu
2. Economic principles of optimizing mixed nuclear and non-nuclear electricity systems   46
   L. Gouni
3. Nuclear and non-nuclear electricity: decision-making for optimal economic performance   67
   T.W. Berrie

NUCLEAR ENERGY AND THE POWER PLANT INDUSTRY   87

4. Manufacturing industry and nuclear power   88
   P.C. Warner

URANIUM PRODUCTION AND DISTRIBUTION   115

5. Prospects and concepts   116
   R. Williams
6. The balance of supply and demand   145
   T. Price

NUCLEAR FUEL CYCLE SERVICES   161

7. The structure and economics of the nuclear fuel cycle service industry   162
   A.J. Hyett

THE ECONOMICS OF ADVANCED CONVERTERS AND BREEDERS — 191

8. Assessing the economics of the liquid metal fast breeder reactor — 192
   A.A. Farmer
9. The case for developing the liquid metal fast breeder reactor in France — 234
   J. Baumier and L. Duchatelle
10. The economics of the CANDU reactor — 255
    L.G. McConnell and L.W. Woodhead

NUCLEAR POWER IN OPERATION — 273

11. The comparative costs of nuclear and fossil fuelled power plants in an American electricity utility — 274
    G.R. Corey
12. The comparative costs of nuclear and coal-fired power stations in West Germany — 288
    D. Schmitt and H. Junk
13. The economics of final decommissioning and disposal of nuclear power plant — 303
    W.H. Lunning

NUCLEAR POWER AND ELECTRICITY TARIFFS — 315

14. The implications of nuclear energy for electricity tariffs — 316
    J.M.W. Rhys

NUCLEAR POWER IN LESS-DEVELOPED COUNTRIES — 333

15. The practical problems and issues pertaining to nuclear energy in less-developed countries — 334
    L. Kemeny

THE MACROECONOMIC ROLE OF NUCLEAR ENERGY — 361

16. Energy, economic growth and human welfare — 362
    S.H. Schurr
17. Nuclear energy as an instrument of economic policy — 371
    L. Thiriet
18. Long-term equilibrium effects of constraints in energy supply — 381
    L.G. Brookes

CONCLUSIONS — 403
L.G. Brookes and H. Motamen

Index — 409

PREFACE

When we first contemplated a book on this subject we were faced with a number of options: (a) to write it all ourselves, which would have had the merit of internal consistency and continuity of style; (b) to produce a collection of existing papers, which would have given us expert views in the various sub-fields of the economics of nuclear energy and would have put us in the position of knowing from the start exactly what the authors' contributions would be; (c) to commission contributions from individual specialists, chapter by chapter; or (d) some combination of these options. We settled for the last: we have written some of the material ourselves, have obtained permission to use some existing papers that seem to us to be valuable contributions to the subject, and have been fortunate in persuading a number of eminent people in their fields to produce papers especially for the book.

This has given us a great deal of work and taken up more time than we planned for but we believe the result justifies this time and effort. It enabled us to design a structure for the book from the outset, recognizing that there are several aspects to the economics of nuclear energy – especially if we take a broad view of what is embraced by the word 'economics'.

We did not see the book as, in primary intention at any rate, a contribution to the nuclear debate. We did not see it, in other words, as an instrument for proving the economic case for nuclear energy – although we recognize that some of our authors might have strong feelings. We were keen to produce something broader in scope and more thoughtful than simple comparisons between the costs of nuclear and fossil fuelling of power plants, although we recognize that such comparisons must play a part in any account of the economics of nuclear energy.

We thought that by careful structuring and selection of material we could do a good deal to remove misunderstanding about the economic role of nuclear energy. If we have succeeded in that, then perhaps we have, to some extent, made a contribution to the debate. If so, we hope we have avoided the more usual adversarial approach.

We believe that the book will appeal to a fairly wide readership – from students at all levels to members of the public who are sufficiently interested in these matters to want to learn more about them. To this end we asked contributors to couch their contributions in terms that intelligent non-economists could follow. Some of the chapters are concerned with commercial as much as economic concepts and others touch upon political implications at the borderline between politics and economics. The book does not (apart from a fairly brief section of the introduction) embrace the externalities of nuclear energy – safety and environmental considerations, for example.

We are most grateful to those authors who have allowed us to use their papers or have written papers especially for us. The method of production that we chose has inevitably meant fairly lengthy delays between the initial receipt of the earlier papers and final publication. We apologize to those authors whose work has been held up in this way. We believe, however, that what they have to say is as valid today as when it was first written.

London L.G. BROOKES
November, 1982 H. MOTAMEN

ACKNOWLEDGEMENTS

The editors gratefully acknowledge a contribution from British Petroleum towards the cost of preparing this book for publication. British Petroleum have made this contribution in the interest of public discussion and are in no way responsible for the content of the book.

CONTRIBUTORS

J. BAUMIER
Department des Programmes, Commissariat à L'Energie Atomique, France

T.W. BERRIE
Armitage Norton Consultants

LEONARD G. BROOKES
Economic Consultant (ex-Chief Economist, UK Atomic Energy Authority)

GORDON R. COREY
Financial Consultant and retired Vice-Chairman, Commonwealth Edison Company, Chicago, US

L. DUCHATELLE
Department des Programmes, Commissariat à L'Energie Atomique, France

A.A. FARMER
Chief Technical Manager, Technical Services and Planning Directorate, Northern Division, UKAEA

L. GOUNI
Directeur Adjoint, Electricité de France

A.J. HYETT
Imperial College, London University

HERBERT JUNK
Energiewirtschaftliches Institut, Cologne University, Germany

LESLIE G. KEMENY
Senior Lecturer, School of Nuclear Engineering, University of New South Wales

WILLIAM H. LUNNING
Ex Central Technical Services, Northern Division, UKAEA

LORNE G. McCONNELL
Vice President, Production and Transmission, Ontario Hydro, Canada

HOMA MOTAMEN
Lecturer in Economics, Imperial College, London University, and Editor of the quarterly journal *Energy Economics*

TERENCE PRICE
Secretary General, The Uranium Institute

JOHN M.W. RHYS
Chief Economist, UK Electricity Council

DIETER SCHMITT
Energiewirtschaftliches Institute, Cologne University, Germany

SAM H. SCHURR
Deputy Director, Energy Studies Center, US Electric Power Research Institute

CHAUNCY STARR
Vice Chairman, US Electric Power Research Institute

LUCIEN THIRIET
Department des Relations Publiques, Commissariate à L'Energie Atomique, France

PHILIP C. WARNER
Director of Corporate Engineering, Northern Engineering Industries Ltd

RICHARD M. WILLIAMS
Adviser, Uranium, Energy, Mines and Resources, CANDU

L.W. WOODHEAD
Ontario Hydro, Canada

OLIVER YU
US Electric Power Research Institute

To Joyce and Brian

INTRODUCTION

THE ROLE OF NUCLEAR POWER IN ENERGY SYSTEMS

Nuclear energy forms part of the world energy and economic system. It is as much complementary to other energy forms as it is a substitute for them. It is, moreover, at present overwhelmingly used for the production of electricity (its role in civil and naval marine propulsion is not, for example, dealt with in this book). Some large submarines are propelled by small special purpose nuclear reactors. One or two experimental nuclear powered surface ships exist. Some experts believe that if large combined electric power and district heating schemes are adopted, nuclear plants may be the cheapest energy source. But all these will remain minor applications for a long time to come.

To understand how nuclear energy fits into national energy and electrical systems one needs to know something about the decision-making criteria that are appropriate for such systems. The first chapter, by Starr and Yu, is not directly addressed to the economics of nuclear energy. It does, however, set the scene for the student of this subject. It brings out the different roles of capital-intensive and less capital-intensive forms of production, establishing that the problem is not simply one of identifying the method that is capable of producing electricity at the lowest average cost and then employing it throughout the system. It is central to the problem that electricity cannot, in general, be stored. Electricity must be generated instantaneously to meet variations and uncertainties in both demand and supply (which may be subject to plant failure). Some variations in demand are predictable – for example those between winter and summer and night and day. Some variations in supply are equally predictable – for example when planned withdrawals of plant take place for routine maintenance. There are, however, hour to hour and minute to minute variations that must be met by decisions of system-operating engineers.

It is not difficult to see that the electricity system needs plant with different technical characteristics for different roles. For what is called 'base-load' – that part of demand which remains substantially continuous throughout the whole winter or summer season – it is important that the plant should be robust and durable, capable of operating for a very large fraction of the year, for three or four decades, before requiring replacement. For such plant, however, it does not matter if it takes several hours to raise it to full power (as indeed it does) because its role is to meet a highly predictable and unchanging load. This is the load that is in practice met by large modern nuclear or fossil fuelled stations. These all have the characteristic of robustness and durability but they all take some hours to run up to full power.*

Quite different characteristics are required for the plant that meets the surge in demand when, in millions of households, the electric kettle is switched on for a quick cup of coffee at the end of a popular television programme. For this increase in load the main requirement is for plant that can produce full power almost instantly. Gas turbine driven generators are most often used for this role. They run up to full power in a few seconds and can be started up by remote control. (There is sufficient energy stored in the boilers of the base-load plants to meet the increase in load up to the point where the peak-load plant, as it is called, takes over.) Hydro-electric plants, including pumped storage plants, also have the ability to meet increases in load at very short notice. A pumped storage plant uses electrically driven pumps to transfer water from a reservoir at low level to one at higher level during periods when the demand upon the system is well below the peak. When the load builds up steeply the pumps are switched off (which immediately makes electrical energy available to consumers) and water from the high reservoir is released to flow through water turbines to supply additional electrical energy.

Between these two roles lies a requirement for what is called 'mid-merit' plant. This is the plant that is required to meet predictable variations during the day or the week – between night and day, and between weekdays and weekends. This calls for some plant being operated for only one or two shifts each day. At present this role is met by the older base-load plants, which are almost fortuitously fitted for it. The explanation is that electricity growth rates have been so high that, at any point in time, electricity systems have consisted of plants covering a whole spectrum of sizes, reflecting the level of total electricity demand at the time they were built. The older, smaller plants are fairly easily fitted into a pattern of one or two shift operation. The UK generating boards believe that time will show that the modern, large plants can take on this role. Other alternatives are the development of special mid-merit plant (for example gas turbine plant designed to operate for a larger fraction of the year than typical gas turbine peak plant) or what is called the part-load operation of base-load plant. This entails operating a

---

* It is a popular misconception that it is only the nuclear stations that have characteristics that require them to be operated in this way. The misconception probably arises because the early (Magnox) gas-cooled nuclear plants operated by the UK Generating Boards are subject to an additional limitation – turning the reactor on and off leads to the build up of xenon in the reactor, which interferes with the nuclear fission process. Later nuclear systems are not subject to this limitation.

proportion of the base load plant at a level below its full output capacity, thus providing a margin of power that can be fairly quickly supplied simply by raising these plants to their full output level.

It is somewhat less obvious that plants filling the different roles in the system must have the right economic, as well as the right technical, characteristics. On average, running a plant to the limit of its availability means running it for about 70% of the year – the remaining 30% being taken up with the fairly lengthy annual shutdown for planned maintenance and any unplanned breakdowns. (Some individual plants achieve much higher levels of availability than this: one of the CEGBs Magnox nuclear stations ran continuously for two years before being shutdown for routine maintenance.) Plants that are required to run for over 6000 hours a year (70% of the year) must have low running costs, principally fuel costs. Because their capital costs are spread over the many billions of units of electricity they generate during their lifetimes, a relatively high capital cost per unit of power is acceptable in return for low unit running costs. The reverse is the case for peak plant. Such plant usually operates for less than 5% of the year and, with present designs, the plant has a shorter life than the large base-load stations. Thus, the capital costs are spread over a relatively small lifetime output and a much lower capital cost per unit of power output is sought for them. On the other hand, because the number of units generated each year is low, the relatively high running costs (mainly due to the distillate fuels consumed by gas turbines) are acceptable. In between comes the mid-merit plant, with capital costs that have been written-off and fuel costs that fall between those of modern base-load stations and peak load plant.

These ideas lead to a concept of 'merit order'. Electrical utility operating engineers refer to plant with the lowest running costs as having the highest 'merit', with all the plants in the system ranged in a descending merit order that is a mirror image of their ascending order of running costs. As demand on the system builds up, the available plant of highest merit is brought into operation. This results in the annual pattern of usage shown in Fig. 0.1.

Note that for operating decisions the criterion is variable cost (that part of total cost that varies in proportion to output): overheads and capital costs are irrelevant. Different considerations apply to investment decisions. If the plant with the lowest unit running costs also had the lowest unit fixed costs (including capital charges) these decisions would be simple. Ignoring, for the moment, the technical requirements for peak plant, the decision would always go in favour of one type of plant – the one with lowest fixed and variable costs. In practice the plant with the lowest variable cost tends to have the highest fixed cost – low operating cost calls for highly sophisticated plant whose unit capital costs are likely to be high. A very simple example of greater sophistication was the introduction of low-pressure turbines combined with condensers to make use of the steam exhausted by intermediate pressure turbines. The quantity of electrical energy produced per ton of fuel

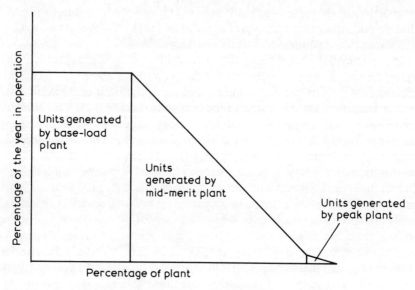

Fig. 0.1 Plant/duration graph (greatly simplified).

consumed was increased, but at the expense of greater sophistication and higher capital cost. The introduction of nuclear-fired plants may be seen in a similar light: the advantages of the very much lower fuel costs obtainable with nuclear plants (about one-third of the fuel costs for coal fired plants) are bought at the expense of an addition to capital costs of 50% or more. It is easy to see that with fixed costs and variable costs moving in opposite directions there is always a level of annual usage at which plants with different patterns of fixed and variable costs break even. The number of units actually produced in a year divided by the number that would have been produced if the plant had run for the full 8760 hours in the year, is called the plant load factor. This leads to the concept of a break-even load factor.

As a first step to understanding the principles applying to investment decisions, let us make a number of highly artificial assumptions:

1. The demands upon the system never change, both in quantity and load factor profile (see Fig. 0.1).
2. The types and costs of plants available remain constant.
3. Plants last for ever and never deteriorate in efficiency.
4. All variable costs including fuel costs remain constant.

It is easy to see that, on these very artificial assumptions, the total cost of running the system, capital charges, overheads and variable costs, is at a minimum when the proportions of each type of plant are such that the boundary between them in the merit order occurs at their break-even load

Introduction

factors. This must be true, for otherwise it would be possible to reduce the total cost of running the system by having more of the plant with lower total unit costs at the expense of the adjacent plant with higher total unit costs. By the definition of the break-even load factor, optimality is achieved when it marks the boundary between adjacent types of plant in the merit order.

It should now be clear that the investment appraisal process to determine the merits of alternative plant investments must in practice be a complex one. It must take account of expectations about future levels of demand and likely future load factor patterns, and about changes in prices of fuels and other electricity production costs. Moreover, electricity utilities are often faced with what seem to be conflicting financial criteria. When assessing and controlling their financial viability they have to take account of the rates at which they can borrow money in commercial and other markets available to them. Similar considerations apply to fixing tariffs to recover their money costs. But when assessing investment proposals they may be required to pay heed to the fact that investment in electrical plant may be a significant element in total national fixed capital formation. This consideration leads to the nationalized generating boards in the UK being required to assess projects on the basis of a real required rate of return that reflects the government's judgement of what the capital resources would earn if put to alternative use. Calculations done on the real (after deducting inflation) rate of return call for all inputs to be expressed in constant money value terms. When this basis is adopted, projected changes in cost have to be given as relative changes (relative to the general price level).

Other complicating factors are: (a) how soon existing plants are likely to become obsolescent (which depends to some extent upon how superior the new plants are) and (b) the degree of assurance sought on the security of electricity supplies. Much more standby plant is needed, for example, if consumers are to be assured that, on the average, their supply will fail in only one day in ten years than if the assurance is only of a failure of no more than one day per month.

With all these complexities it is not surprising that most electricity utilities use computer models of their systems for planning investment strategies and helping them to judge the robustness of their cases for investing in new plants.

L. Gouni (Chapter 2) and Tom Berrie (Chapter 3), analyse the problem of optimizing an electricity system consisting of both nuclear and non-nuclear plants – bringing out the distinction made above between the day to day and hour to hour decisions (which plant to bring on line next) and the capital investment decisions (the answers to the questions: How much plant in each category should be ordered? and When should it be ordered?). Gouni concentrates upon the nature of the economic decisions and the economic principles involved. Berrie discusses the problem of forecasting, pointing out that although forecasts seem not to be biased (they are just as likely to be too high as too low) forecasting errors are disturbingly high and are not getting smaller as time goes by. He also discusses the relative merits of different approaches to economic optimization and decision making.

It should be clear from these contributions that the decision between ordering a nuclear plant or a non-nuclear plant for an electricity system is not the simple one of deciding which one produces electricity at the lowest average cost per kWh. Given that the capital cost of nuclear plant is higher than for non-nuclear plant, while the relationship between running costs is the other way round, the decision about which to order at any point in time depends upon: (a) how near to optimality the system is at the moment, (b) expectations about future demand both as regards quantity and 'peakiness' and (c) predictions of future capital and running costs of the different plants. It should be evident from this that the arguments between the proponents of different types of electrical plant are often wrongly based. It is not a question of whether we should have all of one type of plant or all of another. Arguments about relative costs are really about how much nuclear and how much non-nuclear plant would together produce the optimal system.

NUCLEAR ENERGY AND THE POWER PLANT INDUSTRY

This part provides the second example of the several viewpoints from which the economics of nuclear energy can be seen. There is not, in practice, a separate nuclear energy industry. Even in the United Kingdom where, nominally at any rate, a single nuclear power plant corporation exists, this body is not in practice a producer of nuclear hardware. It is primarily an architect/consulting engineering organization. The civil, mechanical and electrical engineering that goes to make up a nuclear power station is provided by constituents of the engineering industries, for whom nuclear energy is only a part of their business. The situation is substantially the same in other countries with nuclear power programmes, even though some of them (e.g. Sweden, France and Canada) have national corporations responsible for nuclear power plant production whilst in others (e.g. the United States and Germany) no national corporations exist and nuclear plant production is entirely in the hands of the private power plant companies.

One implication of this situation is that the view of the nuclear energy industry as a monolithic body is largely a mistaken one. If the power plant industry were not producing nuclear plants it would be producing additional non-nuclear plants instead. This is not to say that no nuclear energy industrial lobby exists. It clearly does, in the shape of such bodies as the British Nuclear Forum in the UK and the Atomic Industrial Forum in the USA, but the activities of such bodies should be viewed against the actuality of a nuclear energy industry which is more of a concept than an institution.

Philip Warner (Chapter 4) discusses this situation and its implications from the point of view of a senior member of the power plant industry. Taking an analogy from control engineering he draws attention to a mismatch between the time constants of the decision-making process in the electrical industry and the establishment of appropriate levels of production in the power plant industry. Left to itself the electricity supply industry would like to be in the position of tailoring its orders closely to its predictions about the level of electricity demand and the obsolescence of existing plant. The industry would also like to be able to cancel, postpone or bring forward orders whenever there are signs of predictions being falsified by events. At worst it may be quite impossible for the power plant industry to make frequent changes of course in this way and, at best, it is likely to be costly, with productive capacity in the power plant industry geared more towards meeting peaks than towards average demand for power plants. If plant costs are to be kept down and the most efficient use made of resources, patterns of ordering must have respect for the time constants of the power plant industry and must acknowledge that the additional costs of failing to do so may substantially exceed the modest cost of relegating old, relatively inefficient, plant to a standby role a little sooner than a narrowed view of plant obsolescence would have dictated.

Philip Warner also demolishes some of the myths about what are believed to be the problems of changing from one nuclear power plant type to another. Power plant companies will supply any type of plant specified by the customer: it is false to assume that experience with building one type of plant disqualifies that company from building a different type.

URANIUM PRODUCTION AND DISTRIBUTION

Uranium production and distribution provides yet another facet of nuclear energy activity. Again, it is not part of a monolithic nuclear energy industry. In fact it forms part of the worldwide mining and minerals industry. Uranium production is, for the most part, carried out by the same companies who extract nickel, aluminium, copper and a host of other minerals from the earth's crust. Not surprisingly, the oil industry (another extractive industry) has significant interests in uranium production – just as it has in coal production. This is because the oil industry sees itself as an energy industry, ready to diversify into other forms of energy if a market for them emerges, just as the more general mining and minerals industry sees uranium as simply another mineral to be mined as long as a market at a price to justify extraction and investment in exploration exists.

Most of the nuclear power plants in the world today are driven by thermal reactors fuelled by low-enriched uranium. The word 'thermal' has nothing to do with the fact that the reactor produces heat: it is simply the name given to a limited part of the whole spectrum of speeds at which nuclear particles move. 'Thermal' neutrons are those moving at the relatively low speeds at which fission occurs preferentially in reactors fuelled with natural uranium, or with fuel that is only slightly enriched in the isotope ^{235}U. The first thermal reactors used natural uranium (0.7% ^{235}U, 99.3% ^{238}U). The economics of thermal reactors can, however, be improved by enriching the ^{235}U content to a level of 2% to 3%, mainly because power density of the reactor core is thereby increased. Consequently, although there are some natural uranium reactors in the world (e.g. the UK Magnox stations), the great majority use low-enriched uranium. Such reactors require the mining and processing of about 4000 tonnes of natural uranium per 1000 MW of plant during their lifetime. Some of this is needed for the original fuel inventory of the reactor, but most of it arises from the annual consumption of fuel (about 50 tonnes per annum of enriched fuel, requiring about 150 tonnes of natural uranium to be mined and processed).

At present (1982) uranium accounts for only about 15% of the cost of electricity generated by a nuclear power plant. The incentive to develop reactors that economize in the consumption of raw uranium is therefore limited at present. It exists mainly because of uncertainty about the ability of the world's uranium producers to provide the fuel required in the future by a much expanded nuclear component of the world's electrical power systems. The fast breeder reactor ('fast' because it relies upon the fission of uranium atoms by rapidly moving neutrons) is capable of an improvement, by a factor of 50 or thereabouts, in the heat energy produced per tonne of natural uranium consumed. It does this by converting the inert ^{238}U into fissile plutonium. Indeed, it is capable of producing more new fissile atoms than it consumes, thus providing a balance that can be stockpiled to build up an inventory for a new fast breeder reactor.

Substantial improvements in utilization can also be obtained by the use of what are called 'advanced convertors' – thermal reactors designed to be especially economical in the use of fissile atoms and to offer some degree of 'breeding', more usually with thorium rather than ^{238}U as the fertile material.

Uncertainty about the role of advanced convertors and breeders in future nuclear power systems introduces a corresponding uncertainty into the planning and prospects of the uranium industry. This uncertainty is mitigated by the fact that it would take a long time to introduce sufficient advanced convertors and breeders to make any substantial reduction in projections of uranium requirements over the next several decades. Indeed, Terence Price (Chapter 6), argues that the prospect of plants offering much higher utilization rates is an advantage to uranium producers rather than the reverse because it gives electricity producers confidence in nuclear energy as a long-term option.

Richard Williams (Chapter 5) explains the principles applying to the assessment of uranium resources and rates of production. He explains that resources are usually classified on two dimensions – degree of assurance

Introduction

about their existence and cost of production. He draws attention to some failure in the past to take account of all the costs when assessing prospects and forecasting prices, and a tendency among both government planners and electricity producers to take too simple a view of the adequacy of reserves, sometimes totalling up resources in the different categories as if there were no distinction between them. Williams warns that some of these resources may, for economic or other reasons, never be exploited and, among other things, that extending into higher cost resources at any given mine may simply prolong the life of that mine without raising production rates; it is, of course, the production rate that is all important in attempts to predict the balance between supply and demand.

Anxiety about future supplies of uranium, coupled with over-simple views about the assessment of resources, leads to a somewhat paradoxical situation in which electricity producers demand a higher degee of assurance about future supplies than they have ever been given for other fuels, but are ready to accept a less than adequate explanation of the extent of resources. In his contribution Williams discusses some of the measures of resource adequacy that are usually adopted.

Terence Price (Chapter 6) goes more deeply into the demand side of the equation, describes the historical and political factors that have influenced the present structure of the uranium market and past pricing patterns and assesses the future situation.

At the time of writing (1982) the uranium market is in some disarray. This arises for a number of reasons:

1. Both the level of investment in nuclear plant and the level of utilization have fallen short of expectations because of world recession.
2. Stockpiles are high because it was at one time felt that the price trend would be steadily upwards so that no risk was entailed in stockpiling. There are also special factors: a speaker at the 1982 conference of the Uranium Institute drew attention to the effect of a US Department of Energy ruling that required electrical utilities to contract for uranium supplies before they were actually needed.
3. The relaxation of federal government restrictions upon the export of uranium from the US, combined with the discovery of high-grade uranium deposits elsewhere, has led to much relatively high-priced American uranium becoming available at a time when much lower-priced uranium from other sources has come into the market.

The effect of these and other factors is that several speakers at the 1982 Uranium Institute Conference reported the existence of a somewhat chaotic market, with low spot market prices due to aggressive selling from stockpiles. Paradoxically, much selling is by utility users from these stockpiles, whilst much of the buying is by uranium producers, finding themselves able to meet contract obligations more cheaply by buying uranium on offer from

10   *Introduction*

utility stockpiles than by producing it themselves. The purchaser, having to accept uranium sold to him under contract in this way, may then sell to a fourth party.

Inventories are at levels equal to several years annual consumption and production is running well above the level of current consumption, even though much uranium is changing hands at prices below the costs of production of some of the lower-cost producers. Projections of longer-term demand point to the need for a higher level of exploration now, whilst market conditions provide a strong disincentive to investment in exploration. The advice of experienced mining economists is that governments and inter-governmental organizations should resist the temptation to intervene in the market: there are unfortunate examples in the history of commodity markets of such intervention destabilizing markets even further. There is, however, a serious problem of getting new exploration effort under way in good time, and even in maintaining exploration teams intact in the present state of the market.

NUCLEAR FUEL CYCLE SERVICES

The nuclear fuel cycle service industry is an industry in itself. To some extent its activities could be likened to oil refining, since it takes a crude raw material and refines it into a finished product – nuclear fuel elements. Moreover, just as the oil refining industry produces several outputs, from heavy fuel oil to gasoline, so the nuclear fuel cycle industry provides more than one output – both as regards product and services. The comparison breaks down here, however, because, for the nuclear fuel cycle industry, the outputs are almost all links in a single chain. The similarity only extends to the fact that the individual outputs of the nuclear fuel industry are often separately marketed – though sales always take place between different constituents of the industry or are made to the electricity industry.

Figure 0.2 shows the various links in the nuclear fuel cycle chain together with typical quantities. The key to the diagram is that all the quantities are related to 1 tonne of finished uranium oxide fuel going into a nuclear power plant. The example is of a second generation thermal reactor, so the fuel is low-enriched (between 2% and 3% ^{235}U) oxide (UO_2) fuel. One tonne of this fuel produces about 250 million units of eletricity. To put it in perspective, it takes about one hundred thousand tonnes of coal to produce the same quantity of electricity in a coal fired power plant. However, this comparison understates the energy content of uranium. Not all of the fissile material (the ^{235}U) in the fuel element is consumed in the first 'pass' and, as will be seen in Fig. 0.2, a substantial part of the material removed from the reactor in the used fuel element (60%) is returned to the enrichment plant for re-enrichment and use in new fabricated fuel elements. Moreover, the irradiation by neutrons that the ^{238}U receives in the reactor converts some of it to plutonium, which is itself a fissile material which can be used either in thermal reactors or fast breeder reactors (see later in this chapter).

On the other hand it takes, typically, 2000 tonnes of uranium ore to produce the 4 tonnes of uranium ore concentrate (U_3O_8, known as

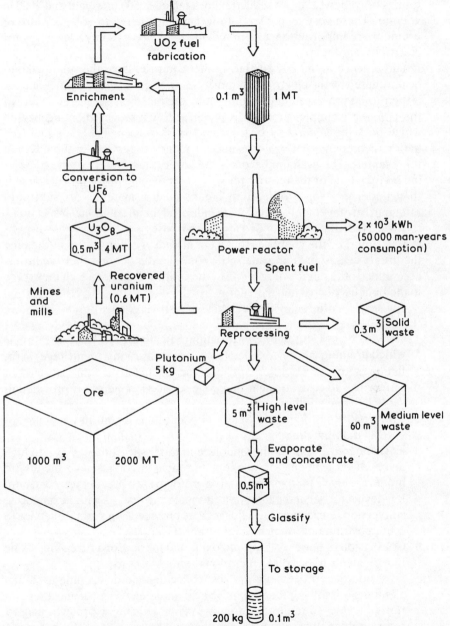

Fig. 0.2 Nuclear fuel cycle. MT, metric tonnes; m³, cubic metres. Values are typical and approximate only. (Reproduced by permission of the UKAEA.)

'yellowcake') that is needed to produce one tonne of finished oxide fuel. This is still a lot less than the one hundred thousand tonnes of coal needed to produce the same amount of electricity, and uranium has the further advantage that the material actually transported from the mine is the 4 tonnes of 'yellowcake' – a reduction by a factor of 500 in weight and 2000 in volume. These very high energy-to-weight and energy-to-volume ratios of uranium are important factors in making it a low cost source of energy for electricity production.

The practice in the UK and most other countries is for the yellowcake to be purchased by the electricity authority or some government agency and for it then to be processed under contract by the fuel cycle service supplier. In the UK, for example, uranium is purchased by a procurement agency on which both the generating boards and British Nuclear Fuels Limited (the national company that provides nuclear fuel cycle services in the UK) are represented. The uranium becomes, however, the property of the generating boards. One of the reasons for special arrangements for procurement is that, under the Non-Proliferation Treaty, nuclear materials are subject to safeguard arrangements that include independent international inspections, to ensure that material intended for civil purposes is not diverted to weapons production. For the countries that were already weapons states at the time the treaty was signed the machinery is voluntary. For other countries, acceptance of the machinery is one of the conditions under which they share in the benefits of civil nuclear energy.

The services offered by the fuel cycle industry are:

1. Conversion of U_3O_8 into UF_6 (uranium hexafluoride). This is the form in which uranium can be enriched (by gaseous diffusion or centrifuge) in the fissile isotope ^{235}U. Perhaps because of imbalance in capacity or differences in costs, nuclear fuel cycle service suppliers may provide this service to one another.
2. Conversion of some yellowcake into uranium metal and canning for natural uranium reactors.
3. Enrichment of uranium hexafluoride in the fissile isotope ^{235}U – to high levels of enrichment for weapons purposes or for research factors, to moderate levels for experimental or prototype fast breeder reactors, or to low levels for second generation thermal reactors. As with hexafluoride conversion, enrichment services may be provided as a separate service to other governments or electricity authorities.
4. Conversion of enriched UF_6 into oxide and fabrication into canned oxide fuel elements for second generation thermal reactors.
5. Provision of reprocessing services: collecting and de-canning used fuel elements, dissolving them in nitric acid and then using chemical separation processes to remove plutonium, generated as a by-product, unused but reusable ^{235}U for recycling to the enrichment plant, ^{238}U and waste

products – mainly 'fission products', the material produced when uranium atoms are fissioned.
6. Stockpiling of plutonium and ^{238}U for future use in fast breeder reactors (as explained above, this material remains the property of the owner of the original uranium).
7. Manufacture of plutonium–^{238}U fuel elements for experimental and prototype fast breeder reactors.
8. Provision of waste management services (storage and, eventually, disposal). All these services are charged separately. Together they constitute about two-thirds of the total cost of nuclear fuel (the remaining third being the cost of the yellowcake).

The proportion of yellowcake going to the enrichment plants is very high, so much so that Terence Price (Chapter 6) has suggested that predicting the inputs to the enrichment plants is one way of predicting the demand for uranium. This may seem an oblique way to compute requirements. It follows from the fact that enrichment services are highly capital intensive. They are therefore usually supplied under very long term contracts (to ensure continuity of work). This introduces a substantial buffer between the requirements for yellowcake and the requirements for finished fuel elements by power plants, making the input to the enrichment plants a better indicator of demand for yellowcake than is the demand for finished fuel elements. Adrian Hyett of Imperial College, London, produced the relevant chapter on the basis of his own researches into the industry, (Chapter 7).

THE ECONOMICS OF ADVANCED CONVERTORS AND BREEDERS

The role of these next generation power plants has been mentioned above. They are capable of increasing the utility of uranium fuel – the fast reactor by a very large factor. However, just as the optimization of a mixed nuclear and non-nuclear electricity system is a complex process, so is the optimization of a nuclear component that includes both thermal and fast reactors.

It could be argued that, in principle, the process is the same. Just as nuclear stations have higher capital but lower running costs than non-nuclear stations, so are fast reactors likely to have higher capital costs but lower uranium requirements than thermal reactors. At first glance it might be thought that, just as in the nuclear/non-nuclear example, there is a breakeven load factor above which fast reactors have the edge and below which thermal reactors are more economic. The problem is not, however, quite as simple as this.

First, it is hard to say what the capital cost of a developed fast reactor will be. Its power density is greater than that of a thermal reactor and it needs no moderating material to slow the neutrons down to 'thermal' speeds. For a given output, therefore, its overall size should be less. But the use of a liquid metal (sodium) coolant (the currently preferred technology) involves two stages in transporting the heat from the reactor to the steam generators. First, the radioactive sodium which passes through the reactor core goes through an intermediate heat-exchanger to pass heat to a second sodium loop, which in turn conveys heat to the steam generator. This is one of the principal sources of additional cost. Longer experience with building fast reactors may bring their costs down, but the indications are that they are likely to cost significantly more per kW than thermal reactors.

Other coolants are a possibility – helium or carbon dioxide for example – but little work has been done on these technologies and there is virtually no prospect of a gas-cooled fast reactor emerging as a practical competitor with the liquid metal cooled fast reactor in, say, the next 15 years. Nevertheless, gas-cooled fast reactors have their adherents and may one day be available as a source of low-cost electricity. Their power densities are likely to be lower, and the size of the reactor core therefore larger, than for sodium-cooled reactors but they may gain some advantages from the simpler cooling technology that they use.

Given higher capital cost than for thermal reactors, liquid metal (sodium) fast reactors have to demonstrate that they offer advantages in terms of lower fuel costs (or greater assurance of fuel supply) if they are to compete with thermal reactors.

There is no doubt that they would consume much much less yellowcake. Indeed, for a very long period, they could utilize the ^{238}U stockpiled as a result of reprocessing thermal reactor fuel; and, for some considerable time, countries that have stockpiled plutonium from the same source need look no further for initial inventories of fissile material. The question that most needs to be answered is how much more costly than for thermal reactors is the fabrication and reprocessing of fast reactor fuel. Present indications are that the total fuel cycle costs of a developed fast reactor might be as high or higher than those of thermal reactors at current uranium prices. For base-load supplies, however, fast reactors could probably already produce electricity as cheaply as coal-fired stations at the coal prices that European consumers at any rate are likely to face. Moreover, the fuel prices of fast reactors, since they depend so little upon the cost of the fuel resource itself, are highly robust against future movements in world fuel prices. A continuing shortage of fossil fuel would be bound to impinge upon the price of uranium, thus affecting the economics of thermal nuclear power stations along with those of oil and coal fired power plants, but the running costs of fast reactors would continue to be dependent almost solely upon the costs of operating the fabrication and reprocessing technologies. However, the larger the component of fast reactors the greater their influence upon uranium prices, so, if fast reactors come to form a substantial part of electricity systems, we are likely to see an equilibrium in which fast and thermal reactors operate side by side like nuclear and non-nuclear stations do now – at least until the capital costs of the fast reactors come down much nearer to those of thermal reactors.

The claim made on behalf of fast reactors that they greatly extend the energy obtainable from uranium reserves (by a factor between 30 and 50) is often made as though it speaks for itself, justifying the adoption of fast reactors regardless of how their electricity production costs compare with those of thermal reactors. This justification for them is sometimes referred to as 'the resource case for fast reactors'. Economists would argue, however, that a special resource case only exists if electricity producers are prepared to put some price upon security of fuel supplies. If they were prepared to assert that their fears about the future availability of uranium for thermal reactors were such that they would willingly pay an 'insurance' premium of 10% of the total cost of electricity production, in return for greater assurance about fuel availability, this would justify adopting fast reactors even though their unit costs of producing electricity were 10% greater than those of thermal reactors. If low-cost uranium reserves were immensely abundant and well distributed, so that no electricity producer felt himself at risk of being short of uranium, there would clearly be no special resource case for fast reactors.

Another often-employed argument is to say that fast reactors will become economic when the uranium price reaches a particular (fairly high) level. This is the uranium price at which fast-reactor electricity production costs would break even with those for thermal reactors at present or predicted capital and running costs, other than uranium costs. This is a somewhat two-edged argument. In the absence of any very large new finds of low-cost uranium, future uranium prices are likely to depend upon the total number of nuclear power stations in the world and the proportion of fast reactors among them. High prices imply an unsuccessful future for fast reactors with little influence upon the world uranium market. This argument therefore constitutes a 'catch-22' situation for fast reactors: they will not be wanted unless uranium prices are high, but they will only be high if very few people adopt them. It would seem that for fast-reactor development to be justified, the developer must satisfy himself that the development potential is such that they are likely to become the preferred – because the cheapest – way of producing electricity, even at relatively low uranium prices.

It should not be too difficult to establish this claim as justified. As stated above, a fast reactor built today with present technology, could probably produce electricity at acceptable costs. For the future, two features about fast reactors offer scope for lowering costs. The first is the high-power density of the reactor core. Fast-reactor designers are likely to exploit this feature by progressively reducing the overall dimensions of the plant, thus bringing down the capital cost. The second is the high energy density of fast-reactor fuel. In the present state of technology this is offset by the greater complexity and greater cost of fabrication and reprocessing of fast-reactor fuel compared with that for thermal reactors. With greater familiarity with the fuel cycle, the high energy density of fast-reactor fuel will become an increasingly important factor in fast-reactor costs.

Once the electricity production costs for fast reactors fall below those of thermal reactors it becomes possible to contemplate a much larger multiplication factor than 30 to 50 for the extension of the energy content of uranium resources offered by fast reactors. The cost of the uranium itself accounts for such a small part of the total fuel cycle costs of fast reactors that very much lower grade uranium ores can be considered as viable sources of fuel. It may not be an exaggeration to think in terms of a thousand-fold increase in the energy available from uranium resources.

We have seen how the different cost characteristics of fast and thermal reactors may result in the economically optimal system being one in which both types are present. During the build-up stage for fast reactors, when they are penetrating existing nuclear systems of thermal reactors, there is an additional reason for interdependence. A large (1300 MW) fast reactor needs an inventory of about 4 tonnes of plutonium. About 3 tonnes of this are in the reactor at any point in time, with the remaining 1 tonne being elsewhere in the fuel system – at the fuel fabrication plant or the reprocessing plant, or simply held until the activity level has sunk to a level that is acceptable to the reprocessing plant. As pointed out earlier in this section, sufficient plutonium stocks exist to provide inventories for the initial programmes of fast reactors. For later programmes there will be two sources of plutonium for inventories: plutonium recovered by continuing to reprocess fuel coming out of thermal reactors, and that obtained by reprocessing the fuel and 'blanket' of the first fast reactors. (The blanket consists of ^{238}U placed in and around the core of fast reactors and which, as a result of capturing neutrons, becomes converted into plutonium.) Much of the plutonium obtained by the second of these two routes is needed to meet the on-going fuel demands of the existing fast reactors themselves, thus saving 4000 tonnes of new uranium in the lifetime of each fast reactor, compared with what would be required for a corresponding thermal reactor. However, each fast reactor is capable of producing a little more plutonium than is required for its own on-going needs – enough to provide a complete inventory for a new fast reactor after a period of about 13 to 20 years. (This period – the length of which is to some extent under the control of the designer – is referred to as the 'doubling time'.) Doubling times of this length are quite acceptable for a complete system of fast reactors, growing only as fast as electricity demand, but are too long for the percentage rate of growth of demand of the fast reactor component of the nuclear system when it is in the process of penetrating the system, growing faster than the total system. It is at this stage in the development of fast reactors that the additional supplies of plutonium obtained by reprocessing thermal reactor fuel are

Introduction

required, adding a technical interdependence to the economic interdependence mentioned above. Ultimately, the fast reactor system becomes self-supporting and it can then, by design and fuel management optimization, be made to supply only as much plutonium as is required to keep the system in step with electricity demand.

Fast reactors present fascinating problems in micro-economics. For example, the total plutonium inventory of the system is reduced if the time between removing fuel and irradiated ^{238}U from the reactor and the extraction of generated plutonium is kept short, but the complexities of reprocessing are increased if time is not allowed for the activity of the irradiated fuel elements to decline. This is just one example of the trade-offs encountered in the micro-economics of nuclear energy. A. A. Farmer, of the Northern Division of the UKAEA gives an account of such micro-economic studies along with his assessment of the economics of fast reactors as a whole (Chapter 8). Baumier and Duchatelle of the Commissariat à l'Energie Atomique provide a French assessment of fast reactor prospects which shows them as having great economic promise (Chapter 9).

Farmer devotes some time to comparing the economic prospects of fast reactors with those of the CANDU reactor – a thermal reactor developed in Canada which employs pressurized heavy water both as a coolant (to extract the heat from the reactor) and as a moderator (to reduce the speed of the neutrons to 'thermal' levels). Heavy water absorbs far fewer neutrons than ordinary water and this enables the CANDU reactor to employ a smaller inventory of uranium (even though it uses non-enriched uranium) than other types of thermal reactor, and also results in a greater surplus of neutrons to convert fertile ^{238}U into fissile plutonium. If this plutonium is extracted from the used fuel and employed instead of ^{235}U to produce new fuel, the overall effect is for the CANDU reactor to improve very substantially upon the ability of ordinary thermal reactors to extract heat from a given quantity of yellowcake. The improvement in fuel utilization is not nearly as large as that achievable by fast reactors, but proponents of CANDU reactors (and certain other thermal reactors with semi-breeding properties) argue that because capital costs and reprocessing costs are lower than for fast reactors, they produce cheaper electricity than fast reactors are likely to do over a fairly wide spectrum of possible future uranium prices.

Some concepts of CANDU reactors and other high fuel-utilization thermal reactors employ a different approach to the partial supply of their own fissile material. Thorium-232 is, like ^{238}U, a fertile material. When thorium-232 captures neutrons it may be converted into ^{233}Th which, like ^{235}U, is a fissile material. By including pellets of thorium-233 in fuel elements operators of CANDU reactors and other semi-breeders can greatly economize in demands for new uranium.

Lorne G. McConnell and L. W. Woodhead of the Ontario Hydro Electricity Utility have given an exposition of the economics of CANDU reactors in this section (Chapter 10).

NUCLEAR POWER IN OPERATION

There is now sufficient experience of nuclear power plants for some assessment to be made of their economics in normal routine operations. Unfortunately, some studies have employed statistical techniques to produce spurious results. Problems of analysis arise because of the very rapid rate of exploitation of nuclear energy. It may surprise some readers to hear this rate referred to as 'rapid', but compared with other technologies the advance from initial conception to exploitation on ordinary hard commercial criteria has been remarkable. It is less than 40 years since the fission reaction was first demonstrated on a laboratory scale and not much longer than that since Lord Rutherford, one of the most eminent nuclear physicists of his time, said that nuclear energy would never have any practical application. Yet today nuclear energy accounts for about half as much electricity as hydropower (which was first exploited over 100 years ago) and, when all the stations at present under construction are completed, it will account for as much electricity as hydropower. Consequently, we have in the world today (1982) over 150 GW of nuclear power (about three times the total electricity power system in the UK or France), with the very earliest nuclear station installed still operating alongside the latest ones. In the meantime, there has been an increase by a factor of 10 in unit size and substantial changes in engineering practices and safety requirements. This has also been a period of large and erratic changes in money values. All these factors cast doubt, to say the least, upon the validity of studies which use statistical techniques (regression analysis) to come to conclusions such as: (a) small plant are more reliable than large ones or (b) nuclear plants are less reliable than coal-fired plants. Practical experience shows that most power plants (nuclear or non-nuclear) become more reliable as they get older, and that, in general, innovatory departures exact penalties in the shape of unexpected costs, and teething troubles in the early stages. Large plants may seem less reliable than small ones simply because they are newer and fewer. For these reasons we have not been tempted to include such statistical exercises at this stage in the life cycle of nuclear energy. Instead we offer two, admittedly semi-anecdotal, accounts of experience with coal- and nuclear-fired stations in the United States and Germany. The first is by Gordon Corey, based on his experience as Vice-Chairman of the Commonwealth Edison Company (an electricity utility based in Chicago), Chapter 11, and the second is by Dr Dieter Schmitt of the Institute of Energy Economics at the University of Cologne, Chapter 12. We have also included a paper by W. H. Lunning (Chapter 13) which outlines the various stages in the decommissioning of nuclear plants at the end of their useful lives.

NUCLEAR POWER AND ELECTRICITY TARIFFS

It is often overlooked that the characteristics of energy sources have an influence on the balance of economic activity. The alliance of coal with the steam engine was associated with the 'railway mania' of the 19th century, when railway systems were developed rapidly all over the world, revolutionizing inland transport. Cheap, Middle Eastern oil has been associated with the rapid spread throughout the community of the pleasures of personal mobility and the growth of the petrochemicals industry. Some writers have drawn attention to the role of electricity in the location of industry and the pattern of production in the United States. It would be surprising if the availability of low marginal cost electricity from nuclear stations did not have some effect on the way goods and services are produced, and upon consumer choice. Electricity tariffs are likely to reflect both the lower average cost of electricity production that is possible with nuclear energy and the different relationship between capital and running costs compared with fossil-fuelled plant.

In Chapter 14, John Rhys, Chief Economist of the UK Electricity Council, explains the theory of electricity pricing, pointing out that in highly competitive markets price tends to equate to marginal cost – the cost of producing the last unit of production. He points out that economic theory requires this to be so if economic resources are to be allocated to maximize the benefit to consumers.

Electricity is not subject to the competitive forces that would, by themselves, bring about pricing at marginal cost; it is produced, for the most part, by public and private monopolies. In order to optimize consumer benefit from electricity, therefore, it is necessary to construct tariffs deliberately so that they reflect marginal cost. For a capital-intensive system of production, like the electricity system, it is necessary to take account of incremental capital costs in meeting increases in demand for electricity, and this has led to the concept of 'long run marginal costs' as a principle for setting electricity tariffs.

Rhys explains that it is not sufficient to see the tariff-fixing process as simply a means of bringing prices into line with costs. It is better seen as a way of bringing production cost structures into harmony with the preferred patterns of consumption of electricity consumers.

In practice, there is likely to be much interaction: the very different cost structure of electricity from nuclear plants, compared with that from fossil-fuelled plants, will, when nuclear plant forms a substantial proportion of the total system, lead to much greater differentiation between the cost of the marginal unit of electricity supplied at times of peak demand, and the similar cost at 'off-peak' times. When tariffs come to reflect this greater differentiation we can expect changes in the patterns of both domestic and industrial electricity demand. The process, to which John Rhys refers, of bringing production cost and consumer demand into harmony is likely to include changes in production practices by industrial consumers and in the type of plant that they install.

Some electricity consumers have expressed disappointment at the apparently very small influence of nuclear energy upon electricity prices. The explanation is that, in most countries, the nuclear component is not large enough for nuclear stations to be the marginal suppliers at any time – they are always on 'base-load'. This is not to say that these stations are having no effect at all upon electricity prices: their lower costs of production lead to reduced average electricity costs compared with those for an all-fossil system, and this is bound to have an effect upon the total expenditure upon electricity by all categories of electricity consumers. When the nuclear component of electricity systems is much closer to its economic optimum, nuclear stations will be meeting the marginal demand for a significant part of the year, and the impact of the nuclear stations upon the pattern of prices charged for electricity will then be much more evident. The most obvious effect will be to produce a substantially larger differential between on-peak and off-peak electricity prices.

Some analyses show liquid fuels to be associated, therm for therm, with higher levels of economic activity than are solid fuels, and show electricity to be superior to both liquid and solid fuels in this respect. We could therefore possibly be in for some pleasant surprises on the quantities of energy required to support a given level of output, as well as expecting to see some changes in the nature of the output, if the energy systems of the world lean more towards electricity as its price becomes more attractive compared with the price of other fuels.

NUCLEAR POWER IN LESS-DEVELOPED COUNTRIES

There is some tendency to dismiss nuclear energy as a source of relief to the problems of the developing countries. This view is often based upon the assumption that the technology is beyond their ability to absorb, that the quantum size of nuclear plants is out of scale with their requirements and that it would make more sense to offer them simpler, labour-intensive technologies (sometimes called 'alternative technologies) that would fit more readily into simple agrarian societies.

Before we examine this view we should first note that if expanding the world's nuclear system results in downward pressure upon the world fuel market, it would provide much-needed relief to the present difficulties experienced in developing countries in paying for the fuel they need. The escalation of what is sometimes called 'the banana price of oil' to developing countries has been much greater than the escalation in the dollar price with which the developed countries have been grappling. However, when we come to look at what nuclear energy offers directly to the developing countries we see that it may turn out to be more than is generally supposed. André Giraud, one time head of the French CEA and subsequently Minister for Industry, has even suggested that the larger part of the future market for nuclear power may be in the developing countries.

First, the paternalism towards the developing countries, that leads to the view that what they need is to be helped to stay in something like their present shape, may be misconceived. The countries that have done most to relieve their poverty are those, like Taiwan and South Korea, which have made a good start in developing an industrial side to their activities. The rich countries owe their wealth to the high levels of output per capita associated with industrial forms of production and some of the developing countries resent the suggestion (however well meant) that they should not be too ready to 'encumber' themselves with the evils of industrial and urban ways of life.

Some of the developing countries – the richer oil-producing states in particular, but also poorer countries like India – already have nuclear plants built or under construction. For them, the finer points of balancing electricity-system requirements against unit size of plant may not loom very large. Investment in the poorer developing countries has been held up because of shortage of electricity. Moreover, they may see their electricity demands primarily in terms of what they need to encourage the growth of industry (or supply what existing industry they may have), with supplies to domestic consumers, especially in rural areas, as a secondary consideration. Their ideas on standards of reliability of supply (which are linked to quantum size of plant) may be very different from those in the developed countries. Moreover, power plant projects are among those for which aid from international organizations is most readily forthcoming.

The oil-producing countries constitute a special case. Countries that produce oil tend to be heavy users of gas, because it is produced in association with oil but is nowhere near as easy to export. In the past, associated gas (i.e. gas extracted in association with oil) has been virtually a 'free good'. Much electricity in the oil-producing developing countries has, therefore, been from gas fired power plants. A number of factors have led to a change in the situation:

1. Home demand is growing.
2. Gas is becoming more valuable as an exportable resource – by pipeline and in the form of liquified natural gas.
3. The amount of gas produced has fallen sharply as OPEC's oil production has been cut back from about 30 million barrels per day to 17.5 million barrels per day.

Thus, gas is no longer a free good and some of the oil-producing developing countries have actual or projected shortages.

One alternative would be to use some oil for the production of electricity (and some countries have done that) but in general the oil-producing states, quite rightly, continue to see their oil, their black gold, as their major source of future income and they are reluctant to burn it in power plants, even though the export market for it is severely depressed. Moreover, although most of the oil-producing states have aspirations towards industrial development, the period since 1973 (when their stocks of investable funds soared) has been an unpropitious one for industrial expansion – because of world recession, which is itself not unconnected with the rapid build up of the OPEC monetary surpluses. The result is that what economists call 'the opportunity cost' of capital invested in nuclear power plants (the value of the alternative investment forgone) is very low. There is, moreover, the psychological attraction of having a tangible capital asset in the shape of a nuclear power plant in preference to a depreciating paper asset in an American bank, which, in practice, may be the only alternative investment. In this light it is easy to see why nuclear power may be attractive to countries rich in oil and gas resources.

In Chapter 15, Leslie Kemeny analyses the many factors that have to be taken into account by a developing country contemplating a nuclear programme and by the developed countries assisting in such a programme. He uses Australia – usually categorized as a developed country, though having many of the features, in this context, of a developing country – as an example to illustrate his points. Australia is an especially interesting case because of her possession of large uranium reserves and the possibility that she may eventually become the largest exporter of uranium outside the Eastern Bloc. He provides details of current plans for nuclear power development. He mentions 'leasing' as an interesting solution to the problem of providing nuclear fuel cycle services to developing countries – a solution which would dovetail with Australia's own future role as a large supplier of nuclear fuel. Leasing is already practised in Comecon countries. The user simply leases nuclear fuel, extracts the energy available from it and returns it to the supplier. The user is thus relieved of all the problems of providing himself with enrichment, fuel fabrication and reprocessing facilities. These are all supplied by the lessor. The system would also provide users with some relief to funding problems. In the ordinary course, the initial fuel inventory is treated for accounting purposes as part of the initial capital plant and is funded accordingly. Under a leasing arrangement the user would hire the initial fuel inventory under the same arrangements as for replacement fuel.

ENVIRONMENTAL, SOCIAL AND POLITICAL FACTORS

No economic analysis of any technology would be complete without some consideration of the economic 'externalities' – the costs and benefits that do not fall directly upon or accrue directly to the direct user of the technology. Some of these costs and benefits are very real and tangible. If, for example, the pollution of an estuary effects the local fish population and reduces the income of local fishermen, we have an example of a cost that does not fall upon the pollutor himself. Similarly, if the substitution of a smokeless technology for a smoky one results in an increase in the crops from local farms we have an example of a benefit that does not accure to the direct employer of the technology.

Other costs and benefits are much less tangible and more difficult to evaluate. Some objectors to nuclear energy have, for example, argued that the spread of civil nuclear technology cannot be dissociated from the spread of nuclear weapons, and others have argued that the safeguards against misuse of nuclear energy or against terrorist action may impinge upon civil liberties. Whether the objectors are right or wrong about this, the strength of the objections has in some countries introduced a very direct cost penalty upon electricity undertakings in the shape of very costly delays both to the initial consents to start building and to construction itself, when objectors have been successful in obtaining injunctions to halt construction.

We have not commissioned a paper on this aspect of nuclear power economics. It would, no doubt, have been fairly easy to have obtained a contribution from one of the leading nuclear power critics, which would have meant a departure from our policy. Unfortunately, the author that we had identified as the best man to write on this topic – his analysis would have been thoughtful, critical and, most importantly, objective – was unable to spare the time within the deadline that we had to set.

It would, in any case, have been very difficult to do justice to this subject in a single paper. Even if we confined ourselves to the strictly economic externalities the task is not simply to compare the direct benefits of nuclear energy (in the shape of lower electricity costs immediately and relief from constraints on the supplies of traditional fuels in the future) with disbenefits like the introduction of an additional source of ionizing radiation to those already existing in both developed and undeveloped societies. One has to consider the benefits and disbenefits of nuclear energy against the benefits and disbenefits of the alternatives to it. The benefits of having coal- or oil-fired electric plants are very similar to (though, as John Rhys has shown, not identical to) those of having nuclear fired plants. The external economic costs are much more difficult to compare with one another. There are some similarities: like nuclear plants, coal-fired plants also release miniscule amounts of radioactive materials into the environment. But they also release large quantities of chemical pollutants, with which there is no parallel. And whilst the transport of nuclear materials to and from power stations calls for special safeguards, the quantities are so much smaller than the corresponding quantities in the case of coal-fired plants that it is hard to make a sensible comparison between the environmental impacts in each case – a few truck loads per annum of material that needs very special provisions for safe transport in one case (especially for the material coming away from the power plant) against millions of tons of fuel and millions of tons of more familiar waste products in the other case.

The comparisons become more difficult still when renewable sources of energy are brought in. The visual, and very probably the acoustic, impact of the equivalent in wind generators to one large nuclear power station would be massive; and the environmental and ecological impact of a system of wave-power generators – like a procession of supertankers nose-to-tail a few miles offshore, for a distance of hundreds of miles – is obviously something that it is extremely difficult to assess.

Perhaps it is, after all, better to leave these very much broader elements in the comparison of power-producing systems to works specially devoted to the subject.

THE MACROECONOMIC ROLE OF NUCLEAR ENERGY

No single energy source is necessary, though some are important in the sense that managing without them may cause severe difficulties – poverty or worse, but economic and social systems are sufficiently adaptable to survive the absence of one or other energy option. Some proponents of nuclear energy who argue that the world cannot manage without it are therefore, in the opinion of the editors, just as vulnerable to charges of over-statement as those nuclear objectors who claim that abandoning the nuclear option would entail no economic cost. Our view is that nuclear energy offers an opportunity not unlike the opportunity offered by the advent of cheap Middle Eastern oil, and failure to seize the opportunity implies forgoing substantial benefits, just as failure to have exploited cheap Middle Eastern oil would probably have resulted in a poorer world over the last few decades and a poorer world today than the one we know.

The role of nuclear energy may also be viewed as providing some relief from the hardship resulting from a tighter world energy market, but it would be going too far to suggest that the world cannot survive without that relief. Some of the misconceptions on both sides arise from seeing energy as always at the receiving end of economic activity. Those who would abandon nuclear energy, project into the future the low economic and energy growth rates that are themselves in no small way due to problems of energy supply and conclude that the world has a long way to go before it needs some large new supplement to existing energy sources. On the other side, projections of desirable economic growth rates are often associated with established relationships between energy and economic activity to produce forecasts of large energy gaps that, from the options available, it is claimed nuclear energy alone is potentially abundant enough to fill. In practice, there is no such thing as an energy gap. In the end, energy price reaches a level that brings supply and demand into balance. A balance at a high price, implying severe restrictions upon energy supply, would almost certainly be associated with a poor world – belying both the economic growth rates projected by the supporters of nuclear energy, and the claims that no economic cost need be associated with abandoning nuclear energy made by its opponents.

What must not be overlooked, however, is the very important role that rising supplies of low cost energy coming on to the markets of the world has played in stimulating economic activity since the beginning of the industrial revolution. Carlo Cipolla, writing in the early 1960s, argued that there had been only two climactic events in the whole of man's economic history – the invention of farming 8000 years BC and the invention of the steam engine in the 18th century AD. In Cipolla's words the second development provided mankind with mechanical slaves fed with inanimate energy, relieving man from the severe constraints imposed by the limitations of human and animal muscle power. This view of fuel energy as having a fundamentally important role to play in economic systems is shared by many other writers, but is not universally accepted among economists. Many of them argue that if heat engines had never been invented, the world might have been just as rich today, but enjoying different economic satisfactions. Economists who argue in these terms see an increase in the price of energy as no different in principle from an increase in the price of, say, copper in terms of the effect upon economic systems. Others argue that whilst an increase in the price of any commodity produces economic setbacks by creating bottlenecks and by pricing some activities out of the market, an increase in the price of energy damages the production process itself.

There are two other ways in which the distinction between those who attach special importance to energy and those who do not may be seen. Those who attach no special importance to energy argue that an increase in price will be followed by the substitution of capital and/or labour for energy in the production process and by changes in output and consumption away from energy intensive goods and services. Those who see a special role for energy would argue that substituting capital for energy means some move away from investment that is complementary to energy consumption towards investment that is in substitution for it, and that substituting labour for energy means a step back towards a more labour-intensive, poorer society. The argument is bedevilled by the relative price changes that would take place if there should be very large changes in the structure of economic output. It may be extremely difficult to make any meaningful comparison between the total value of the economic satisfactions of an energy intensive society, endowed with cheap abundant fuel, and one in which energy was a scarce expensive resource.

The second way in which the two schools differ is an extension of the first. Those who see no special role for energy tend to make their analyses with the aid of the macroeconomic production functions and identities that are important planks in macroeconomic theory. Because energy is only an intermediate good in the production sector of the economy, special strategems have to be adopted to analyse the effect upon the macroeconomy of a change in energy prices; and the form of analysis most commonly used works on the basis that the economy will gravitate towards a full employment equilibrium after equilibrium has been disturbed by a large energy price increase. Such analyses tend to lead to the result that the effect of an increase in energy costs is to reduce economic activity by an amount approximately equal to the increase in the total cost of energy.

Because macroeconomic analysis does not ordinarily provide room for market relationships, such as are normally depicted by movements in supply and demand curves, it is difficult in this type of analysis to distinguish between a sharp energy price rise that is due to a shift in the demand function (perhaps because the economy has suddenly become endowed with some new resource that stimulates economic activity and produces an increase in demand for fuel at all prices) and one which is due to a shift in the supply function (when, for example, a powerful energy cartel demands a higher price at each level of output). In the first case higher energy price is associated with a higher level of economic activity and with no change in the availability of energy to the economic system whilst in the second case higher energy price is associated with a lower level of economic activity and with a reduction in the availability of energy to the system. In the second case, it seems plausible to accept the possibility that a reduction in the availability of energy might lead to an equilibrium at less than full employment of the main factors of production – capital and labour. The problem that the world faces today is one of a reduction in the availability of energy to the economic system: for any given price, a smaller quantity is placed on the market than hitherto. For those who see some special role for energy in economic systems and who see the risk of economic damage in a switch of a significant proportion of capital resources from energy complementarity to energy substitution, the argument that the high price will bring its own solution (by bringing about substitution of capital for energy and bringing into the market previously over-costly sources of energy) is unconvincing.

The foregoing is about fundamental long-term relationships between energy and economic activity and about the long-term effects of a change in the availability of energy (in the shape of a higher price at any given level of production and consumption). There is much less disagreement between economists about the short- and medium-term effects of large energy price rises. These are seen as great disturbances to the world's monetary and banking systems and sharp changes in the pattern of international indebtedness that lead directly or indirectly to a reduction in the level of overseas trade. The immediate problem for individual oil-importing countries is that they must export a higher proportion of their output in order to import their

Introduction 27

oil requirements. If the oil-producing countries were able to absorb, in the shape of imports, the goods and services previously consumed by those workers in consuming countries who must now devote some of their efforts to producing more goods in return for oil it would be possible to maintain employment in the consuming country. In practice the oil producers are not able to absorb all the goods and services that workers in the consuming countries can no longer afford to buy themselves. The result is a two-pronged movement in which there is a general rise in price levels throughout the world (reflecting the large increase in the international money supply) and a fall in the demand for goods and services in the consuming countries (in response to the demand deflationary effect of higher prices without a corresponding increase in money incomes).

In addition to the problems for world trade of a large increase in balance of payments deficits among oil consuming countries there is also a world banking problem. This has become especially acute since the second oil crisis in 1979. The poorer countries of the world have been enabled to continue buying the oil they need by borrowing from the banks in which the oil producers have placed the large sums of money that they have earned from the new oil prices. These deposits are largely of a short-term nature, which leaves the banks in their traditional position of 'borrowing short and lending long', because the poorer countries are not in a position to repay their loans at short notice. The assets of banks have been traditionally large enough to cope with short-term problems caused by their lenders withdrawing their deposits. However, when the world was struck by the second oil crisis, bank reserves were already under strain; and, although the second round of price increases was smaller proportionately than the first, it was larger in absolute terms. The result has been that the world banking system has been placed under very severe strain.

In Chapter 16, Sam Schurr discusses the problems of a reduction in the availability of energy to the world economic system pointing out that energy has been the constraint reliever *par excellence* in the past, and that the world has not really come to terms with the problems created when energy itself becomes a constraint. His paper is not directly addressed to the relief that nuclear energy might be able to give: it is a wise and balanced discussion of the role of energy generally in economic systems.

In Chapter 17, Lucien Thiriet offers a thoughtful analysis of the effects of the increase in oil prices upon the world economy. He traces the changes that the oil price rise combined with floating exchange rates has brought about – price inflation and demand deflation existing side by side. This, combined with the balance of payments problems of oil-importing countries, means that neither devaluation (in an attempt to stimulate external demand) nor internal inflation (in an attempt to stimulate home demand) are any longer available as stimulants to the national economies of energy importing countries. Without the increased competition in the energy market and the reduced dependence upon imported fuels that nuclear energy can provide, Thiriet sees a rather bleak future of continuing erratic upward movement in imported oil prices prolonging the economic malaise of the seventies.

Thiriet concentrates upon the effects upon the world monetary and trading system and upon the economies of oil importing countries of increasingly costly imported energy. Leonard Brookes, Chapter 18, attempts to identify some of the long-run fundamental relationships between energy and economic activity, including the long-run effects of attempts to manipulate world energy prices. With the aid of a simple illustrative world model, he considers the role of substitution of other factors of production for energy (which is what improvements in the efficiency of energy use amount to) as a solution to the world energy problem and concludes that the part that might be played by this type of energy conservation may be commonly overstated. The interrelationship between energy and economic activity (with increased availability of energy stimulating economic activity and higher levels of economic activity creating an increased demand for energy) may produce a strongly equilibrium-seeking system, with attempts to raise the energy price working themselves out more in terms of inflation and reduced world output, than in any substantial improvement in the efficiency of energy usage. This may be because the tendency of the new real energy price to final domestic and industrial consumers to settle at a level not greatly different from its original price (even after a large initial nominal price hike) provides only a limited incentive to substitute other factors of production. Sharp price hikes would therefore seem to jolt the world economic system into an inferior equilibrium, and one reason for this may be that the value of energy at any point in time depends upon the structure of the world economy and the state of technology. Attempts to make sharp changes in the energy price only result in the real price, through the medium of inflation, converging fairly quickly to a price reflecting the value of energy to currently structured systems and current technologies. Moreover, the lower rates of investment that would occur in a world damaged by sharp changes in the prices of world-marketed fuels might slow down the historic rate of improvement in energy efficiency so as to offset any improvement stimulated by higher energy price. In such a situation, any new large injection of relatively low-cost energy is desirable as a way of bringing downward pressure upon the world energy market, reducing both the likelihood of, and vulnerability to, arbitrary changes in the price of world-marketed fuels. Nuclear energy provides just such an option.

THE ROLE OF NUCLEAR POWER IN ENERGY SYSTEMS

1

C. Starr and O. Yu
THE ROLE OF CENTRALIZED ENERGY IN NATIONAL ENERGY SYSTEMS*

1.1 INTRODUCTION

Energy is an essential input to the well-being of any society, and energy consumption is closely related to its production of goods and services. Thus, every society seeks an energy supply system which is most appropriate for its current needs and its future growth. This chapter addresses the electricity supply sector, and reviews the mix of factors which determine the optimum choice among the spectrum of available systems.

1.1.1 Criteria

What is the role of economic criteria in making energy system choices? Either explicitly or implicitly a major criterion for choice is to maximize the difference between the benefits achieved and costs incurred. Much depends on the point at which benefits and costs are measured, and to what extent externalities – both positive and negative – associated with energy use are taken into account.

At the highest level of decision making, which may be termed societal, the costs and benefits need to be all embracing. Negative externalities such as environmental damages need to be considered on the cost side. But so also must the positive externalities, such as energy's impetus to economic and social development, be taken into account among the benefits. The latter set of considerations has been important in the past in the United States in relation to rural electrification and river basin development, of which the outstanding example is the Tennessee Valley Authority. Developing economies need to pay attention to both categories of externalities, although it is fair to point out that those countries which have already experienced successful development have paid far greater attention to the positive externalities of energy use, if only because concerns about such

---

*Presented at Conference on Systems Aspects of Energy and Mineral Resources. Cosponsored by the International Institute of Applied Systems Analysis and the Resource Systems Institute of the East–West Center, Laxenburg, Austria, 9–14 July, 1979.

matters as environmental damage had not assumed their present-day importance.

The analytic difficulty introduced by externalities is that they are difficult to measure; in some instances, even impossible. Because their quantification poses great difficulties, they are frequently omitted from quantitative assessments, and decisions are reached more on judgmental grounds, most often political and social judgments. This is necessary because of the lack of acceptable quantitative measures, a condition which, in turn, reflects a lack of satisfactory understanding of the underlying processes.

Another level of decision making is concerned with what might be termed the implementation of societal decisions. At this level a more narrow set of quantifiable economic calculations is brought into play to choose among pragmatic options, and it is to this level of decision making that the bulk of our paper is directed.

Traditionally, most electricity supply systems have been shaped by economic analyses seeking the most effective use of key resources. A regional electricity supply system is usually a combination of equipment, for production at several centres and distribution to many users. The traditional job of the electricity-system planner has been to evaluate the trade-offs of various alternative systems subject to the constraints imposed by the costs of capital, labour, materials (including fuel), and external specifications of reliability. Usually, the development over time of most electrical networks has tended towards larger plant size and greater interconnection, as representing an economic optimum mix evolving from growing and changing electricity demand characteristics.

1.1.2 Substitutability

Among the many energy supply options is the substitutability of other energy forms for electricity, as well as substitutability among various electricity supply systems. Thus, electric space and water heating can be replaced by natural gas or passive solar systems. The electricity supply spectrum ranges from large central power stations feeding distribution networks to small individual units, such as diesel generators or photovoltaic solar units with energy storage. This spectrum of supply alternatives reflects a corresponding spectrum of characteristics which *cannot* be described by a single parameter, such as 'centralization'.

If the term 'centralization' is intended to convey the existence of a central institution, operating unit, or manufacturing unit which can dominate the performance of an electricity supply, then most alternative systems have such centralized components. The conventional central stations, and the utilities that operate them, are the most obvious manifestation of this. They exist as a result of technical, economic, and institutional forces intended to achieve an economic optimization in the use of resources. At the other end of the system spectrum are those devices which appear to provide electricity self-reliance, such as small diesel generators, or the small solar-photovoltaic unit with battery storage. These individual units can be made in standard model sizes, and represent small physical and economic packages to the user. However, their availability depends on large-scale centralized manufacturing and skilled maintenance services – and in the case of diesels, a centralized network of fuel supply.

Thus, the spectrum of electricity options differs not in the existence of some 'centralization', but in the nature and location of the centralized facilities and institutions involved. These differences are most significant, from the consumer's standpoint, in the number of competitive sources from which they permit the users to choose. Thus, utility services based on an area franchise are non-competitive, and are administratively controlled by public agencies. Individual electric generation devices and their maintenance, on the other hand, do permit competition, and in this sense would be comparable to the automobile and its ultimate dependence on a centralized supply and distribution network for both equipment and fuel. To the consumer, these differences in degree of control may be of considerable importance in his psychological evaluation of self-reliance values.

1.1.3 Performance objectives

The traditional optimization of economic efficiency in the use of resources embodies an assessment of both the supply-system design and the demand-system characteristics. On the supply side, the principal factors are the economy of scale and reliability of supply. On the demand side, there is both a reliability requirement and a scale effect, which is termed the 'load factor'. The load factor measures the time dependence of electricity use by the consumers; it is quantitatively the ratio of the average power used in a specified time interval (day or year) to the peak power the consumers used during this interval. Thus, a variety of interconnected consumers using power at different times would raise the load factor, resulting in spreading to the individual consumer a smaller fraction of the cost of the capital investments. As regards reliability, the greater the requirement, the more costly the redundancy and quality required. This is often achieved by system interconnections. The absence of low-cost electricity storage technology makes these factors particularly important. Thus, the consumer's behaviour (load factor and reliability) tends to foster system interconnections.

1.2 ECONOMIC AND TECHNICAL DETERMINANTS OF ELECTRIC POWER SYSTEM DEVELOPMENT

In a previous study, Yu [1] has examined the major economic and technical determinants in the development of electric power systems. The major findings are summarized in this section.

The role of centralized energy

Yu first identifies two fundamental motivations for centralization in the electric power system.

1. *Energy production efficiencies of system sharing* from the following sources:
   (i) Efficiency in capacity utilization due to load diversity.
   (ii) Saving in system capacity due to reserve pooling.
   (iii) Increase in productivity due to labour specialization.
2. *Economies of scale in power equipment* from the following sources:
   (i) Capital costs.
   (ii) Operation and maintenance costs.
   (iii) Fuel delivery costs.

Examples of these scale economies are given in the Appendix and shown in Figs 1.1, 1.2 and 1.3.

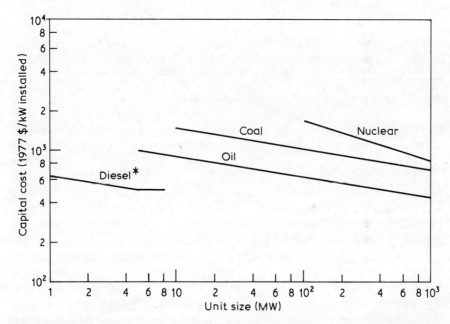

Fig. 1.1 Economies of scale in generating unit capital costs.* Due to technical constraints, scale economies for locomotive diesel units are limited to 5 MW. Sources [11, 12, 15].

Yu further identifies the major factors affecting the cost of electricity.

1. *Equipment characteristics*
   (i) Equipment capital cost.
   (ii) Equipment life.
   (iii) Generating size(s).
   (iv) Generating unit heat rate.

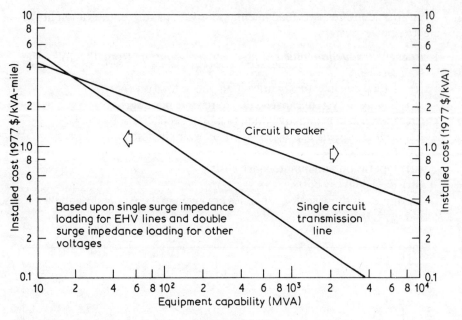

Fig. 1.2 Transmission equipment costs. Source [11].

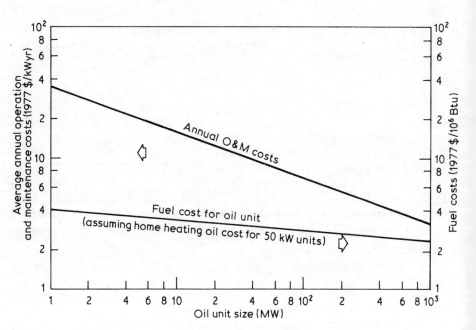

Fig. 1.3 Scale economies in operation and maintenance costs and fuel delivery costs. Sources [13, 14, (adjusted to 11)].

2. *Load characteristics*
   (i) Annual peak load (L).
   (ii) Annual load factor (f = annual kWh consumption/($L \times 8760$)).
   (iii) Load density or distance between neighbouring load centres.
3. *Resource availability*
   (i) Capital (expressed in terms of the fixed charge rate).
   (ii) Labour (including technical capabilities and managerial skills).
   (iii) Material including fuel.
4. *System constraints*
   (i) Reliability (as determined by the integrated load requirements).
   (ii) Environmental.
   (iii) Institutional.

Yu then applies electric utility system planning methods to analyse the economics of alternative power system development. (The study does not endeavour to address the sociologic, environmental and institutional issues, except to the degree that they are reflected in capital and operating costs.) To keep the analyses tractable, the study uses an average generating unit approach which, in fact, assumes that all generating units in a power system are identical. This approach tends to over-estimate the electricity production cost for large power systems. This is because in large systems there are more opportunities to minimize production costs through a mixture of generation units, so that various portions of the load with different load factors can be served individually by units of appropriate size, cost, and performance. Therefore, this approach is conservative in analysing the economic benefits of centralization in the electric power system.

1.2.1 Optimal average generating unit size

The optimal average generating unit size that minimizes electricity cost subject to reliability constraints is determined by a balance between the scale economies in the generating equipment capital costs and the generating capacity reserve requirements. The scale equation for the generating equipment cost, K, is given by:

$$K = K_o s^{-a} \qquad (1.1)$$

where K_o is the nominal per kW capital cost at 1 MW, s is the generating unit size in MW, and a is a scale factor. For a typical loss of load probability of 1 day in 10 years, the capacity reserve requirement, R, expressed as a fraction of the system peak load L, has been empirically determined in the study to be:

$$R = 0.58 \, s^{0.6} \, L^{-0.4} \qquad (1.2)$$

where s and L are in MW. Using 0.15 for the scale factor a in Equation (1.1), a commonly-used estimate for fossil fuel units, the optimal average generating unit size s^* has been derived as a function of the system peak load:

$$s^* = 0.4\ L^{2/3} \qquad (1.3)$$

where s^* and L are in MW. This result, shown in Fig. 1.4, agrees well with empirical observations of the average unit sizes in isolated systems obtained from references [2 and 3]. It appears from the correlation in Fig. 1.4, that the general trend of increasing generating unit size with increasing system load is common among different regions, regardless of their individual differences and state of economic development.

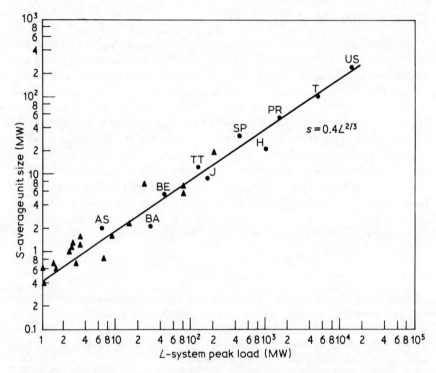

Fig. 1.4 Average generating unit size (S) as a function of system peak load (L), where ▲ is Alaskan Villages, AS is American Samoa, BA is Barbados, BE is Bermuda, H is Hawaii, J is Jamaica, PR is Puerto Rico, SP is Singapore, T is Taiwan, TT is Trinidad and Tobago and US is a large US electric utility. Sources [2, 3].

Using units of optimal average size, electricity generating costs (i.e. the sum of levelized generation capital costs, operation and maintenance costs and fuel costs) for alternative generation technologies in power systems with a fixed annual load factor ($f = 0.6$), have been computed and are shown in Fig. 1.5.

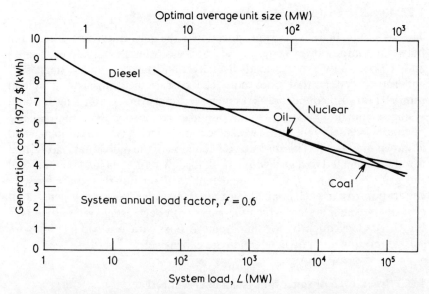

Fig. 1.5 Electricity generation cost with optimal average unit size (new plant cost basis).

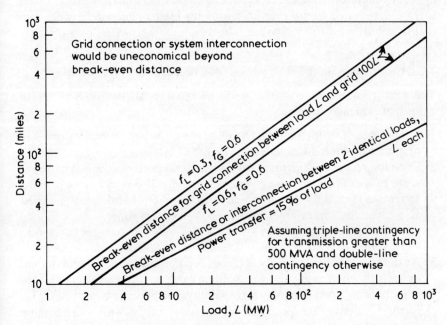

Fig. 1.6 Estimated break-even distances for grid connection and system interconnection.

1.2.2 Break-even distance for grid connection

The economic benefit of grid connection is the generation cost differential between a large power system, that has the advantage of scale economies and a potentially greater load diversity (load factor), and a small self-generating system that does not. The cost is the transmission system required for grid connection. Assuming optimal average unit sizes for both systems, the break-even distance for grid connection, i.e. the distance beyond which self-generation would be economical, has been computed and is shown in Fig. 1.6. This distance is a function of the individual load L (that is to be connected to a grid with a peak load of $100L$), its load factor f_L and the grid load factor f_G. It is important to point out that the electricity generation cost at each point along a break-even curve is the self-generation cost of the small system which increases with decreasing system load (see Fig. 1.5). Consequently, for small remote loads with low load factors, both self-generation and grid connection are expensive.

1.2.3 Break-even distance for system interconnection

Finally, by trading off the generating capacity reserve savings against the additional transmission requirements, the break-even distance for interconnection between two power systems, each with load L and both using the optimal generating unit size, is estimated to be approximately proportional to $L^{1/2}$. Estimated break-even distances for interconnection between two identical systems with 15% power transfer are also shown in Fig. 1.6.

1.2.4 Conclusions on the determinants of power system development

The major conclusions of the study by Yu [1], all in agreement with empirical evidence, are:

1. Economies of scale in capital costs cause the average generating unit size to increase with load growth (Fig. 1.4).
2. Even for a fixed load factor, as the system load increases the economies of scale make capital-intensive generation technologies increasingly more attractive (Fig. 1.5). In other words, economies of scale in capital cost also cause the capital intensity of a power system to increase with load growth. Further, the shift to lower cost fuels also encourages the shift to capital-intensive generation.
3. As the load grows, it rapidly becomes economical for a power system to expand by connecting the surrounding loads and interconnecting with neighbouring power systems for reliability and load factor purposes. This further increases the potential for interconnection with more remote systems (Fig. 1.6). Consequently, the economically attractive degree of interconnection of a system increases with load growth.

In summary, the study concludes that economies of equipment size, efficiencies of capacity sharing, and load growth are the major driving forces of centralization in the electric power system. Furthermore, it identifies load density, load factor, capital availability, fuel availability, labour availability, and reliability requirements as the key components for determining the degree of interconnection, the unit size, the system capital intensity, and the technical sophistication of the supply technologies in a power system.

Although the study has been conducted mainly for US systems using conventional technologies, the method of analysis is applicable to other types of power technologies in any region.

1.3 REGIONAL SOCIO-ECONOMIC CHARACTERISTICS AND POWER SYSTEM DETERMINANTS

Based upon the work of Liu and Anderson [4], indicators of regional socio-economic characteristics were compared with the key determinants of power system development. Some major relationships are briefly discussed below.

1.3.1 Load factor

Population growth, population concentration, and economic development increase the load factor. In general, urban areas and developed regions have higher load factors than rural areas and less developed regions. However, the key regional characeristic for load factor is the degree of industrialization. Data from Association of Edison Illuminating Companies [5] indicate that in the United States, residential customers have load factors between 10% to 30%, small commercial and industrial customers (those with annual consumption less than 20 000 MWh) between 15% to 50%, and large commercial and industrial customers between 30% and 80%. Therefore, the system load factor increases with the proportion of large commercial and industrial loads.

1.3.2 Capital availability

Capital availability in most regions is constrained by the level of economic development. However, regions well-endowed with exportable resources, such as the OPEC countries, have substantial capital resources. The fraction of economic output used for gross capital formation tends to increase slowly with increasing per capita gross domestic product (GDP). Data from the United Nations [6] suggest that typical capital formation rates are 5% to 15% of the GDP in the poorest countries (those with GDP per capita of less than $100 US) and 15% to 30% for the developed countries. Starr [7] has shown that capital availability could be the most important constraint to energy system development in the less developed regions.

Table 1.1 Regional classification of the values of key system determinants and a parametric analysis of appropriate power systems

Typical region	Key system determinants					
	Load density	Load factor	Capital availability	Fuel availability	Labour resource	Reliability requirement
1. North America, urban	H	H	H	H	H	H
2. Japan, urban	H	H	H	L	H	H
3. Kuwait	M	M	H	H	H	H
4. Switzerland	M	M	H	L	M	H
5. Indonesia, urban	M	M	M	H	M	M
6. India, urban	M	M	M	L	M	M
7. North America, rural	L	L	M	H	M	M
8. Japan, rural	L	L	M	L	M	M
9. Indonesia, rural	L	L	L	H	L	L
10. India, rural	L	L	L	L	L	L

Typical region	Estimated characteristics of appropriate power system					
	Degree of inter-connection	Typical system peak load (MW)	Average unit size (MW)	Capital intensity	Technology sophistication	Electricity cost ($/kWh)
1.	H	20 000	300	H	H	5
2.	H	10 000	200	H	H	10
3.	M	500	25	M	M	5
4.	M	500	25	H	H	5
5.	M	1 000	40	L	M	10
6.	M	1 000	40	M	M	15
7.	L	50	5	L	M	10
8.	L	50	5	M	M	20
9.	L	0.01	0.01	L	L	25
10.	L	0.01	0.01	L	L	30

Potential values of system determinants

	High	Medium	Low
Load factor	0.6–0.8	0.4–0.6	0.1–0.3
Capital availability (fixed charge rate)	0.075–0.15	0.15–0.20	0.20–0.30
Fuel availability (oil cost in $/MBTu for 500 MW units)	1.25–2.5	—	2.5–5.0
Reliability requirement (percent reserve margin for 1000MW system with 40MW units)	22–33	11–22	0–11

1.3.3 Reliability requirements

In general, regions with higher levels of economic prosperity often require greater electric power service reliability and availability. The degree of industrialization also has a major effect on reliability requirements. Data from Bhavaraju and Billington [8] show that the direct cost of power interruption is greatest for industrial users, averaging about $7 per kWh. Commercial users are the second most affected, averaging about $4 per kWh, and the residential users are the least affected averaging about $0.8 per kWh.

1.4 APPROPRIATE POWER SYSTEMS FOR DIFFERENT REGIONS – A PARAMETRIC ANALYSIS

Based upon the relationships developed in the previous section, different combinations of regional characteristics can be used to broadly indicate the potential values of key determinants of power system development. Some examples are given in Table 1.1. By using specific values for the system determinants and applying the method of analysis in Yu [1], the character-

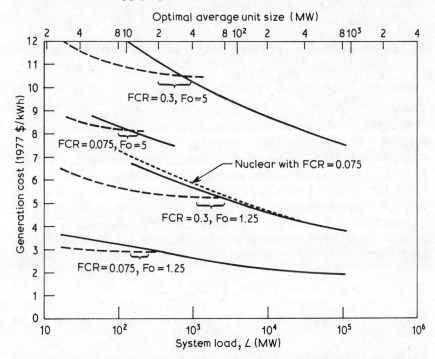

Fig. 1.7 Generation costs for alternative power systems (high load factor case), where ———— is diesel, ———— is oil steam, FCR is fixed charge rate, Fo is oil cost in $/MBtu for 500 MW units and f is 0.1.

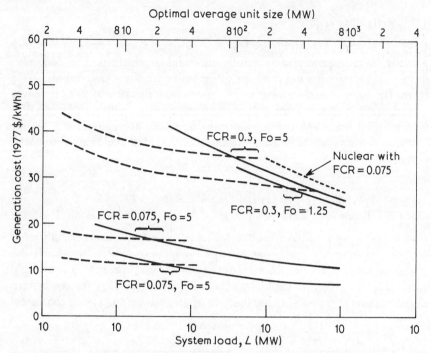

Fig. 1.8 Generation costs for alternative power systems (low load factor case), where ———— is diesel, ———— is oil steam, FCR is fixed charge rate, Fo is oil cost in $/MBtu for 500 MW units and f is 0.1.

istics of the power systems that economically match the regional characteristics have been estimated and are also given in Table 1.1 Computational results using extreme values for the system determinants are shown in Figs 1.7 and 1.8 to provide an indication of the ranges of these system characteristics. The computed values for the average unit size and electricity cost for the appropriate systems are in general agreement with actual data [2] and estimates from independent sources [9, 10]. This agreement appears to indicate that economic optimization has been the dominant criterion for electric power system development in all regions of the world.

1.5 CONCLUSIONS

Examination of these results leads to the following conclusions for the present study:

1. The ideal degree of centralization in a power system, i.e. the level of interconnection, the size of generating units, the capital intensity and the sophistication of the technology, tends to increase with such factors as

population concentration, advancement of economic development, increasing degree of industrialization and a high level of fuel availability.
2. A mismatch between power systems and regional characteristics can have disastrous effects on the cost and availability of electricity to these regions. For example, highly-interconnected, capital-intensive, technically-sophisticated large-scale power systems for sparsely-populated, economically-underdeveloped, resource-deficient agricultural regions; or conversely, isolated, unreliable, backyard type micropower systems for densely-populated, well-developed industrial regions.

In summary, there is no one system that is universally applicable to all regions; there is no one answer to the question of the appropriate degree of centralization.

APPENDIX: COST FORMULA AND INPUT ASSUMPTIONS

The cost of electricity supply, C, may be expressed in terms of its major components as follows:

$$C = \text{Fixed costs} + \text{Variable costs}$$
$$= \{[(1+R)K+T+D]r+M\}/(f\times 8760)+(F+V)$$

in \$/kWh, where:

R = Generating capacity reserve margin (fraction of the system peak load)
K = Average generation capital cost (\$/kW)
T = Average transmission capital cost (\$/kW)
D = Average distribution capital cost (\$/kW)
r = Annual fixed charge rate (fraction per year)
M = Average annual fixed operation and maintenance cost (\$/kW/yr)
f = Annual load factor (%)
F = Average fuel cost (\$/kWh)
V = Average variable operation and maintenance cost (\$/kW/yr)

The fixed charge rate is dependent on, among other elements, the cost of capital, the equipment life, and the tax requirements. A detailed discussion of the fixed charge rate is given in [11].

The following assumptions are based upon data from [11, 12, 13, 14, 15].

A. Equipment costs (1977 dollars)

1. Generation

Technology	Automotive (diesel)	Locomotive (diesel)	Oil (steam)	Coal (steam)	Nuclear (steam)
Size range (MW)	0.001–0.05	0.05–5	5–1500	10–1500	100–1500
Equipment life (yr)	5	10	30	30	30
Fixed charge rate	0.3	0.2	0.15	0.15	0.15
Capital cost:					
Constant K_o (\$/kW)	200	640	1240	2190	6510
Scaling factor a	0.15	0.15	0.15	0.165	0.30
Heat rate (Btu/kWh)	10 000	10 000	9500	10 200	10 400
Fixed O&M (\$/kW/yr)	$15 s^{-0.15}$	$35 s^{-0.35}$	$35 s^{-0.35}$	2.5	2.9
Variable O&M (\$/lWh)	—	—	—	4.2	0.7

s is the unit size in MW.

2. Transmission

(i) Average total transmission plant capital cost is assumed to be \$150 per kW of generating capacity.
(ii) Single surge impedance loading (SIL) is assumed for above 220 kV transmission, and double SIL is assumed for lower voltages. The SIL is approximately equal to (transmission voltage)^2/400 ohms.
(iii) Triple-line contingency is assumed for transmission of 500 MVA or greater, and double-line contingency for lower levels of transmission.
(iv) Specific capital cost assumptions:
    Single circuit transmission line, including right-of-way:
        $23 L^{-0.66}$ \$/kVA-mile,
    Circuit breaker: $9.5 L^{0.35}$ \$/kVA,
    Transformer: 4 \$/kVA,

where L is the equipment loading capability, which lies between 1 MVA and 10 000 MVA.

3. Distribution

Average total distribution plant capital cost is assumed to be \$150 per kW of generating capacity.

B. Fuel costs (1977 dollars)

  Oil (including storage): $4 s^{-0.075}$ \$/10^6 Btu,

where s is the unit size in MW, and the formula yields a nominal oil cost of \$2.5/10^6 Btu for 500 MW units;

        Coal: \$1.0/10^6 Btu
        Nuclear: \$0.5/10^6 Btu.

REFERENCES

[1] Yu, O. S. (1979) Economic and Technical Determinants of Power System Development. *Proceedings of the Joint Power Generation Conference*, Institute of the Electrical and Electronics Engineers, New York.
[2] Electrical World (1973) *International Directory of Electricity Suppliers*, McGraw-Hill, New York.
[3] Electrical World (1978) *Directory of Electrical Utilities*, McGraw-Hill, New York.
[4] Liu, B. and Anderson, C. F. (1979) Industrialization, Energy Requirements and the Quality of Life: An International Comparison. *Proceedings of the Silver Anniversary Conference of Society for General System Research*, London.
[5] Association of Edison Illuminating Companies (1973–1977) *Reports of the Load Research Committee*, New York.
[6] United Nations (1975) *Yearbook of National Account Statistics 1974, Volume III, International Tables*. New York.
[7] Starr, C. (1977) Energy System Options. *Proceedings of the Tenth World Energy Conference*, Istanbul.
[8] Bhavaraju, M. P. and Billinton, R. (1979) Cost of Power Interruptions – A User's Viewpoint. *Proceedings of 1979 Reliability Conference for the Electric Power Industry*, American Federation of Quality Control, New York.
[9] World Bank (1975) *Rural Electrification*, Washington, DC.
[10] Desai, B. G. (1978) Solar Electrification and Rural Electrification: A Techno-Economic Review. *Proceedings of International Solar Energy Society Congress*, New Delhi.
[11] Electric Power Research Institute (1978) *Technical Assessment Guide*, Palo Alto, California.
[12] Alexander, T. (1978) The Little Engine that Scares ConEd, *Fortune*, December 31, pp. 80–84.
[13] Aschner, F. S. (1974) *Planning Fundamentals of Thermal Power Plants*, Wiley, New York.
[14] US Department of Energy (1979) *Monthly Energy Review*. DOE/EIA/005/1(79), Washington DC.
[15] US Energy Research and Development Administration (1974) *Power Plant Capital Costs, Current Trends and Sensitivity to Economic Parameters*. WASH 1345, Washington DC.

2

L. Gouni

ECONOMIC PRINCIPLES OF OPTIMIZING MIXED NUCLEAR AND NON-NUCLEAR ELECTRICITY SYSTEMS

In this chapter, an attempt will be made to show how and why, viewed from the economic angle, nuclear energy and electricity systems supplement each other, since the former requires large size facilities, and the latter provide already existing networks for the supply of all users. Consequently, it is primarily through the electric vector that the rational development of the nuclear industry may be ensured.

Section 2.1 sets forth the essential rules for economic calculation. In Section 2.2 we discuss the competitive factors among final-use forms of energy in regard to utilization, and we attempt to show how nuclear energy transmitted through electricity systems may meet such terms. Finally, Section 2.3 deals with, and specifies the characteristics of, electricity systems based on nuclear energy and, in particular, the rates to which they lead.

2.1 THE PART PLAYED BY ECONOMIC CALCULATION IN THE DECISION-MAKING PROCESS

For the decision-maker, economic calculation constitutes a valuable help, in particular, since one of its functions is to seek the 'best' solution. However, such calculation has its own concepts and requirements, and, by being oblivious of the corresponding constraints, one may be easily led to ambiguities, albeit to errors. Consequently, one must be clearly aware of the possibilities afforded by economic calculation and of its limitations. In particular, calculation can only be effected through quantification of actual facts. Hence, economic calculation becomes illusory when no measurement can be made or the efficiency of measurement is inadequate to solve the problem posed.

It is indeed because of the existence of important and highly complex options allowing for the measurement of numerous factors that economic calculation plays an important part in the decisions taken by the electricity supplier.

2.1.1 Main rules governing economic calculation

(a) Criteria and constraints

There is a possibility of choice only when several options are available and the resulting solutions to be contemplated can be ranked. Only then can one refer to an economically optimum system. However, this requires pre-determined criteria, or one criterion. In the present state of the art, and despite the avenues opened up by multi-criteria analysis, economic calculation hardly favours multiple criteria, and becomes operational only within the scope of a single criterion, accompanied by constraints if need be. In practice, it is not always easy to assimilate all the various motivations underlying choices and to discern the most fundamental one. When choosing a car, its performance data on the highway, and its handling ability in town, its comfort and its price, its aspect from an aesthetic viewpoint and the specific prestige which it will confer upon the owner are all taken into account. Which is the actual criterion? At which point do we set the difference between constraints and the criterion of choice?

Even in cases where economic calculation is operational, the difficulty of defining the criterion does not constitute a side issue. Thus, in developed countries, lacking energy resources, the energy policy objective is often a compromise between a (quantified) expenditure and a self-sufficiency constraint more difficult to apprehend. Efforts are made to secure the required supply for the country at the lowest cost, while avoiding being over-dependent upon foreign countries energy-wise. This latter requirement is essentially political. For instance, a payment made to a group of united oil-producing countries does not bear the same significance as a payment made to a foreign coal supplier who does not belong to a trade organization.

It is obvious that ranking of the solutions depends on the selected criterion. The less costly solution in terms of total cost is not generally the most advantageous from the viewpoint of capital expenditure, number of jobs created, or fossil-energy savings. Thus, strangely paradoxical situations may result from playing upon the criterion. However, the superiority of the 'total cost' criterion needs to be stated. Indeed, insofar as a price system providing for equivalence between the scarcities of various products exists, the total cost criterion affords the most acceptable synthesis of an extensive number of factors for decision-making. It is a criterion of this type which is adopted by industrial companies, and, in particular, electric utilities. Through prices, the aggregation of numerous transactions is effected by calculation.

The aggregation of future years' expenditures must also be ensured since the decisions taken bear on several decades. Aggregation is achieved through discounting. D_n expenditure for year n is reduced to the value of year 0 through application of the following formula:

$$D_n^0 = \frac{D_n}{(1+a)^n} \qquad (2.1)$$

where a equals the average discounting rate of the future. The expenditures of the various years may then be added up.

Conversely, there are rules whereby from a discounted total expenditure, an annual expenditure may be worked out as an input to the calculation of unit cost. Those rules are quite strictly defined; unfortunately, the accounting procedures used in practice are generally quite different.

(b) Need for computational conventions

The introduction of the discount rate a shows that computational conventions need to be specified. In particular, the question should be asked whether the computation to be made should reflect the advantages accruing to the whole country, or be limited to the financial advantages accruing to the electricity supplier. The two concepts are not similar; and the fact that electricity supply is a public utility service which, *de facto* or *de jure*, constitutes a monopoly or which enjoys a franchise or other special privileges makes the distinction all the more important.

Hence, the conventions adopted for economic calculations must be determined through a dialogue between the electricity supplier and the representatives of the national community. These conventions include:

1. The method of allowing for inflation; it is generally convenient to reason in constant currency which does not mean that the price of the various products do not evolve over a period of time – the results stated in this chapter are based on the assumption that currency shall remain constant.
2. The choice of the discount rate a (which is here assumed to be equal to 9%).
3. The assumptions made about prices of imported goods and materials (oil products, for instance).
4. The treatment of domestic taxation; such taxation, in principle, does not constitute a burden for the country. However, it plays an important part in the consumer's behaviour.

It may be noted that, in numerous countries, national accounting is not based on constant currency, so no allowance for inflation is made. It follows that, if such national accounting conventions are adopted, production costs for capital-intensive systems are generally undervalued. Consequently, such calculations favour, for instance, hydro or nuclear power plants, compared to fossil fuelled thermal plants.

Economic principles of optimizing mixed systems

(c) Technological choices when the consumption level is predetermined

The unit cost of a product depends upon consumption which, indeed, determines the supply structures, including production-unit size and the characteristics of the distribution system. Conversely, in regard to both level and distribution, consumption depends on the prices of goods sold to users. Consequently, there is a close interrelationship between consumption and the supply system. Unfortunately the evolution of electricity demand in relation to its numerous explanatory factors, and, in particular, the prices relating to electric power and other forms of energy (cf. Section 2.2.1) cannot be mathematically expressed. Hence, the calculations of electric utilities are generally calculations bearing on technological choices, where future consumption has already been determined and is the subject of several assumptions.

2.1.2 The conditions for practical economic calculations for electricity systems

In practice, the economic calculation to which the electricity supplier resorts is highly complex by reason of the technical characteristics specific to electricity (cf. Sections 2.3.1 and 2.3.2) on the one hand, and time-constraints (decisions involving several decades) on the other hand. Consequently, such calculation uses rather large-scale mathematical models, each of them being adapted to a specific purpose, while complying with the proper management requirements of the overall system. Although the specific structures of such models cannot be discussed, a few comments may be useful.

The models concerned are dynamic, rather than static optimization models. Indeed, the options to be considered involve numerous related factors and, furthermore, the optimum decision to be taken at a given moment is also affected by earlier decisions, short- or long-term. An especially interesting example is furnished by considering the development of fast-neutron power plants, where the nuclear fuel is itself a byproduct of thermal neutron power plant operation. Technical and chronological linkage between the fast reactor programme and the thermal reaction programme must bear not only on equipment but also on fuel. For problems of this nature, the so-called 'instantaneous snapshot', which consists in comparing two costs assumed to have been defined at a given time, is hazardous. It cannot produce optimum time-linkage for management and equipment decisions. Such linkage is achieved through dynamic optimization, employing optimum control language (maximum principle of Pontryagin for instance) which systematically singles out: *state variables* characterizing the system at instant t, and *control variables* whereby the system may be acted upon at each instant t.

These models also lead to seeking systematically the economic meaning of the intermediate variables which the mathematical calculation of the optimum associates with each constraint, whether such constraints be expressed as equalities or inequalities. Such intermediate variables are construed in terms of cost, and if one has some idea of the cost of the constraint, its weight can be incorporated into the objective function. Thus, a dialogue is established between physical elements (which cannot be directly assessed) and an assumption bearing on the price of these factors. For instance, a 'deficiency value' (outage cost), which is far more operational, may be substituted for a physical risk constraint (cf. Section 2.3.3(a), a case of power determination).

Mathematical calculation of the optimum expressed as equalities (squaring of derivatives to zero) leads to highly fruitful results, subject to the important reservation that it be kept in mind that such results apply only to the optimum. Thus:

1. The variables relating to the state equations of an optimal control problem represent the 'use value' of a marginal decision (an additional equipment item or a quantity of water which was not run down from storage).
2. The economic depreciation is then perfectly determined: it is the difference between the value in use of an equipment item over a two-consecutive-year period.
3. The economic evaluation of the anticipated construction of an equipment item upon which there is no constraint is nil.

Finally, owing to the ever-growing uncertainties about the future, on the one hand, and to the characteristics specific to electricity (cf. Sections 2.3.1 and 2.3.2), on the other hand, calculation techniques allowing for randomness must be applied. Introducing stochastic phenomena makes the search for the optimum much more difficult. Given current means, and in order to reduce the scale of the models and to reach concrete results, one is led to make decisions of the following nature:

1. Either model dynamism must be reduced; or
2. Randomness must be reduced; or
3. Reality must be simplified, certain data must be aggregated, which involves a loss in the quality of the representation of actual facts.

It must be emphasized that to reach decisions about the complex electricity systems of developed countries, such models are essential. Optimal planning cannot be determined without using them. Some means of presenting results in a readily understood form must then be established. For such purpose, simplified procedures are developed, but these procedures must be considered only as means of interpretation, since they rest on the assumption that the overall problem has already been solved. The calculation set forth in Section 2.3.3(b) is an example of such simplified presentation. In reality, this procedure, which grossly over-simplifies actual facts, can only

Economic principles of optimizing mixed systems

give a mere idea of the operations involved in comparisons made by using large-scale models.

2.2 NUCLEAR ENERGY, ELECTRIC POWER AND ENERGY MARKET

2.2.1 Flows of energy from resources to final use and the consumer's viewpoint

The possibilities of evolution of the energy market constitute a vital factor for the development of the nuclear energy market. Indeed, in the choice of energy, customers have the last word; and the number of such customers is substantial, running for each developed country to tens of thousands in the industry sector, and to millions in respect of commercial, institutional and domestic buildings.

An essential problem consists in establishing linkage between primary energy – the energy found in nature – and the very numerous users. Often, and this applies to nuclear energy, production is centred in very large units, whereas each user's consumption is comparatively very small. The highly complex systems for energy processing, transmission, and distribution consist of many steps, constituting the energy flow, and each step contributes added value.

The user is not interested in energy *per se* but in the service which he gets from it. This service requires not only energy purchase expenses, but also substantial plant maintenance and labour expenditures. Hence, a comparison between forms of energy calls for the definition of complete energy supply processes such that services rendered by each form at the end point are identical (including product quality, comfort or extent of pollution). It calls also for the determination of all the expenditures to be incurred in order to reach its final stage in each case. Efficiency, in the physicist's sense, may allow for some ranking of the possible solutions. However, if the total expenditure verdict is different, there is little case for the user to allow himself to be lured by technical considerations (cf. Section 2.1.1(a)).

The example set by railway electrification is a good illustration of such considerations. Despite the fact that the technical efficiency of the electric engine indisputably exceeds the efficiency of the internal combustion engine, all railways were not (and probably will not) be electrified. This is because of all the items that go to make up the total cost of providing rail travel, energy costs – at least when considering average traffic lines – do not weigh the heaviest. They account for less than 20% of the total. Labour and maintenance costs (which favour electric traction) and investment costs (which favour diesel traction) are far more important items.

A comparison between electricity and conventional fuels used for heating purposes, whether in respect of industry or households, should be analysed in similar terms. The determination of a satisfactory decision cannot be made solely by considering the price of the energy purchased.

Let us, for instance, work out the annual expense for heating purposes (heating of premises and water), for an approximate 80 m² apartment in Western Europe, in a temperate climate area. In the following table, a comparison is made between two identical apartments, thermally insulated at the optimal level, one of which is oil-heated (domestic fuel oil) and the other electrically heated (Joule effect). Assume the country concerned launched a substantial nuclear programme in 1974, whereby electric power prices will decrease after 1980 (cf. Section 2.3.5), whilst oil product prices will escalate steadily. The expenses for three years are split up as follows*:

1. 1972, that is prior to the oil crisis.
2. 1980, by which time only a few nuclear plants will have been commissioned.
3. 1990, the year for which it has been assumed that the nuclear plant construction policy bears fruit, whereas oil product prices for domestic use continue to escalate regularly at an approximate rate of 4% per annum, in constant currency.

The expenses, readjusted to allow for inflation, are expressed as indices, the reference index being the total expense for the electric solution in 1980.

Table 2.1 Annual heating and hot water expenses for a new apartment (with optimum insulation). Index 100 in 1980 for electricity

	1972		1980		1990	
	Oil products	Electricity	Oil products	Electricity	Oil products	Electricity
Energy purchases	47	51	66	72	81	70
Maintenance	11	5	11	5	11	5
Depreciation of plant						
Heating apparatus	30	16	25	12	12	12
Thermal insulation	—	18	8	11	11	11
Total	88	90	110	100	125	98

Table 2.1 illustrates that, even for uses where the energy expense is substantial, energy expense is not the only factor to be taken into account:

* This calculation is grossly simplified since it does not take into account the discounted sum of future year expenditures.

Economic principles of optimizing mixed systems

1. Even before the 1973 crisis, electric energy, although close to being competitive, required a thermal insulation investment, the burden of which was alleviated through far lighter heating equipment than that of the 'oil product' solution; the latter technique did not resort to insulation owing to the low price of fuel.*
2. The energy price escalation modified the whole picture: in 1980, all forms of heating were backed by insulation; the electric solution has a very marked advantage, because it is only partially affected by the oil price escalation.
3. In 1990, the cost advantage of electricity will be even greater since, through the commissioning of nuclear power plants, the real price of electricity will be reduced, whereas oil product prices will have continued escalating, calling for greater insulation.

This example shows how, with price escalation of hydrocarbon fuels and the increased competitiveness of nuclear generated electric power, new markets will open up for electricity. It may henceforth be substituted for conventional fuels for numerous heating purposes.

However, let us revert to the beginning of the electronuclear process in order to explain the difficulties resulting from the necessity of linking large-size production tools to users with an average small consumption rate.

2.2.2 Nuclear reactors and energy consumer linkage

Energy is rarely consumed where it is produced. Between the production site and the place of consumption, there generally intervenes a substantial added value for transmission and distribution, which varies according to many factors. The result is that the user pays a price which is quite considerably higher than the production price. Since highly complex phenomena are involved, they are quite frequently ignored and their omission gives rise to a great number of errors and considerable misunderstanding.

Just as for the other forms of energy, the economic characteristics of nuclear energy are substantially affected by transmission. The main factors can be summarized as follows:

1. In a nuclear reactor (of the types which will be in industrial service during the next 30 or 40 years), energy is released in the form of heat. Heat, like electricity, is not a tangible entity. It requires a physical support – a vehicle – for its removal, its transmission and use. The physical support which is generally resorted to because it is cheap, is water, or rather steam;
2. Moreover, the cost of nuclear energy production is highly sensitive to the

---

*If thermal insulation technology, as a substitute for oil products for residential and domestic purposes, had been implemented long before the oil crisis, the results achieved would have been significant.

size of the reactor*: in view of the magnitude of the investment, the improvement of neutron effectiveness with size and the essential safety measures, production should be centred in a limited number of units, as large as possible. By contrast, the geographical location and size of the energy utilization centres follow different rules. In fact, very few energy consumers have a consumption sufficient to absorb on their own the production of an economically sized reactor.

3. Hence, appropriate linkage must be ensured between two different structures. The solution would be easy if heat transmission were simple and its cost a subsidiary matter. However, apart from certain strict requirements (short distance and high consumption) such transmission is difficult. The easy transmission and storage of heat in containers cannot be effected. If it is contemplated, as in the solutions presently under consideration, to resort to a continuous line between producers and users, the fixed cost portion of transmission is heavier than the production cost by about 80 to 90%. It depends basically on the quantity transmitted and on the geographical location. It also depends upon the regularity of transmission, and on the annual utilization factor. At present, the most ambitious projects hardly exceed 20 kilometres (15 miles);

4. Thus, in the competition between nuclear energy and conventional energy forms, it may be predicted that the promptness with which a substantial delivery coupled with a linkage system suitably and regularly used can be achieved constitutes an essential factor. The time, quantity and regularity-of-utilization factors bear heavily on the search for acceptable prices. The distribution network must be brought to a fairly high utilization level as soon as possible. One must then ask whether the conventional forms of energy which can already benefit from their own distribution system warrant such a time limit;

5. In short, if, for nuclear energy, the transportation cost of the fissile material is quite negligible the transmission and distribution arrangements for the generated energy loom very large.

The need to find a solution for these difficulties leads in fact to imagination and research efforts, which the general public is often unaware of. Naturally, the vectors which came to mind at the outset are the electric system (since we know how to convert heat into electricity) and water, which is the ordinary means of heat transmission. Consequently, during an initial phase, nuclear energy will be transmitted by these two means. However, other more ambitious, much longer-term projects, for which success is uncertain are being considered in many parts of the world. They all consist of seeking a

---

*For an electronuclear reactor operating at base load, the cost per kWh is worked out as follows, by reference to an index of 100 for a 130 MWe reactor:

130 MWe	100
360 MWe	55
900 MWe	40

better vector for nuclear energy than the 'water–steam' solution. They generally involve, besides a nuclear reactor, a very large central unit producing a fuel-fluid which can be transported (e.g. hydrogen production, or the production of methane or methanol from coal). Unfortunately chemical processes are confronted with the difficulty that industrially achievable nuclear boilers provide temperature conditions which are not high enough for satisfactory chemical reaction rates.

2.2.3 A marriage of convenience: nuclear energy and the electricity system

The foregoing paragraphs have explained why and how the electricity transmission and distribution system is called upon to become, in future, the first and most important linkage between the nuclear reactor and the energy consumer. Indeed:

1. The electricity supplier, whether willingly or reluctantly, had to adapt to a form of energy which cannot be stockpiled (cf. below Section 2.3.1). This is a heavy constraint which requires the existence of a continuous line between producers and consumers. However, this constraint has resulted in producer solutions which are precisely the reasons for which the linkage system between large production units and numerous consumers already exist. In particular, in industrially developed countries with an evenly-distributed population, the electricity supplier has for a long time been availing himself of interconnection between large production units, in order to ensure the best management of the system (cf. Section 2.3.2(a)). The existence of such networks and their increasing returns to scale currently make the electricity vector very much the front runner within the range of possible transmission technologies. Current steam-power systems cannot exceed urban size.
2. Moreover, in most developed countries, the annual growth rate of electricity demand, and the need to replace old units, warrants the annual construction of several large-scale nuclear units with the assurance of suitable on-load operation. The time allowed to reach economically efficient operation is here very much reduced, whereas it is dreaded in other distributed energy systems.

Given these realities, one is led to the conclusion that electricity is the choice intermediary for the necessary substitution of nuclear energy for hydrocarbon fuels in the many countries lacking oil resources. Electricity may thus replace fossil fuels for uses which did not traditionally fall within the scope of its purposes.

2.3 ECONOMIC CHARACTERISTICS OF ELECTRICITY SUPPLY

We shall discuss in this section how the economic characteristics of electricity supply may, for nuclear energy, be dissociated from the price of conventional fossil fuels (coal and oil).

2.3.1 An exceptional technical constraint for a high-quality product: no storage

Electricity energy is high-quality energy; a comparatively high price is paid by the consumer, but he finds substantial advantages in electric appliances and equipment. This characteristic, which is important in regard to competition (cf. Section 2.2.1), is nevertheless in no way exceptional. On the other hand, the fact that electricity cannot be stockpiled, without being a unique feature in the world of economics, creates an exceedingly cumbersome constraint, and also, though electricity has many other characteristics (e.g. the coexistence of active energy and reactive energy), the main concern for the management of a costly system arises from the fact that electric power cannot be stored. A continuous line must exist between the producers and the consumers, and one cannot rely upon a product reservoir which would resolve the outage problems of the plants for maintenance or demand fluctuations at little cost. On the contrary, the production–transmission–distribution tool must be able to meet the vagaries of demand at every instant. Thus, the electric utility is tied to the 8760 hours of the year, and to the mandatory distinction between power (kW) and energy (kWh), which is not well understood by the public.

2.3.2 Two main consequences on the structures

The fact that electric power cannot be stored entails numerous consequences in practice. We shall only consider here the two most important consequences affecting structures: strong technical interconnection among production centres and the virtual lack of international electricity exchanges.

(a) Interconnection between production centres

In the absence of storage capacity, the need for stand-by facilities, whether production plants or transmission lines, assumes a very different significance from that which obtains in most other industrial sectors. For electricity we cannot think in terms of a reserve capacity which permits several weeks' outage (as for maintenance, for instance) for an increase in only a few percentage points (3 to 5%). On the contrary, the idea of an isolated unit

Economic principles of optimizing mixed systems 57

must be given up and the doubling of equipment or at least the existence of stand-by units must be systematically envisaged. For a nuclear unit the depreciation and the fixed plant and personnel costs represent approximately 80% of the total electricity production costs. If the unit must be doubled to provide standby power, the production cost will be increased by 80%, not by a few percentage points.

The problem is solved at the lowest cost, in the case of countries of limited geographical size and rather uniform consumption density, through the creation of a high-voltage transmission system which ensures not only the transmission of electricity from generating centres, but also interconnection with the centres. This applies to all Western European countries, and to all areas of the world where consumption patterns are similar. If we consider the historical evolution of the various countries we see that the development of all electricity systems followed the same pattern, once an adequate economic level was reached.

The reasons for this pattern of evolution are complex. They ultimately result from the interconnection network being able to provide standby systems between units and to meet contingencies for both demand and supply. The more comprehensive the interconnection, the fewer the number of additional plants needed to cope with such contingencies. It is true, on the other hand, that transmission lines must be built. But given that the investment in a 1000 MW production unit corresponds to the investment required for 3200 km (2000 miles) of very high-voltage lines, the choice is quite obvious. Moreover, cost factors, other than capital, favour interconnection and offer great scope for meeting the load at lowest operating cost. Likewise, the available production during slack hours warrants – through pumped storage schemes – the creation of an hydraulic reserve capacity which avoids the daytime operation of old thermal units, which are costly in terms of fuel.

(b) Little international exchange

Long distance (more than 1000 km) electric energy transmission raises sensitive technical problems, so that transmissions of such nature are effected only exceptionally; for instance, when electricity generation, because of natural conditions, is located far from the consumption centres. This is the case for certain very large hydroelectric schemes. However, whenever possible, the preference goes to long-distance fuel transmission. This is certainly the case for nuclear fuel since its transportation and its storage are both relatively cheap.

Thus, within average size areas (approximately 1000 km in diameter), due to the characteristics of electric power, very specific technical and administrative structures must be set up: the system must be intelligently conceived to provide satisfactory customer service and assistance between production centres and, moreover, a virtually monopolistic organization for electricity sales. At the same time electric power exchanges among countries are exceptional, worldwide.

This constitutes an essential difference from fossil fuels, especially petroleum, which are the subject of substantial international trade. At present, approximately 50% of the petroleum consumed throughout the world travels 12 000 kilometres between production site and place of consumption. As a result, we have petroleum structures and a petroleum world price system which assert themselves on, and conflict with, the domestic scene. Electricity knows no such constraints. There is no world price for electricity. In short, electricity retains a national dimension more compatible with the maintenance of local equilibria (e.g. the level of employment or the safeguard of home currency).

2.3.3 An approach to economic calculation in respect of electric utilities

The description of the complex models used by an electric utility to define optimal planning and management cannot be dealt with in this paper. Using a simplified presentation, an attempt will be made only to show how problems are posed and results are expressed. For reasons of simplification, distinction need be made between two parts:

1. Determination of the total capacity required: the demand being determined (cf. Section 2.1.2(c)) the necessary supply must be calculated, allowing for contingencies.
2. Determination of the nature of the equipment and plant to be built, total capacity having previously been defined: this involves an optimization computation.

(a) Determination of total overall production capacity

Since the electricity producer does not have the flexibility afforded by a storage capacity, demand, at each instant t, must be met by equipment which is incomparably more costly than storage facilities. Concurrently, calculation of the overall production capacity must resort to probability computation and allow for two essential concepts:

1. Demand and production randomness: indeed, if the reasoning is based on average data, there is a 50% chance that the overall production thus determined will be inadequate.
2. The convention of an allowed deficiency risk. If absolute warranty for meeting demand were required, some production units would be used only exceptionally (a few hours for a twenty- or thirty-year life duration) and, consequently, the cost of the warranty would be much too high –

especially for a public utility service, which is expected to show a proper cost consciousness.

If C_t is the consumption at instant t, P_t the available overall production capacity, the production facilities would be inadequate if $C_t - P_t < 0$. Consequently, if K_t is the allowable deficiency factor at instant t^*, P_t is defined as:

$$\text{Probability } (C_t - P_t < 0) \geq K_t \qquad (2.2)$$

C_t and P_t are random functions, the main random aspects being:

1. For C_t, errors on forecasts and temperature fluctuations.
2. For P_t, fluctuations of hydraulic inflows; unforeseeable breakdowns of any production units and of the transmission system; uncertainties as to new power-plant construction times; various contingencies.

If those uncertainties can be represented by Gaussian laws, characterized by their means and standard deviations, then the function $C_t - P_t$ is also a Gaussian random variable, the mean and standard deviation of which can be calculated. For fixed C_t, it is then possible to determine P_t; satisfying Equation (2.2) by an iterative calculation.

Table 2.2 Determination of capacity

France in 1985 (winter 85/86)
Critical period: day hours of work days of average winter week (from December 15 through March 15). Allowed deficiency risk for such period: 5%

Estimated consumption (energy)	340 TWh
Power output to be guaranteed	
Averge power to be supplied during critical period hours	52.8 GW
Supplemental power required to allow for hazards and an allowed 5% risk	8.6 GW
Total	61.4 GW
Average contribution of plant already existing or decided upon (including equipment failures)	
Hydraulic	12.5 GW
Conventional thermal	25.1 GW
Nuclear	26.8 GW
Total	64.4 GW
Outright allowance for factors not taken into consideration (in particular, the network): 4%	2.6 GW
Effective contribution of the overall system $64.4 - 2.6 =$	61.8 GW
Surplus of power output $61.8 - 61.4 =$	0.4 GW

---

*It has already been said (Section 2.1.2) that a deficiency value, far more workable for calculation purposes, may be substitted for this physical risk.

Through analysis of the 'Probability $(C_t - P_t < 0)$' function, throughout the year, one or more groups of hours during which such probability is highest can generally be determined. If such a group of hours is comparatively stable over a period of time, and in regard to assumptions, it may then be considered as sufficient to define the necessary capacity from Equation (2.2) and the method becomes operational. For instance, if the production system consists only of thermal plants, the condition in Equation (2.2) will be met for all the hours of the year, provided it has been ascertained for the group of heaviest-load hours of the year. This no longer applies when the power-generating system includes hydraulic reservoirs whose function is to alleviate the difficulties of the busiest hours, but which may no longer be available for other difficult periods. In this case, a much more detailed analysis must be used in order to arrive at an acceptable risk balance among the various difficult periods of the year. The most representative period may then be a specific hour or set of hours, in a vulnerable week or season, making due allowance for equipment maintenance requirements.

Table 2.2. is an example of capacity calculation for the year running from 1 July 1985 to 1 July 1986 in the case of France, where hydraulic facilities are quite substantial.

(b) Determination of the overall production structure: application of optimization rules

The total capacity having been determined for each forthcoming year, the contribution of each production technology must be specified. Production with minimum discounted total expenses will then be considered: the choice is made through optimization.

The problem is divided into two parts:

1. Determination of operating conditions at each instant by assuming that the equipment has been defined. To obtain the lowest operating costs, in the absence of constraints*, equipment should be brought on line by reference to operating costs – lowest cost always being given preference. Operating costs are basically costs proportional to the energy generated, that is, to the fuel costs.
2. Determination of investments leading to the lowest total expenditures (investments, labour, operation) taking into consideration the annual operating costs which were previously computed.

To give a clear image of the computation, the annual load curve of demand is often represented by ranking the average hourly loads in decreasing order: the simple curve thus obtained is called a load duration curve. It is convenient for easy comprehensive purposes, but it has a serious drawback in

---

*For a study covering a hydroelectric unit with limited storage capacity and its own maintenance requirements, such constraints play a fundamental part.

Economic principles of optimizing mixed systems

that it provides no information about the chronological order of the changing patterns of demand.

Figure 2.1 shows that the supply curves of each production technology implemented are also simple. For the different hours considered, taking into account the age of the equipment and maintenance operations, each technology supplies a certain capacity, and comes on line in order of operating costs. For instance, given only two technologies, nuclear power plant and oil-fired thermal power plants, then nuclear plants, whose fuel cost is low, operate the lower end of the diagram, while conventional thermal power plants cover the upper part of the diagram. The line separating the two parts of the diagram is an 'almost' horizontal line and the ladder steps reflect the rules of the scheduled shutdowns of nuclear plants.

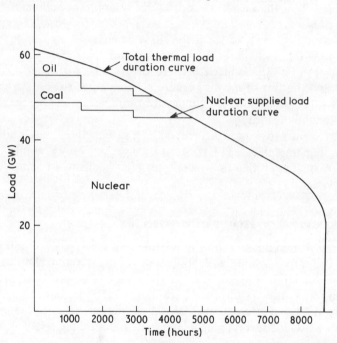

Fig. 2.1 The supply curves of oil, coal and nuclear production technologies (France, 1990).

Thus, it may be seen that to define the sharing of the installed capacity between the two categories of power plants, nuclear and oil-fired thermal plants, the annual U duration, corresponding to the operating limits of the two categories of plants, need be found for each year: the 'last' nuclear power plant, and the 'first' oil-fired plant.

In simple terms, it may be stated that, at the optimum, an equality is ascertained showing that, on the dividing line of the diagram, there is no difference between the total cost of a nuclear kW and the total cost of a thermal kW. If C means the total cost, U is hence defined, each year, by the following equation:

$$C_{nucl} = C_{thermal\ oil}$$

C being explained as follows:

$$C = \frac{aI}{k} + F + Ue$$

where: a is the discounting rate, I is the cost of the installed kW for a new plant, k is less than 1 and is the ratio between the installed capacity and the capacity actually available during the year concerned, F represents annual fixed costs other than capital expenditures (manpower and maintenance), and e represents the operating costs (fuel) per kWh produced, U represents the energy generated for an available 1 kW capacity.

In actual fact, whenever a technology (for instance, the oil thermal technology) must be reduced because of the advent of a new (nuclear) technology, the term I must be taken as being equal to zero for the technology in recession. Such technology, through its mere existence, benefits from an economic rent.

Finally, as already stated in Section 2.1.2, such methodology is not applicable to the complex cases of Western Europe; it is only presented here in order to convey an idea of how optimization is effected over a period of several years.

2.3.4 Nuclear energy economic characteristics

The foregoing comments show how difficult it is to speak of unit cost per kWh in order to assess the profitability rates of the various technologies; unit cost depends on the capacity level at which the kWh is produced in the load diagram, or on the U utilization period of the thermal power plant, which provides the capacity concerned. However, in practice, such thermal plant does not exist, for several plants intervene to ensure the relevant supply in relation to various events. The unit cost concept then loses its meaning if its definition is not carefully specified.

If some appreciation is to be gained of nuclear power profitability compared to conventional thermal plant, the cost of the kWh for an electricity supply at the *base of the diagram of demand* (that is, 8670 hours per annum)* might usefully be considered. Table 2.3 summarizes the results when the other assumptions are as follows:

1. Currency assumed to be constant (with readjustments for inflation).

---

*Distinction must be made between the demand diagram and the supply diagram. It is obvious that a given plant cannot operate without interruption throughout the year.

Economic principles of optimizing mixed systems

2. Costs are given in indices, index 100 being the estimated cost of the nuclear kWh in 1980.
3. The cost involved is a discounted average cost covering the life duration of the power plants with a 9% discounting rate.
4. Fuel prices are discounted average prices allowing for future escalation of fuel price (for instance, 2.5% per annum for liquid fuel with the 1980 projections as a base).
5. The costs given are those which were forecast early in 1972 (prior to the petroleum crisis), 1975 (after the crisis), and 1980; the corresponding nuclear plants cannot be put into operation until several years later (5 years in 1972, 10 years in 1980).

Table 2.3 Projected costs of the nuclear (PWR) and conventional thermal kWh. (Power plant operating at the base of the load diagram, constant currency, in indices with 100 = nuclear power cost in 1980)

Date of projection Date on which the plant becomes operational	Jan. 1972 1977	Jan. 1975 1980	Jan. 1980 1990	
Nuclear kWh				
Capital expenditure	32	36	50	
Operating costs	12	15	19	
Fuel cost (including processing)	15	19	31	
Total cost A	59	70	100	
Conventional thermal kWh	Oil	Oil	Oil	Coal
Capital expenditure	19	21	33	38
Operating costs	10	14	18	21
Fuel costs	31	93	197	106
Total cost B	60	128	248	165
Ratio of A to B costs	0.98	0.55	0.40	0.60

From Table 2.3 the following comments can be made:

1. From 1974, the nuclear kWh is highly profitable, as compared to the oil-fired kWh. It is approximately 50% less expensive; its total cost is even very much less than the cost of the fuel alone to produce the fossil-fired kWh. Compared to the cost of the coal-fired kWh, the cost of the nuclear kWh is 40% less. Naturally, as explained above, the advantages will lessen for equipment not operating at the base-load portion of the diagram, and there is a limited utilization period for which coal becomes a better proposition than nuclear energy. It should also be noted that before 1973, the cost of the nuclear kWh was substantially equal to the cost of the oil-fired kWh.
2. The cost of the nuclear kWh breaks down into 50% for capital expenditure and only 15% for uranium which, for Western Europe, must be

regarded as only partially imported. On the whole, the cost of the kWh contains approximately 85% of national added value, corresponding to technologies indigenous to industrially developed countries. By contrast, the cost of the conventional thermal-fired kWh involves 65% (in the case of the coal) to 80% (in the case of petroleum) of fuel costs, which for many Western European countries are a foreign exchange cost. The fast reactor will produce an even larger differential because the portion of the fuel in the cost of the nuclear kWh will then amount to only a few percent.
3. With such a cost structure, even if uncertainties about the operating conditions of nuclear reactors remain, the future cost of the nuclear kWh is known with far greater accuracy than that of the petroleum or coal-fired kWh. The price of fossil fuels, especially hydrocarbon fuels, on the other hand, may be subject to hazards of a political nature which nobody can assess (cf. Section 2.3.2(b)).

2.3.5 Implications for electricity generation conditions and electricity rates

The cost advantage of nuclear plant should result in those countries which adopt optimal planning substituting nuclear electricity for conventional plants for all base-load production. In the long term, the borderline between nuclear energy and coal will be around $U = 4000$ hours In the case of France, where nuclear programmes have been very substantial since 1974, the electricity production pattern is expected to evolve as shown in Table 2.4.

Table 2.4 Electricity production pattern in France expressed as a percentage

	1975		1980		1985		1990	
	Capacity	Energy	Capacity	Energy	Capacity	Energy	Capacity	Energy
Hydroelectric	38	34	32	25	25	19	22	14
Conventional (thermal):								
Coal	25	19	24	27	17	18	12	9
Hydrocarbon fuels	30	38	23	25	15	6	10	4
Nuclear	7	9	21	23	43	57	56	73
Total	100	100	100	100	100	100	100	100

Provided the rate structure reflects unit cost, this pattern has two major implications for electric power rates:

1. First there will be a progressive but marked divergence of the average cost per kWh from the fossil fuel average cost: it is obvious that when the nuclear production portion represents 57% of overall production, while

the conventional thermal portion represents only 24% (which will be the case for France in 1985), the average cost, having regard to the divergence in cost referred to for basic production, will be substantially affected. Hence, while fossil fuel world prices will continue to increase, the kWh price, after having reached a 'peak' when the first nuclear power plants were commissioned, will drop to lower levels. Figure 2.2 shows the evolution of the price per kWh and per 1000 kcal to large industrial users in France in constant currency in indices (base $100 =$ price of the kWh in 1980). It shows how during the 1980–1990 period, the price differential between the kWh and fossil fuels will assume a scale completely different from that prevailing prior to 1970.

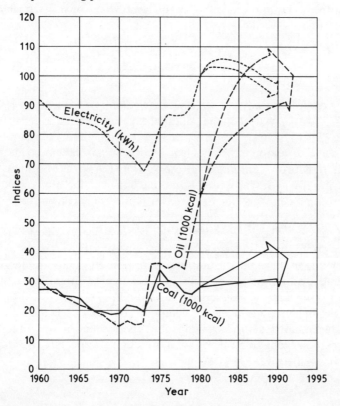

Fig. 2.2 Price evolution of energy delivered to large industrial users (about 100 000 toe/year). The case of France. Constant currency. In indices; base 100: price of the electric kWh in 1980. Note: (1) in regard to electricity, this is the price of the electric kWh; and (2) in regard to fuels, this is the price per 1000 calories (HHV).

2. Secondly there will be an increasing spread between rates at different times of the day and periods of the year. Let us examine, for instance, the case of the day hours of winter time and summer time: the marginal cost

of many summer hours (not all of them) includes only fuel costs. On the other hand, the marginal cost of winter hours includes, in addition to fuel costs, fixed equipment and labour costs. The very different cost structure of the nuclear kWh and of the fossil kWh (see Section 2.3.4), will produce a very marked change in the marginal cost ratio. Thus, in a system with an optimum balance between nuclear and conventional production, the marginal cost ratio of the summer time kWh (marginal nuclear cost) to the winter time kWh (marginal coal or oil cost) is likely to be of the order of 10. On the contrary, in a system where the nuclear generated kWh plays no part, the ratio approximates to 3. In short, when fixing electricity tariffs for a system that includes nuclear generated power, the user should be informed that there will be many hours in the year when very favourable rates are available.

2.4 CONCLUSION

Given a completely new relationship between future kWh and conventional fuel rates, we can reflect on the potential electricity market referred to under Section 2.2. The changes in rates now taking place as a result of the progressive development of nuclear plants, within the electricity systems, provide reassuring guarantees for users, while the prices of traditional fuels and especially hydrocarbon fuels, closely linked to the uncertainties of the international scene, are due to increases at rates which no one can assess any longer.

The impact on the allocation of the energy market among the various sources of supply will be very substantial, since electric power should assume an ever larger share in consumption in response to nuclear energy expansion. The absolute cost of electricity should tend to diminish as the size and number of nuclear plants increase and from the productivity of the electric network. In short, the nuclear–electricity combination is the ideal one for developed countries who lack indigenous energy but who wish to continue to develop their technological potential, to retain a respectable position in the world of tomorrow.

3

T. W. Berrie

NUCLEAR AND NON-NUCLEAR ELECTRICITY: DECISION-MAKING FOR OPTIMAL ECONOMIC PERFORMANCE*

3.1 INTRODUCTION

Forecasting the values of a large number of variables is at the heart of all planning, and planning must always, therefore, involve some uncertainty in its outcome. The electricity supply sector is one of the most difficult for which to plan because of the long 'load-time' needed, from obtaining the go-ahead to order the plant from all concerned, to the completion of construction. However, one cannot say that, because of this, planning in such a sector bears little relationship to planning either in (say) the manufacturing sector or in one's own every-day life. Such a viewpoint would show a schizophrenic attitude to planning in the electricity supply sector. 'Spot' pricing described later in this chapter, helps to 'normalize' the power system planning process.

Planning means making predictions about: the long-term and the short-term; the quantifiable and the unquantifiable; the predictable and the unpredictable; the palatable and the unpalatable. Also, whether one is conscious of it or not, one is aiming at a standard of reliability of out-turn. Normally it requires extra capital to be spent on equipment, generators, transmission lines, distributors, etc., to improve the standard of reliability. The question always is: is this extra capital worth it? Planners must try and answer this question quantifiably; mathematically if possible, but at the very least, empirically. The setting of an economic optimum level of reliability is an integral part of planning; this is followed up later on in the chapter.

3.2 POWER SYSTEM PLANNING: THE PLANNING CYCLE

To hold a balanced perspective, the planning process in any sector must be regarded as a cycle [2]. Although a number of events are usually going on

---

*This chapter owes a lot to the notes prepared by the author for the Intercollegiate M.Sc. course in Electrical Engineering, London University, 1980 and used by permission of Imperial College, London, UK. For further information the reader may also find it profitable to consult [1].

simultaneously, for an electricity system there is an easily recognizable planning cycle with decision points at (1) load forecasting; (2) setting the size of the planning plant margin to cover both uncertainty and obtain a given standard of reliability; (3) project selection from a number of alternative projects; (4) project justification of the chosen project(s); (5) project design and construction; (6) project operation; (7) ex-post project evaluation; (8) project demolition. At each of these decision points, there is a 'learning curve' which ensures planners making better decisions in the future. The important function in the planning cycle of ex-post evaluation [3] has only recently been fully recognized. Among other things it enables lessons learned to be codified, evaluated and digested for future use.

In the electricity sector the planning cycle is broken into at the load-forecasting stage, i.e. on the 'demand' side. In industrial sectors the equivalent break is often made on the 'supply' side, by calculating the costs of production for making various quantities of goods, putting on a 'mark-up' for profit and overheads to get prices, deducing the size of the market at these prices, recalculating the costs of production for the likely volume of sales, readjusting the prices accordingly and thus closing the loop. The electricity sector doing things 'the other way round' often confuses the industrial economist working for the first time in the electricity-supply sector.

Because planning is a cycle, somehow one must take into account the supply side when dealing with the demand side and vice versa. This is especially true when determining the economically optimum-system reliability standard in electricity planning; the point is expanded later.

3.2.1 Load-forecasting period

In this chapter we use the normal definitions of the electricity supply sector. 'Demand' or 'power' is measured in kW (1000 watts) or MW (1 million watts). It assumes that somewhat theoretical concept of a maximum demand made on the electricity system, which is associated with plant capacity. 'Consumption' or 'electrical energy' measured in kWh or MWh has a time dimension. It relates to plant output or total production of electricity over time. Both factors cannot exist on their own. Their combination is defined by the electricity system or plant 'load factor' or 'utilization factor'.

It is dependent upon the type of electricity supply system in question and the type of planning problem whether the short, medium or long-term forecasts are the most important. Interconnected electricity systems often have a long-established average rate of growth in load, e.g. the doubling every ten years of the electricity system in England and Wales up to the late 1960s. Such systems are broadly speaking in an 'equilibrium of growth', taking one year with another. For this type of system the medium-term (5 to 7 year) forecasts of the system maximum demand (kW) and the electrical

Decision-making for optimal economic performance

energy consumption (kWh) are the most important for planning. Usually the system has reached a sufficient size and integrated structure that large base-load, conventional steam or nuclear generating plants are being added continually to the system. Such plant takes 5 to 10 years to go from selection to operation, given sociopolitical delays. In the case of power systems which are small, but growing rapidly, as in developing countries, short-term forecasts for the next two or three years ahead are the most important, for this is the lead-time for the installation of the type of generating plant being installed on such systems, i.e. gas turbines and diesel plant. When, as is becoming true in many developing countries, hydroelectric power projects are important, then the medium-term forecasts must be extended a few years, to ten years and beyond, to accommodate the lead-time for hydroplant. Also, developing countries do not have the dense transmission network of developed countries, and local geographical variations in demand assume much more importance. Scale affects, e.g. a single large new load, may be a significant part of demand growth.

When considering the economic justification for a plant-extension programme (see later) some approximate load forecasts are needed for the long-term, i.e. for (say) every five years ahead, over the economic life of the plants, in the plant extension programmes. This means making load forecasts for about 15 to 20 years ahead for systems containing mainly conventional and nuclear thermal plant, and as much as 50 to 60 years ahead for systems with appreciable hydroelectric plant. 'Spot' figures for even later years, in each case, may be useful to obtain perspective.

3.2.2 Uncertainty

Load forecasting introduces one of the greatest uncertainties into the planning cycle. In many planners' minds, errors in load forecasting simply grow larger the further ahead the forecasts are being made, starting from small errors for next year. A good example of this was shown by the author himself in 1967 [4], for England and Wales, reproduced again here as Table 3.1. This table has been brought up to date since 1967 by several authors, both for England and Wales and for other countries.

The only truly comprehensive study of load-forecasting errors which the author can find was done by the World Bank in 1969–70 [5] for a wide variety of countries. It estimates for these errors their 'random' (probably unforeseeable) and their 'systematic' (probably foreseeable) components. After looking into forecasting methods and the data used in a number of cases in developed and in developing countries it suggests steps that can be taken by the planner to improve the accuracy of load forecasts. The paper also gives one approach to calculating the degree of risk involved when making a load forecast. Readers interested in such detail should consult reference [5].

The World Bank study shows that, taking into account equally all the

Table 3.1 Errors in forward estimates of maximum system demand (in standard weather conditions) in England and Wales expressed as a percentage of the estimates. + indicates over-estimate, − indicates under-estimate. Reproduced by permission of the *Electrical Review* [4]

Year of estimate	Number of winters ahead to which estimates relate							
	1	2	3	4	5	6	7	8
1947	−3.4	−3.3	−7.1	−8.8	−5.4	−6.0	−6.4	−8.5
1948	−3.6	−8.0	−9.4	−6.3	−7.3	−8.0	−10.3	−14.5
1949	−6.6	−8.6	−4.8	−4.9	−4.9	−7.5	−11.0	−11.9
1950	−6.3	−1.9	−1.5	−1.2	−2.7	−5.7	−6.8	−8.7
1951	+2.1	+3.2	+3.0	+1.2	−2.2	−2.6	−3.8	−6.6
1952	0	+1.2	+0.6	−2.8	−2.6	−4.4	−6.6	−9.8
1953	−0.6	−2.4	−4.5	−4.8	−6.5	−8.6	−12.3	−16.4
1954	−1.8	−3.9	−4.2	−5.4	−8.1	−11.3	−15.9	−16.1
1955	−2.0	−1.8	−3.5	−4.8	−7.8	−11.3	−14.9	−17.7
1956	+0.8	−0.1	−1.1	−4.2	−7.3	−10.7	−13.3	−12.2
1957	−0.4	−1.0	−3.8	−6.8	−9.9	−12.4	−11.4	−11.2
1958	−1.0	−3.2	−6.2	−8.9	−11.3	−10.2	−9.8	−10.5
1959	−2.4	−5.0	−7.2	−8.8	−6.9	−5.6	−5.1	−1.6
1960	−1.6	−4.2	−6.1	−4.6	−4.0	−3.6	0	
1961	−0.7	−1.4	+0.8	+2.6	+3.4	+7.1		
1962	+0.7	+3.6	+5.8	+7.3	+11.5			
1963	+4.2	+7.8	+10.4	+15.6				
1964	+3.7	+6.8	+12.4					
1965	+1.0	+5.4						
1966	+5.0							
Mean error	−0.6	−0.9	−1.5	−2.7	−4.5	−6.7	−9.1	−11.2
Standard deviation (%)	3.1	4.7	6.1	6.7	7.2	8.5	10.5	12.5

countries examined, percentage errors in both kW and kWh forecasting have a 'normal' distribution around an overall mean of zero. This means that forecasts, taking one country with another, are just as likely to be too high as too low. One interpretation that can be put on these findings is that forecasting is done in an unbiased way, in the sense that it has a normal distribution. Another interpretation, however, is just as possible but much more worrying, i.e. in the degree of randomness of the possible errors. This invites the question: why spend much time on forecasting, bearing in mind the likely level of uncertainty? However one interprets the overall findings, the errors in some individual cases can be frighteningly large [5] given the numbers and the likely circumstances. Also, the same orders of magnitude of errors seem to apply throughout the range for making of forecasts of 1 to 10 years (and probably further) ahead, for both large and small electricity supply systems, and for systems in all rates of growth.

Unfortunately, from the evidence of this World Bank paper, it would appear that if a forecast is high (or low) in the short-term it is just as likely to be high (or low) in the long-term. The prospects for deliberately and systematically reducing errors over a long-term period are likely to be higher

Decision-making for optimal economic performance

than those for reducing errors in the short-term forecasts, just because we have more time to learn to make the necessary forecasting adjustments. It is vital whenever possible to quickly address and correct 'systematic' errors, i.e. errors which, in principle, can be reduced.

One would expect that those new to forecasting would make the worst forecasts and that they would make better forecasts given time and experience. The World Bank paper indicates that this is indeed true and that forecasting errors in developing countries are in general about twice those in developed countries. More ex-post evaluation of the reasons behind making load-forecasting errors is needed.

It is somewhat alarming that, in general, forecasting accuracy does not appear to be improving with time. That is, uncertainties introduced by forecasting in the 1950s seem to be no greater than those introduced in the latest figures examined, by forecasting in the first half of the 1970s. This again brings up the question raised before: why spend a lot of time on forecasting? Each individual planner must answer this question for himself but, as long as load forecasting remains in the planning cycle, the best estimates for load growth must be made.

3.2.3 Probability and sensitivity analysis in planning

The outcomes from the planning cycle must be tested against uncertainties introduced by the main parameters. This can be done by a simple sensitivity analysis [6], i.e. changing each parameter in turn by a specified amount to examine the effect on the outcome from the planning cycle. Some combinations and permutations in changes of parameters are usually worth trying. Should a particular case in question warrant it, and if the data is known, probabilities can be placed [7] upon whether any particular piece of data will have a particular value. Probability distributions can then be used for the outcome from the planning cycle and the uncertainties covered in this way.

As variation in load forecasts is one of the most important sensitivity analyses which should be done for power system planning, it is also imperative to carry out appropriate sensitivity tests when putting together the load-forecasting process itself, i.e. whilst all assumptions made during the forecasting process are fresh in the mind.

3.3 POWER SYSTEM RELIABILITY

3.3.1 Standard of reliability: planning plant margin

It is convenient to deal with the 'planning plant margin' at this stage in the planning cycle. The question of what degree of uncertainty to allow when deciding upon how much spare (i.e. the amount of generating-plant capacity (kW) over and above the level of the system maximum demand) generating plant capacity to install, is still usually decided upon by a mixed analytical and empirical approach. The degree of uncertainty can be defined in a number of different ways, e.g. the number of hours (or years) in ten or one hundred when:

1. All load (kW or kWh) will not be met at full system voltage and frequency (also giving how many kW or kWh will then be lost if possible).
2. All load (kW or kWh) will not be met even at a technically acceptable safe, but reduced, voltage and frequency (also giving how many kW or kWh will then be lost if possible).

The larger the margin of spare capacity built into the system at the planning stage, often called the 'planning plant margin', the greater the investment required but the less the risk of failure to supply all the kW or kWh required. The smaller the 'planning plant margin' the smaller the investment required but the greater the risk of failure to supply all the kW or kWh required.

In most cases of electricity planning it is still usual to deal with the question of what planning plant margin is required somewhat empirically, by considering as benchmarks (1) either the standard of *actual* performance of meeting the *actual* load in the past which has been considered 'satisfactory' or not by consumers and (2) whether they would be prepared to pay for a more reliable system. If it is considered that the past performance has not been considered 'satisfactory', then an adjustment upwards to the planning plant margin is indicated. (This does not mean necessarily that the planner should always make an upward adjustment, see Section 3.3.2.) If it is considered that past performance has been considered too 'satisfactory', i.e. there has been obvious over-capacity for a few years and the consumers would like a reduced reliability with a reduced tariff, then a downward adjustment in the planning plant margin is indicated. It is worth noting that it may not be known *why* the performance has been considered more or less than 'satisfactory'.

It is usually possible, however, to consider analytically in what proportion any shortcomings in the standard of 'satisfaction' has been due to uncertainty introduced by:

1. Generation and network plant outages, either 'planned' by maintenance schedules or 'forced' by failure in service; the 'loss of load probability' due to forced outages can usually be estimated and quite sophisticated methods are available [1].
2. Late construction of generating and network plant.
3. Errors in load-forecasting, for whatever reason.*

*In the U.K. and some other countries, where climate plays a large part in the actual amount of electricity demanded and consumed, it is customary to make forecasts for 'average' climatic conditions. Even although this practice is used throughout the year, in the U.K. the standard climatic condition is known as 'average cold spell'.

4. Unforeseeable changes in water storage for hydroelectric plant.

The economic optimum planning plant margin is determined by comparing the marginal cost of lost production of industry and commerce, plus the leisure costs of households, whenever load is shed, with the marginal cost of meeting that load, i.e. the marginal cost of adding plant capacity so that the load is not shed.

In the 1980s about 12% to 15% of public sector investments in developed countries, and approaching 20% to 25% of public sector investments in developing countries are expected to be for electricity supply. The scarcity of investment capital, and in the case of developing countries especially foreign exchange, indicates the need to develop a more rigorous framework in the economic optimization of electric power sector reliability.

Most economists deal with this subject under public-utility pricing, because the problem is looked upon as being within the compass of 'welfare economics'. The planning engineer does not see it like that. As mentioned at the beginning of this section, he sees the subject of determining the optimum planning plant margin as adding something to the load forecast because the total capacity of plant planned to be installed must be equal to the load forecast *plus* the optimum planning plant margin: we can accommodate both the economist and the planning engineer by saying that:

1. Economic efficiency is optimized when price is set equal to the marginal supply cost – the economist's test under 'welfare economics'.
2. Plant capacity and, therefore, the planning plant margin and the reliability standard are optimized when the marginal supply cost of adding to the plant capacity equals the marginal consumer outage costs avoided by adding to the plant capacity – the planning engineer's test.

3.3.2 An economist's approach to reliability

A new economist's approach to determining the optimum reliability of an electricity supply system is taken by Mohan Monasinghe [8] of the World Bank. Monasinghe's approach helps to get over one of the fundamental problems encountered by all electricity economists, namely, how to treat electricity as a good or a service which has a value *per se* to the consumer. Putting a number on this value has tended to defeat all concerned although many attempts have been made. Electricity shortage has a cost measured in economic terms primarily by a reduction in consumer surplus, i.e. a net reduction in a consumer's (including his business) well-being.

Monasinghe gets over many of the traditional problems by treating electricity as an intermediate product, i.e. it has no consumer surplus (direct consumer satisfaction) of itself. The cost of reductions in voltage and frequency and complete loss of supply are then measured directly in terms of the loss in goods or services which are caused by those. By this means the main difficulty of all electricity sector economists appears, at first sight anyway, to be avoided, i.e. the difficulty of having to use consumers' proven willingness-to-pay for electricity consumption as a measure of benefits forgone by an electricity supply failure. Normally costs of outages will greatly exceed such proven willingness-to-pay for most electricity usages. Yet many planners still use the present level of the retail tariff to calculate lost revenue as a minimum estimate of electricity outage costs.

The problem of how to take into account outage-time in outage costs has always been a difficult one to deal with. Planners sometimes estimate outage costs by the product of average value-added per kWh of electricity supply, times the kWhs lost in outages during working periods. This not only ignores the likelihood of making up any lost production later in the industrial sector, normally the only sector considered in detail, but the calculations made are seldom tested empirically in practice. Because of the latter cause alone, such outage costs are heartily disbelieved by most planning engineers.

Monasinghe produces a convincing list of 20 principal industrial types of electricity consumer, whose outage costs have been worked out on the assumption that electricity is an intermediate good and, therefore, the outage cost has the net value of the loss in the final good, after allowing for make-up time, as its main criterion. He has a more unconventional approach to households than he has to industry, by treating them as having productive outputs in the form of (1) home utility, (2) cooking and (3) leisure. There are very few substitutes for electricity which are associated with leisure in the evening. The main 'opportunity cost', i.e. alternative-use-of-time cost and therefore the outage cost, is thus the cost of time lost. This should probably be measured at the household's net income earning rate during the hours that the household is at leisure. One instinctive worry of this approach to the planning engineer is that it gives a value for the outage cost considerably higher than the one he is accustomed to, i.e. lost revenue derived from the existing consumer tariff; nevertheless it is a better approach than his.

Taking his own approach Monasinghe optimizes the long term power system development programme somewhat differently to the planning engineer, at least in the arithmetic; he optimizes the *total* costs, i.e. the normal electricity supply system costs, (as taken by the planning engineer) *plus* the outage costs. This 'internalizes' the calculation for the optimum standard of supply and the optimum planning plant margin mentioned above, and it brings this rather cumbersome criterion into the calculation of what is the optimum plant extension programme, which will be described later in this chapter.

3.3.3 Plant and load balances

The object of making generating-plant extension programmes is to equate demand and supply at an optimum economic standard of reliability. The specific points of time ahead at which they are planned for being equated, depends mostly on the lead-time needed to bring a new generating plant into commission from the date that the investment decision is made. This is usually longer than the lead-time for transmission and distribution plant and is three to ten years, according to type, e.g. gas turbine, diesel, steam, nuclear or hydro. This lead-time includes the time taken to obtain the necessary financial and administrative approvals.

We put together generating plant and load balances to get from the existing situation to that required for the equating of plant and load at some point(s) of time in the future. The balances at this stage on a large system will be purely numerical, i.e. they will not give much consideration to the type and size of plants in the balance as such things have only a marginal effect on the balances, except in the case of developing countries when indivisibilities are important. The balances may need to be adjusted for these factors in any case at a later stage in the planning cycle. On medium and small systems due attention must be given to plant type and size at the stage of forming the first balances, because such effects are not marginal; systems build-up can be affected fundamentally by the particular type and size of the additional plant. There may be some years when the plant and load balance is noticeably over or under par, because of indivisibilities, e.g. the optimum size of additional plants is larger or smaller than that needed to keep the plant and load just in balance.

3.4 POWER SYSTEM ECONOMICS ALTERNATIVE TECHNOLOGIES

Once we know the size of the approximate plant extension programme we can take a first look at the 'plant mix', i.e. the choice of technology to use, expressed in generating plant by size and type of fuel, in networks by voltage level and conductor size.

3.4.1 The basic situation

If we were putting together the mix of generation *ab initio*, all equipment being made from the same vintage of drawings, we would use a quite simple method [9] to choose the optimal. The plants would differ in their investment cost to fuel cost ratio, and in their technologies, but not in their degree of obsolescence. There would be no 'sunk' costs of capital to take into account. On large and medium-sized systems we would take low capital cost, high fuel cost, gas turbines and diesels plus peaking hydroelectric stations (if available) for the peak demand. We would take a fairly large block of highly efficient, low fuel-cost plant to meet the base-load. Finally we would fill in the times between base and peak demands, with a mixture of plant. On a small system the size of the system itself would rule out some types of plant altogether because of the size needed for them to be economic, e.g. large steam and all nuclear plant. Our basic criterion throughout would be to minimize discounted total (capital plus running) system cost throughout the planning period.

We seldom, however, deal with an electricity supply system *ab initio*. The best we can do is to determine what seems likely to be the best plant-mix for some particular year(s) ahead, e.g. when the influence of the existing plant is quite small. This is almost the same as planning *ab initio* and the time ahead can be quite short in developing countries, when the demand for electricity is often doubling every three to seven years. Having taken a view of the optimum plant-mix as seen from that year(s) ahead we can then examine how to get there from the present plant-mix.

An actual plant extension programme takes account of five main things:

1. The present plant-mix is non-optimal because of errors made in the past.
2. The possibility of using existing plant more economically rather than acquiring new plant which is especially suitable for a particular time of system loading as described above.
3. The short term effects of the investment on the financial cash flows of the utility.
4. The investment (foreign exchange and local) requirements necessary to implement the development programme.
5. The economic return on the investment.

Many electricity planners are not required to go further than (2). Although this situation is starting to become more rare, only (1) and (2) are appropriate to this chapter.

3.4.2 Fuel cost savings

Any method of determining the size and composition of a generating plant extension programme uses the concept of 'fuel cost savings'. This is not a simple concept for those not familiar with the economic operation of complex systems. As a result of (1) technological advance and (2) changes in the efficiency of using fuels, generating plant designed today usually has a higher thermal efficiency than a similar existing plant. The new plant's fuel costs per kWh are therefore lower than those of similar existing plants.

For the economic operation of an integrated system, generators are given an 'order-of-merit' for being operated sequentially, i.e. an order in which to be brought onto the system to meet the system load in kW and kWh at any period (Fig. 3.1). One obviously uses the generation with the lowest fuel cost

Decision-making for optimal economic performance

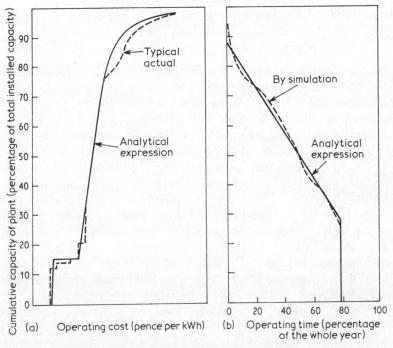

Fig. 3.1 Generating plant capacity/operating cost curve and generating plant capacity/operating time curve. Reproduced by permission of the *Electrical Review* [4].

per kWh generated as much as possible.* This 'merit-order' has nothing to do with the capital cost of the generating plant, for that is a 'sunk' cost which has been committed whether the plant is operated or not. At the top of the merit order are the generators with the lowest fuel costs per kWh generated; a typical merit order would look something like this, depending upon the system in question.

Operate first: Nuclear plant, run-of-river hydro-plant.
Operate second: Modern large coal-fired steam plant, combined-cycle plant, geothermal plant.
Operate third: Modern large diesel, relatively modern and medium-sized coal-fired steam plant, hydro-plant with spillage over a dam.
Operate fourth: Modern large gas-turbine plant, older small coal-fired plant and modern large oil-fired steam plant.

---

* In the case of hydroelectric generation, the amount of energy output in kWh is limited by the water available. Their operation is usually reserved, therefore, for the time near system peak demand, when the generating stations most expensive to run would otherwise be required to operate.

Operate fifth: Medium and small oil-fired steam plant.
Operate sixth: Small gas-turbine plant, large dam capacity hydro-plant with spillage.

Nuclear plant and modern, large, coal-fired plant will normally operate at 'base-load', i.e. as continuously as technical considerations like maintenance allow. Plant with the highest fuel costs per kWh generated will be brought in as infrequently as possible, i.e. only when they *must* be operated to meet a peak demand. Hydro-plant will be operated to make the optimum savings in fuel cost taken over a period, normally a year, in other plant.

The most significant saving made by new plant is the saving made by an integrated change in the merit order. When a new plant is installed it has an effect upon the merit order to an extent depending upon the difference in fuel cost per kWh generated between the new plant and the existing plant. In this respect the introduction of small gas turbine plant will make little or no change to the existing merit order. On the other hand, the introduction of base-load plant will have an effect on the whole merit order, for the new plant displaces the plant which was previously the highest in the merit order, which therefore is called to run for less hours in the year than it otherwise would have run. This effect cascades down the merit order, with all plant below the new base-load plant in the merit order displacing the plant just below it. The latter plant therefore runs somewhat less. The summated result is a saving in total system fuel-cost due to the introduction of the new generating plant, i.e. a summated fuel-cost saving.

For plant designed to run at load factors between base and peak-load, the fuel-cost savings will be less than those for base-load plant. The amount that these fuel-cost savings are less will depend upon the position of the new plant in the merit order. The actual load factor at which a plant will run is thus a prime factor in its economics.

For one year, on even a medium-sized system, the arithmetic of working out the fuel-cost savings by hand is formidable enough; yet assessments of fuel-cost savings are required throughout the economic life of the new project. Also, future plant introduced after the new plant with which we are presently concerned is likely to be even more efficient than the latter. Thus the fuel-cost savings as defined above, made by the introduction of a generating plant today, will tend to decrease over its life due to the influence of future plant, making calculation impossible unless carried out on a digital computer (see later). The computer simulates the merit order loading of all generating plant on the system, theoretically for every day of every year of the economic life of the new plant. In practice, however, it is usually done for 'spot' days and 'spot' years only; the fuel-cost savings for intermediate days and years are then found by interpolation.

The annual fuel-cost savings over the economic life of the plant are usually present-valued at the test rate of discount (as laid down usually by government for calculations done at constant price levels). The present-valued annual fuel-savings are then summed to a total present value, which is then either used directly in any economic comparisons or annuitized over the economic life of the project at an interest rate equal to the test rate of discount. On large power systems it is often permissable to work incrementally, i.e. to consider an increment of kWh of each alternative type of generating plant by fuel. If we ignore load growth (see discussion of climate influence, p. 72), because we are dealing with an increment, then it is the fuel cost savings made by a generating plant which are paramount and *not* the fuel costs *per se*. This is not true for any medium or small power system, and for some large power systems not in growth (see later and reference [1]).

3.4.3 Optimum plant-mix in the background plan

In order to determine the optimum generating plant-mix at any specific point in time, we can look at the ratio of capital costs to fuel costs of various

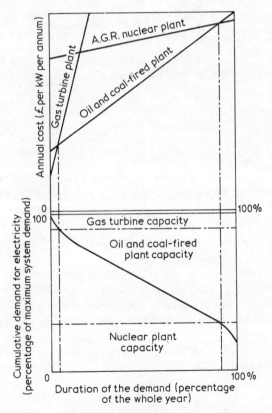

Fig. 3.2 Determining the plant mix. Reproduced by permission of the *Electrical Review* [4].

plant types. This is best done by constructing the diagram shown in Fig. 3.2, which combines the decision on which type of plant is optimum for different load factors of operation with the means of deciding the optimum proportion of any particular plant type in the plant-mix. (Strictly speaking a cumulative plant capacity curve should be used and not a cumulative demand curve, but this point is outside the scope of this chapter.) This gives an optimal plant-mix at the specific point in time. Any generation-plant extension programme must contain the right plant to be added to proceed from the existing plant-mix to this future optimum plant-mix.

If an optimum plant-mix is determined as above for a number of different years ahead, say 10, 15, 20, 25, 30, we can draw up a 'background plan' to show the optimum proportion of each *general* (e.g. gas turbine, oil-fired, nuclear, etc.) type of plant in the mix which should exist on the system at any discreet point in time, at least under present assumptions. A specific plant extension programme is then any *named* plant contained in the first few years of this background plan. Any disagreements between optimum plant-mixes at the discrete points in time due to practical considerations (e.g. plant, once built, must appear in subsequent mixes) can be dealt with by dynamic programming on a computer.

3.4.4 Marginal analysis and robustness

By repeating all or part of the planning cycle (see p. 67) improvements can be made to the background plan. Sensitivity of the plan to different assumptions concerning load forecasts, fuel prices, capital costs, test rate of discount, etc., is an indication of robustness. It is normal to 'refine' the background plan by making small changes in the amounts of the different types of plants by fuel, and possibly by size. Such refining is known as 'marginal analysis'. Any change which reduces the present value of the total (capital plus fuel) system cost over the next (say) 30 years should be counted an improvement and accepted. A model of the electricity supply system is usually needed to do the calculations involved, using one or more of several different types of approach [10].

Some planners believe it is meaningful to carry out the above refining process of the background plan down to a degree of detail equal to introducing, in turn, each particular named generating plant alternative being considered, in order to compare directly the change in total present system cost made by each alternative; the more the present value of the total system cost is reduced by introducing any particular alternative generating plant, the 'better' is that alternative.

3.4.5 The constant load factor comparison method

In some cases a simple outline background plan is all that is needed – for

Decision-making for optimal economic performance

example when comparing total costs per kWh of plants that are direct alternatives to one another, all operating at the same load factors throughout their economic lives. Base values for all costs and the particular load factor which is appropriate come from the background plan. The capital investments for the different alternatives are converted to annual charges, i.e. 'annuities' taking account of economic life and the appropriate rate of interest on capital (this is usually taken to be equal to the test rate of discount). The annual charges for the different alternatives, divided by the total kWh generated throughout the year, are then added to the fuel costs per kWh generated, to give the total (capital plus fuel) costs per kWh for the comparison of one alternative with another. The alternatives can then be ranked in ascending order of total costs per kWh generated.

The basis for providing the alternatives for such comparisons will be the various proven generating plant alternatives on offer today, or likely to be on offer at a common date in the future. These latter are difficult to describe precisely [11] or cost accurately, because of continuing technical change. For instance, when considering now (1982), plant to be commissioned in the late 1980s, planners must ask themselves how far their past experience with respect to capital cost per kW and the increases in fuel prices can be expanded into the future. Moreover, it is recognized to be true that some elements in generating plant design, e.g. with respect to nuclear and the larger conventional thermal plant, have still to be fully tested in practice. An example of the presentation of the costs per kWh at the same constant load factor is given in Table 3.2 for base-load plants. The figures are purely notional, and differences exaggerated:

Table 3.2 Costs per kWh at the same constant load factor for base-load plants

Plant type	Capital cost (units of money/kW)	Fuel cost (units of money/kcal)	Total generation cost at constant load factor (units of money/kWh)
Nuclear 'A'	1050–1130	12.0	2.2–2.4
Nuclear 'B'	940–1020	10.0	2.0–2.1
Nuclear 'C'	810–850	9.0	2.0
Thermal 'D'	550–590	11.0	2.5
Thermal 'E'	550–590	11.0	1.8
Thermal 'F'	550–590	16.0	2.1

The specific difficulties about the constant load factor method are twofold:

1. For the simple method ground rules have to be adopted – including a set of assumptions about the lifetime load factors of different types of plant, say 65% to 70% for base-load plant, 30% to 40% for mid-merit plant and 10% to 20% for peak-load plant. In practice, one may have to decide between a new peaking plant and a new addition to base-load plant: it is

not always a matter of comparing alternatives for the same role in the system. Moreover plants may, during their lifetimes, depart very substantially from the ground rule load factors as new, more efficient, plant coming on line pushes them further down the merit order.
2. The decisions to be taken often depend upon quite small differences (say 6%) between the total generation costs of alternative plants and these differences may be masked by the inaccuracies inherent in simplified methods.

Dealing with difficulty (1) without departing from 'constant load factors comparison' method can often be attempted successfully [12]. When comparing two generating plants, each with different but constant load factors throughout their lives, not too much distortion (as measured against the results obtained from more rigorous methods) is obtained by calculating their 'costs per kWh' at some third constant load factor, chosen by experience, which is intermediate between the two constant load factors of each alternative.

To deal with difficulty (2) without departing from the constant load factor method is more difficult. However, making decisions on the basis of small differences between large numbers when making comparisons between alternative individual generating projects is common to all methodologies. It is always difficult to demonstrate with any degree of confidence that one plant is more economic than another when the difference between their respective impacts upon total costs – for whatever reason – is small. The constant load factor comparison has the virtue of great simplicity and should not be ignored when it can safely be used.

3.4.6 The 'net effective cost' method

A more sophisticated, but still conceptually simple, method of comparing the economic merit of individual alternative generating plants is to use the 'net effective cost' of investment in any new plant. If an incremental approach is valid then, when the value of the annuitized capital charges, taken over the economic life of a new generating plant (plus its associated transmission and distribution), is less than the annuitized total present value of the annual system fuel cost savings made over its economic life by installing it, then the new plant should be installed. Furthermore, the new plant should be installed until its net effective cost becomes positive. Such a test can also give first approximation to the optimum timing of new plant, i.e. by measuring the improvement in net effective cost causal by postponement of the new plant. With respect to judging the most economic alternative generating plant that should be installed for any particular reason, e.g. load growth, plants are ranked in their descending order of negative net effective cost and their ascending order of positive net effective cost.

Decision-making for optimal economic performance

Some factors which influence the validity of whether an incremental approach is valid (see [1]) are:

1. Whether the power system is in a state of steady growth from a large 'base'.
2. Whether plant already installed is very different in size, fuel and technology from the new plant under consideration.
3. Whether introducing a particular plant will seriously disturb the underlying background plan upon which all calculations are based.

It is for the above reasons that the more simplistic methods of constant load factor comparison and net effective costs must be used with great caution, especially with respect to the power systems in developing countries.

3.4.7 Electricity systems in developing countries

Normally, for power systems in developing countries, we must return to comparing different background plans because, in such circumstances, the plan is likely to change fundamentally in accordance with each alternative generating plant being considered; we need a definite different approach to marginal analysis.

The change of approach to the background plan is one of emphasis. Basically it means keeping the first (say) ten years flexible and, from then onwards, leaving the format of the plan in rough 'plant-mix' percentages only (see p. 79). This means that a good deal less arithmetic has to be done than in the net effective cost case; but only a small amount of arithmetic is often warranted under the conditions of uncertainty and the accuracy level of data met with in developing countries.

The change of approach with respect to marginal analysis is somewhat more fundamental. It involves trying each alternative generating plant in turn *as part of* the background plan and testing whether the background plan is improved or made worse thereby, i.e. whether the total (capital plus fuel) cost present-valued over the life of the background plan is reduced or increased. Because of the fluid nature of the background plan in these cases, permutations and combinations of alternative generating plants are permissible, not to say encouraged, for the early part of the plan and care must be taken to ensure that each plant 'sequence' which makes up the background plan carries out the same function, e.g. with respect to the standard of reliability of power system supply.

The marginal analysis is then carried out by refining the background plan itself, rather than by a marginal analysis of an incremental (kW) approach (see p. 80). However, there must obviously be a systematic approach to carrying out the refinement, especially in the light of the testing of the robustness of the optimum solution with respect to uncertainty in the data. Plant sequences can be ranked in accordance with ascending order of present total (capital plus fuel) system cost over the economic life of the sequence.

Strictly speaking, each uncertainty must be tested by sensitivity analyses applied to each alternative generating project and each combination of generating-plant alternatives. In practice many of these tests can be eliminated by inspection. Typical sequences of generating plants to be compared are shown in Table 3.3

Table 3.3 Typical sequences of generating plants to be compared for their economic excellence to meet a given hard forecast

Year	Maximum demand (MW)	Planning plant (MW)	Total plant (MW)	Sequence 'A' installed capacity (MW)	Sequence 'B' installed capacity (MW)	Sequence 'C' installed capacity (MW)
0	500	100	600	600	600	600
1	550	110	660	660 (GT)	660 (GT)	660 (GT)
2	605	120	725	720 (GT)	725 (GT+D)	725 (D)
3	665	135	800	800 (GT)	800 (GT+D)	800 (D)
4	730	145	875	1000 (S)	900 (GT+D)	875 (D)
5	800	160	960	—	1200 (S)	900 (S)
6	880	175	1055	—	—	1055 (D)
7	970	195	1165	1400 (S)	—	1300 (S)
8	1070	215	1285	—	1500 (S)	—
9	1200	240	1440	1800 (S)	—	1500 (S)
10	1320	265	1585	—	1800 (S)	—
11	1450	290	1740	—	—	1900 (S)
12	1600	320	1920	2400 (S)	2100 (S)	—
13	1750	350	2100	—	—	2300 (S)
14	1925	385	2310	—	2700 (S)	—
15	2115	425	2540	2800 (S)	—	2805 (S)
16	2330	465	2795	—	—	—
17	2560	515	3075	3500 (S)	3500 (S)	3300 (S)
18	2800	560	3360	—	—	—
19	3100	620	3720	—	—	4000 (S)
20	3410	680	4090	4000 (S)	4000 (S)	—

In columns 5, 6 and 7, the items in parentheses are the type of plant being added:
GT: gas turbines of sizes (say) 20 MW
D: diesels of various sizes from 5 MW upwards
S: steam plant of sizes, 30 MW, 50 MW, 100 MW, 200 MW.

3.5 TYPICAL OPTIMIZATION GENERATION COMPUTER PROGRAM

A typical optimization computer program is designed to assist the power system planning engineer in achieving an optimum power station planting programme for up to 30 years ahead. The core of the program can be a production simulation facility which calculates the annual running (fuel plus maintenance) cost and reliability of a number of power generation sources on a power system for given patterns of load demand.

Two possible operating modes for such a program are typically (1) to predict the performance of a system for a generation planting programme selected by the user or (2) to invoke the automatic optimization facility which calculates the optimum planting programme on a year by year basis by choosing suitable sizes and types of plant from a list of possible options provided by the user.

The optimum planting programme is that which provides a least-cost solution, subject to specified reliability criteria and, often, environmental constraints. All costs may be discounted back to a chosen year at various discount rates. Reliability is usually assessed either as loss of load probability or planning plant margin.

Data must be assembled for the generating plant and the system load. Generating plant data includes information on capital costs, running costs, reliability, plant life, maintenance scheduling, type of plant, preferred operating modes and environmental factors. It is desirable to be able to (1) model all types of generating plant including thermal, gas turbine, hydro-electric, pumped storage and renewables and (2) alter all data elements on a year by year basis to reflect changes in operating requirements. The load data is frequently entered in terms of annual maximum demands and load factors with additional detail on daily, weekly and monthly variations in power and energy demands.

After data assembly a simulation process is carried out on this type of model. Using both present plant commitments as well as a merit order, the program then simulates the production of electrical power and energy on (say) an hourly basis for each year in question. Using the known load duration curve, plant allocation is carried out to achieve the desired level of reliability. Account is taken of both forced and unforced outages as well as variations in fuel costs and availability of water for hydro-energy.

If an automatic facility is used, shortfalls are met by selecting plant from the list of future planting options. Various alternatives are tried and the least-cost one selected and retained for future years. Alternatively the program may calculate the probable shortfall in energy and loss of load probability for a given planting programme meeting the specified load pattern. A cumulative cost over the planning period is accumulated and discounted to give net present costs.

Loss of load probability in generation reliability evaluation has been calculated by some workers using a straight line peak-load distribution characteristic. This increases the error of either having a higher than normal risk level or a very low risk margin – leading either to inadequate reserve or to over-reserve in the system. The peak-load curve could be modelled as a probability distribution curve defining the probability that the peak-load in a given day will surpass or equal a given value. This technique takes into account the uncertainty in the load forecast. If the loss of load probability curve is shown as a normal distribution, then the mean and standard deviation can provide indices for generation and reserve allocation.

REFERENCES

[1] Berrie, T.W. (1982) *Power System Economics*, Peter Peregrinnes, UK.
[2] Baum, W. (1978) *The Project Cycle*, reprint from *Finance and Development*, World Bank, December.
[3] *Operations Evaluation* (1976), World Bank publication and the *Annual Review of Post-Project Audits*, published each year by the World Bank.
[4] Berrie, T.W. (1967) The Economics of System Planning in Bulk Electricity Supply, *Electrical Review*, 15, 22 and 29 September, and reprinted in *Public Enterprise* (ed, Ralph Turvey), Penguin Economics Classics, 1968.
[5] *Ex-post Evaluation of Electricity Demand Forecasts* (1970), World Bank Economics Department Working Paper no. 79, 18 June. (The reader should note carefully the definition of percentage error used in the paper.)
[6] Pouliquen, L. (1970) *Risk Analysis in Project Appraisal*, World Bank Occasional Paper no. 11.
[7] Reutlinger, S. (1970) *Techniques for Project Appraisal Under Uncertainty*, World Bank Occasional Paper no. 10.
[8] Monasinghe, M. (1980) *The Economics of Power System Reliability and Planning*, A World Bank Publication, March. Published by the Johns Hopkins University Press, Baltimore.
[9] Berrie, T.W. (1967) The Economics of System Planning in Bulk Electricity Supply *Electrical Review 22nd September*, reproduced in *Public Enterprise* by R. Turvey (1968) Penguin Economics Classic.
[10] Turvey, R, and Anderson, D. (1976) *Electricity Economics*, World Bank Publication. Published for the World Bank by the Johns Hopkins University Press.
[11] Posner, M. (1973) *Fuel Policy – A Study in Applied Economics*, Chapter 6.
[12] Berrie, T.W. (1967) The Economics of System Planning in Bulk Electricity Supply *Electricity Review*, 22 and 29 September.

NUCLEAR ENERGY AND THE POWER PLANT INDUSTRY

4

P.C. Warner

MANUFACTURING INDUSTRY AND NUCLEAR POWER

4.1 INTRODUCTION

The role of manufacturing industry in nuclear power is at first sight fairly obvious: it is to manufacture components in a manner which makes sensible use of its resources, which satisfies its customers in terms of specification, quality and delivery, and at a price which enables it to service its borrowings and renew the enterprise. That is the ultimate objective of all its activity.

By the 'manufacturing role' is meant not only the performance of production processes (the physical working of the material so to speak) but all the functions that are necessary for that, both before and after, from conceptual design and development all the way through to installation and setting to work. Thus the phrase 'manufacturing industry' is a shorthand for an activity which includes the engineering of components and often also of systems; and since this is the test of good design, the manufacturer accepts responsibility for the good behaviour of the components in service. It may be taken as evident that engineering and manufacture are not separate.

The theme of this chapter is to look at the implications of being a manufacturer in the nuclear industry. For convenience, the examples have been chosen from the experience of Northern Engineering Industries Ltd, but the points have a general validity.

The manufacturer is involved at two levels. There is the supplier level where the problems tend to be orthodox but take on a special peculiarity when applied to the nuclear market; and there is the different and unique level which might be termed 'nuclear politics', which has a number of aspects including energy policy, choice of reactor system, efficient use of industrial resources, and public acceptability. Something will be said about all these in order to explain how they appear from inside a manufacturing group and what are seen as the important factors. First, however, we need some background.

4.2 THE NATURE OF MANUFACTURING INDUSTRY

The range of nuclear products is wide and many different types of engineering are called upon; the diversity can be emphasized by some examples, starting with the conventional side in order to remind ourselves that well over 50% of the equipment for a power station would be conventional.

You could not tell by looking at switchgear, transformers, or electrical equipment generally what form of power station, nuclear or fossil-fuel burning, they were associated with. Strictly that is true also of a water treatment plant, but by looking carefully at the method of operation you might deduce that you were dealing with a once-through boiler and therefore probably with an AGR. Similarly, with a turbogenerator set you can tell only from details of the feed heating system whether it is for an AGR or a fossil-fuelled station; and of course if it is set like that in Fig. 4.1 which is for a water reactor, the lower steam conditions cause more obvious differences. Figure 4.2 shows a reactor crane: we know at once that this must be nuclear because it is a polar layout, characteristic of light water reactor stations, but the distinctive nuclear engineering is not so much in the shape but in the special structural and seismic features.

Fig. 4.1 An 800 MW turbine generator at Bruce CANDU nuclear power station, Canada.

Fig. 4.2 A 130 t polar crane at Asco PWR station, Spain.

The equipment in the control room (Fig. 4.3) is standard, but it has been assembled in a way that is appropriate to the type of station, in this case an AGR. Some instrumentation is specific, and Fig. 4.4 shows pulse-counting channels for neutron flux monitoring.

The next three illustrations are unquestionably of nuclear equipment: a gag mechanism in its final assembly (Fig. 4.5); one of the AGR boiler units during insertion (Fig. 4.6); and, finally, the internals of a high active waste storage tank (Fig. 4.7). The last is not for a power station as all the others have been, but for a fuel-reprocessing plant.

That gives some idea of the diversity; the reality is of course even wider, and many things have been left out. However, that is enough to show how the spectrum is continuous, going from the wholly conventional to the very specialized nuclear components. There is no step change.

At the nuclear end of that spectrum, illustrated by the gag assembly, the boiler and the storage tank, the configurations are quite special to the nuclear application; but this does not imply a new form of engineering, merely an extension of ordinary principles, and it need not be more difficult. For instance, the environment inside a gas-cooled reactor is not necessarily

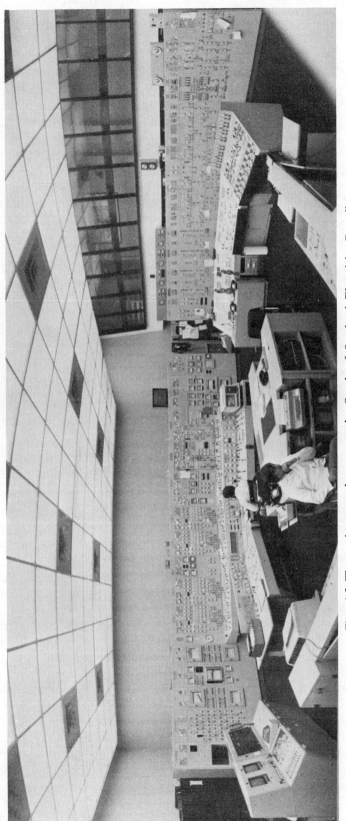

Fig. 4.3 The main control room at the South of Scotland Electricity Board's Hunterston 'B' AGR power station.

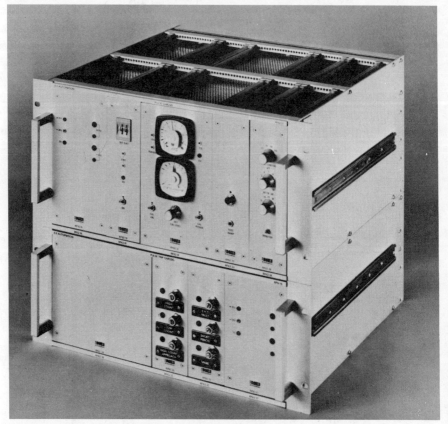

Fig. 4.4 Typical pulse-counting nucleonic chassis for neutron flux monitoring.

harder than, say, inside a coal gasifier, or deep down the leg of an oil rig. People sometimes say that two special things distinguish nuclear plant: the standard of reliability dictated by difficulties of access for maintenance or repair, and the demands of safety. Even these requirements are found in other applications, however, and one could argue that there is perhaps a further group of industries which would do well to adopt them. Again, as we move towards the nuclear end, manufacturing facilities are progressively more dedicated, and at the extreme we find clean conditions shops, special assembly areas, and even (for graphite) special machine shops.

What has been said so far is true of all reactor systems, even though some of the illustrations apply to water reactors, but most of them to gas-cooled ones which is where the UK's experience is strongest.

This progressiveness, this lack of a clear separation of the nuclear manufacturing industry from the rest of engineering (and it is certainly not being suggested that any such separation would be desirable) has certain obvious

Fig. 4.5 Final assembly of a gag mechanism for Hinkley 'B' AGR power station.

corollaries. The first is that manufacturers do not specialize in nuclear work to the exclusion of other things: it is just one among many of their markets and usually quite a small proportion of total business. You cannot say how many people are employed in nuclear manufacturing, and even if you could it would vary from time to time as the work-loads vary.

In short, there is no such thing as the nuclear manufacturing industry, in the sense of a sector that is hived off and separately recognizable. One consequence of that, probably, is a tendency to overlook the manufacturer when the policies of nuclear power are under discussion. This has many facets, to which we will return, but we now have enough background to continue.

4.3 THE CUSTOMER

The organization of the nuclear industry varies from one country to the other and it is important to understand the British system, to which the

Fig. 4.6 AGR boiler unit being lowered into position at Hinkley 'B' power station.

comments in this chapter mainly apply. Leaving out regulatory aspects, the major functions are:

1. The Utility, who in due course will operate the station and market the electrical power, and who therefore is the ultimate customer.

Fig. 4.7 Internals for a highly active waste storage tank.

2. Development of the nuclear system – for the gas cooled systems in the United Kingdom (and others such as the pressure tube Steam Generating Heavy Water Reactor (SGHWR) and the Sodium Cooled Fast Reactor) the developer has been the Atomic Energy Authority. If the system is a foreign one, for instance a light water reactor, then the developer is, in effect, the foreign licensor of such a system.
3. Overall engineering and construction of the power station, including the

architect/engineering. In the UK, that has been undertaken by what were formerly referred to as 'Nuclear Consortia', but now there is a single organization called the National Nuclear Corporation (NNC).
4. Contracts for manufacture and construction: the manufacturer, who is the subject of this chapter, comes under this heading but it also includes the civil engineering contractor and others.
5. The supply of fuel elements, a function which in the UK is undertaken by British Nuclear Fuels Ltd directly on behalf of the Utility.

The fairly wide role of the NNC–and of the Consortia before it–in the engineering of the reactor has included the specification and subcontracting of nuclear components as well as for the conventional plant. At the same time, neither NNC nor the Consortia have themselves been significantly involved in manufacture.

What all this means is that the proximate customer, from the point of view of the manufacturer of components, is the organization performing function (3), namely the Consortia in earlier times and the NNC today. There may be other contracts, especially in the area of repairs and maintenance of existing stations, which are placed in other ways, but that is broadly speaking how things stand for new construction.

4.4 THE ROLE OF THE SUPPLIER

To be successful, it is not enough for a supplier's organization to be efficient internally: there are also certain minimum requirements from the external environment in which he is to work. These may be summarized under three headings:

1. To be dealt with fairly in his search for new orders.
2. A working climate that allows him to perform well in his contracts.
3. In the longer term, reasonable consistency against which to plan and invest.

Those requirements apply to all markets and types of business. In order to see how they work out in the nuclear market in particular, let us take each of them in turn.

4.4.1 The search for new orders

The first measure of a manufacturer's success is his ability to win orders for his products. In nuclear power, as elsewhere, he needs to feel that the tendering conditions will be fair, and the tender appraisal both competent and just. The history of the British nuclear industry has not always been perfect, but on the whole these points have been well understood and well managed.

Years ago, the leading firms on the manufacturing side joined together to form consortia whose main role was the design and construction of nuclear power stations: as we all know, over the years these sometimes became fused, and sometimes even defused, and we have now evolved to a condition where there is only one consortium, the NNC. When there were several, there was little competition between the manufacturers in any individual one: either they had taken good care to exclude their rivals from membership, or where the merging of consortia had frustrated that, working arrangements would be arrived at, for instance, between the two boilermakers in The Nuclear Power Group (TNPG) in the years 1961 to 1970.

When a consortium placed manufacturing orders, fair treatment was obtained partly by the absence of competition at that manufacturing level, which meant that there was no real doubt provided the consortium itself was successful, and partly by the participation of the manufacturers in the management of the consortium through its Board of Directors. In 1974 the consortia were finally and completely merged into a single design and construction organization, now known as the NNC. In fact, competition between the consortia since the mid 1960s had become somewhat hypothetical (under the pressure of other circumstances), and there had been a growing feeling that it might be time to shift competition down to the manufacturing level: now that there is a single design and construction organization, we can have several manufacturers of like capability submitting competitive tenders to the one customer. This raises no particular difficulty for plant which is towards the conventional end of the spectrum; we rely on traditions of impartiality in tender assessment to avoid conflicts of interest.

It becomes more difficult to produce competition between manufacturers as one moves further towards the nuclear end of the spectrum. The problem is mentioned here in general terms without offering a solution, since this would depend on the type of equipment under consideration and a whole range of other circumstances.

4.4.2 Performing the contract

Once he has secured an order, our typical manufacturer wants only to get on and perform the work to the agreed terms, and in such a way that the margin at the end of the job is at least as good as when it appeared in his estimate. There are some very good customers who let him do this, and others who perhaps are not quite as good.

There could be no objection to proper intervention under the contract, say on quality matters, on approving design, or even on calling for changes; nor is there, in practice, much substance in the old complaint that the customer is interfering too much in design matters. Such so-called overinterference may be caused as often by a manufacturer's technical inadequacies (speaking generally) as it is by customer fussiness.

The rule that the 'customer specifies and the manufacturer designs accordingly' would be a fair one if it were not ambiguous; because you cannot rely wholly on the words and on the contract. A good customer is one who understands that and brings to bear his professional engineering judgement to supplement the letter of the law.

The old consortium system, in which major manufacturers were represented on the Boards of their customers, did ensure a responsible management approach to the project which was treated as a common problem to be solved in partnership. I have no doubt at all that this factor played a very big part in the success of the Magnox programme, and for that matter in the success of Hinkley 'B' and Hunterston 'B' which were built largely under that kind of regime. It is a system which is especially important for site questions, where interaction between suppliers is greatest, and it is also particularly effective in times of major crisis; a good example would be the gag vibration troubles experienced by the first two AGRs during commissioning. That demanded a concerted effort by manufacturers and their customer (who had recently been transformed from TNPG into the Nuclear Power Company, a subsidiary of NNC) and, of course, by the two ultimate customer Boards. It was essentially teamwork, and could never have happened successfully if the consortium management had been less enlightened, or if the parties had been encouraged to concentrate on their strict contractual liabilities and to lodge claims before taking any action. Teamwork of this kind is in any case good business because it brings solutions to problems more quickly, puts the plant in commercial service sooner, and leaves everyone satisfied.

In its 30 years of nuclear work, manufacturing industry has been fortunate to enjoy on the whole this kind of professionalism from its customers; it was very evident throughout the Magnox programme, notably in the great fuel element rattling and distortion crisis of 20 years ago which many will remember, and one must confidently expect it to be so in future.

So much for the letter of the law. One may add that manufacturers need to have confidence that their customer himself is determined on the success of every one of the projects he is engaged in. At the moment, the new projects are the two AGRs and the first PWR. Each of them will require wholehearted teamwork, the sort of teamwork just described on gag vibration, together with commitment at the very top.

4.4.3 The longer term

The third of the supplier requirements mentioned above was consistency in the longer term. In the limit, what brings the lowest cost and the highest quality and performance is a 'steady-ordering programme'. To see that demonstrated one need only look across the Channel or at Canada.

It may be retorted that manufacturers should be content with the business they do get, and cannot expect plant to be ordered if it is not required. That is true, but it misses the point. We are not talking of unnecessary plant, but of timing.

The old idea that you can treat the manufacturing sector as an infinite pool, to draw from at will, is valid only if orders are relatively small. That does not always apply as it depends on the component, and if individual contracts are commensurate with the actual throughput of the shops, then irregular ordering will simply put up the costs, and customers are noted for their unwillingness to believe that high prices can possibly be due to their own buying habits. In the limit, there is a degree of haphazardness beyond which the manufacturer cannot cope: he cannot hold his skilled and trained resources, he runs the risk of poor performance, and of course he has great difficulty in carrying through productive investment. Any control engineer will supply the explanation: the manufacturing process has a high time constant. To illustrate this, take the example of the AGR boilers.

Those for Hinkley 'B' and Hunterston 'B' were built, 48 units in all, over the period 1968 to 1972. There was then a long pause until January 1978 when a further two AGR stations were announced and it was possible to

Fig. 4.8 Assembling boiler units for the first two AGRs in a specially built 'clean conditions' shop, 1970.

start bringing back production facilities. In the figures that illustrate this (Figs 4.8–4.18) there are two things to note: the substantial degree of productive investment for this one component, and the technological changes to processes in the twelve-year period. Figure 4.8 shows the special shop built to assemble boiler units under clean conditions in the first AGR programme, with some seven or eight units being assembled. It is interesting to note that when work was finished not only was the shop used for conventional boilers, but the technique shown, of assembling tube banks in the works for delivery to site as packs, which had been developed for the AGR, was then extended to conventional boilers and in particular the convection banks of Inverkip and Littlebrook were done in that way. This shop has now been refurbished and Fig. 4.9 illustrates its use for assembly of fast reactor tube bundles; the AGR will follow.

Upstream in the process, Fig. 4.10 shows one of the earlier bending machines, in which the jacking is pneumatic, but the bending round the dies is by hand. In contrast, Fig. 4.11 shows the new CNC bender. A view of the handling cascades is shown in Fig. 4.12. The first operation is to butt-weld the tubes to form long lengths which are then rolled towards the centre to

Fig. 4.9 View of 'clean conditions' shop after refurbishment, showing fast reactor tube bundles being assembled, 1982.

reach the bending machine. The arms are lifted pneumatically and the tubes roll under gravity. There are fixed head welders on either side, and one of them is shown in close-up in Fig. 4.13. The same shop is pictured earlier (Fig. 4.14) when Littlebrook manufacture was in full swing. The purpose of showing that, as for the clean conditions assembly shop (Figs 4.8 and 4.9) is to bring home the extent of reinvestment implied by product switching.

Fig. 4.10 Earlier type of tube bending machine.

102   *Nuclear energy and the power plant industry*

After bending, for obvious reasons, butt-welds need to be done orbitally – that is the machine moves around the tube. The machines incorporate wire feed, gas supply and drive, and one of the new ones is illustrated in Fig. 4.15.

Spacers (the small metal pieces that hold and separate the tubes) were all welded on by hand on Hinkley and Hunterston, using manual MIG. The new system used mechanical robots: the platens lie horizontally, and the combined gantries and robots give complete two-dimensional coverage. Figure 4.16 shows a robot in action, and Fig. 4.17 shows three gantry systems working in parallel. Finally, Fig. 4.18 shows the stress-relieving furnace; the lid lifts and the platens are fed in on trolleys.

These illustrations give you some idea of the extent of production investment that has had to be reinstated and brought up to date, even though the shops were fully used in the meantime and the experience was incorporated in other boiler work. Lying behind that photographic story is an extensive programme of training and working up, of production and quality planning, and much else. The same tale could be told about many other nuclear components in other works in the UK. That is the significance of a high time

Fig. 4.11 CNC tube bending machine as installed for the new programme.

constant: time to build up, time to run down, time to convert, and always associated with cost, which nationally is irrecoverable.

So, essentially what the manufacturer is asking for is an ordering programme whose rate of rise and fall, and of conversion to other things, is geared to take proper account of the time constants in the manufacturing process. That should be a factor in national energy policies.

4.5 ENERGY POLICIES AND PROSPECTS

No manufacturer can afford to wait in the works for inquiries to come to him. He must go out and do some marketing, not only in order to publicize his capability, but also, and more importantly, to assess what the market may require in future and to equip himself to provide it. There are people to be recruited and trained, designs, production and quality processes to be developed, new technological factors to be investigated in the laboratories, facilities to be renewed, perhaps even outside know-how to be examined and licensed. All this takes time to reach fruition, and goes on continuously

Fig. 4.12 Tube handling equipment.

in any well-ordered enterprise (despite the opinion often expressed nowadays that new ventures, as they are termed, are the prerogative of small firms). The manufacturer must take a long view, not just about the nuclear future which as we saw earlier is only part of his work (often quite a small part) but about energy as a whole and his other markets.

Nearly two-thirds of present UK energy is derived from oil and natural gas, both of which have limited reserves and need to be husbanded. As seen in Table 4.1, the basic picture is the same whether you look at an active year (1979) or at a recession year (1980), whose energy figures were back to 1976 levels. Indeed year-to-year fluctuations are not that important. Incidentally, these usage figures include the oil non-energy feedstocks. The general view is that the UK will cease to be self-sufficient in oil around the mid 1990s, so we have some fifteen years in which to achieve significant substitution of the oil by other sources of primary energy – coal and nuclear power. That is the argument in a nutshell – substitution for oil – and perhaps it can best be developed by anticipating some familiar reactions.

First, what about conservation? However much conservation is pushed, and it should be, it will not make a serious dent in those oil figures. Roughly

Fig. 4.13 Close-up of butt-welding fixed-head machine.

Manufacturing industry and nuclear power

Table 4.1 UK primary energy usage, mtce

Source	Year	
	1979	1980
Coal	130	121
Oil	155	134
Natural gas	71	70
Nuclear	14	13
Hydro	2	2
Total	372	340

speaking it will compensate for the underlying growth in demand. The best known of the conservation studies, the Leach report, achieved just about that result and only succeeded in proving that nuclear power was unnecessary by assuming that there was unlimited oil for imports to the UK even in

Fig. 4.14 The tube shop during conventional boiler manufacture.

AD 2025: hardly a reasonable assumption. Not only are world reserves of oil limited, but the population is rising very steeply and so are its aspirations. Those who think we could rely on oil imports expect us to use our financial strength to get supplies, as we could; but it would be bad economics, would promote misery among developing countries who may have no alternative fuel, could lead to instabilities, and politically would be very foolish; and it might not work. So much for importing oil.

Next, there are the environmental objections to nuclear power, which are well known, and detailed answers are not appropriate to the present chapter. It is noticeable that in Northumberland environmentalists ally themselves in resistance to nuclear power with people in the coal industry, who see it as a threat to their employment; but in Leicestershire, at the Belvoir enquiry, the environmentalists are against coal.

In fact, coal is extremely important. Current annual production at 120–130 Mt needs to be increased significantly over the next 15 years. One of the sad features of the nuclear debate is the way the coal and nuclear people seem ready to form rival factions. A glance at the figures shows how wrong that is: we desperately need an expansion of both. There is something else

Fig. 4.15 Modern version of orbital welder.

Manufacturing industry and nuclear power

that coal and nuclear power have in common: a tendency for their prospects to be judged on short-term demand factors, and for the reality of long-term shortages of oil to be overlooked. This brings us to the last of the reactions we are anticipating, namely, the present excess of electricity generation capacity. Let us note in passing that in other forms of industrial investment excess capacity is not a valid objection, because the proper justification is economical: in a workshop, better and more efficient plant gives better productivity, whether or not there is over-capacity of older less efficient plant. Why should not the same principles apply to electricity generation?

This is a line of reasoning that has much to commend it, and a good economic case can be made for nuclear power. What is more, it appeals to the many people who are at home with fairly direct supply and demand arguments. Unfortunately, its drawback is that the financial assumptions are just as easily challenged, rightly or wrongly, as for instance in the House of Commons Select Committee's report and more recently in the one from the Monopolies Commission. Critics who believe they have successfully challenged the economics may suppose that they have demolished the whole case for nuclear power.

Fig. 4.16 One of the robotic welders in action.

It is better, therefore, to meet this objection of excess generating capacity by looking more carefully at demand. The current habits of consumers are no guide to their late behaviour. Any manufacturer of energy-using equipment knows that few of his customers look ahead 15 years. It is not enough to point at today's excess capacity: you must ask what the capacity and the demand might be in the 1990s. When oil shortages develop, they will translate into high prices for petroleum derivatives, and consumer demand will shift towards forms of secondary energy which are derived from more plentiful primary sources: coal and nuclear. So electricity demand will rise steeply. The oil substitution argument, which is at the primary energy level, translates into an electricity demand argument at the secondary level. Now, even if one nuclear station is started each year from 1981, the extra installed power at the end of the century is only 14 GW or so (nothing started after 1993 would be ready); this might give access to an extra 30 mtce of primary (nuclear) energy, not a large figure relative to our present consumption of primary oil.

We may find it strange that lay commentators generally do not seem to recognize the importance of the long construction time. A recent example

Fig. 4.17 General view of spacer welding facility showing the three gantries carrying the robots.

Manufacturing industry and nuclear power

was the report of the House of Commons Select Committee on Energy which concentrated on the demand estimates for the present decade, very little of which can be affected by new construction decisions. It is also strange, and perhaps a little unfortunate, that this fundamental argument from primary energy is given relatively little prominence in the advocacy of nuclear power. Perhaps it is because of the strong conviction that it is in any case cleaner, safer, and cheaper than other sources of energy and should shine by its own light. However, the need to substitute for oil is a most telling argument because it shows nuclear power as a sheer necessity and not as one among a variety of primary energy sources from which we are free to choose. If only this argument could be stressed more often.

4.6 CHOICE OF THERMAL REACTOR SYSTEM

No review of the role of manufacturing industry for nuclear power could be complete without some reference to the choice of thermal reactor. The controversy in Britain between the AGR and the PWR has lasted a long time. Years ago the rival to the AGR was actually the BWR, a system now

Fig. 4.18 Stress-relieving furnace for tune platens.

not generally favoured and we must all surely be thankful that it was not the one chosen in 1965 for Dungeness 'B'. However, the PWR has had strong advocates in this country for the past ten years or so.

We are now in a lull. The decision has been taken to continue with two AGRs and to build a first PWR, subject to the public inquiry. While everybody gets on with that, controversy is less sharp and we can limit ourselves to a few observations.

First, the present policy will bring out the true characteristics of a PWR in the British context: cost, amount of site work, engineering aspects, construction time, and so on. If any of its claims were exaggerated they will be exposed, because genuine evidence is on the way. Meanwhile the AGR is also building up experience, both from the operating stations and from those now starting their construction. Future debates will have more facts and less myth, and while we wait for those, we can be more relaxed.

Secondly, it is an error to talk as if some manufacturers are better suited to one system, and some to the other – as if there was a contrast between two camps. It is true that for a system like the AGR, which has been around for some time, there would be strong experience and demonstrated ability in a number of firms; but it does not follow that the firms that would be stronger on PWR are different ones. It is a supposition without intrinsic logic, which invites the extraordinary thought that the qualities needed to manufacture PWR components are the opposite of those for the AGR. However improbable it may seem, let up hope that we will not end up in two camps: it would only transfer the argument to outsiders.

Thirdly, investment factors should play a proper part in deciding the mix of thermal systems after the immediate present programme. The AGR boiler production facilities were mentioned earlier and this example could be multiplied across the whole industry, including manufacturers, designers, laboratories and operators.

For the first PWR, the components that would require significant production and technical investment are to be imported. That seems sensible as a broad policy, but it does of course mean that any major investment by the home manufacturing industry for the PWR would be made after that, and it becomes a factor in deciding what system to follow up with after these first three stations. There is an apparent contradiction here which must be properly understood, because it may appear that manufacturers with good experience and hence with substantial investment for its manufacture would be disinclined to launch into PWR production.

In practice, that is not so. Investment is an important factor and it is true that existing investment, whether nationally or in a company is predominantly for AGR components. A manufacturer might therefore express the view that the AGR would seem, at least on those grounds, to be preferable to the PWR. He may even be more directly conscious of it because he can see the scale of investment in his own works.

It will, however, have been written-off. Indeed one of the consequences of the irregularity and uncertainty in orders referred to earlier is that investment must be amortized over the few projects that are definite – and that does push up their cost. So there is not an outstanding debt in some sense to be picked up in future orders. It remains necessary for the customer, in considering the next decision, to take due account of the cost of any further investment that may have to be incurred if there is a change of reactor type.

However, if the customer does opt for something new in spite of that, the manufacturer will cheerfully go ahead and put himself in a position to supply accordingly. The qualities needed are similar. Almost everybody in British manufacturing concerned with nuclear power has good experience of gas-cooled reactors, but good experience of the key components of the PWR is much more limited. In looking for manufacturers for the new PWR, there is the field to choose from: it would be illogical to exclude the people who understand the AGR. The supposition that there are some manufacturers who are good at one system and other manufacturers who are good at another is without validity; in fact, good experience of nuclear work and in particular of pressure parts for one type of plant, is helpful and conducive to success in undertaking pressure parts for nuclear plant on a different system. Indeed, people who have studied the American and other practices have said that in many ways, if you allow for differences in national style and culture, the management of production is very similar. If you look at quality programmes in nuclear manufacture – not limiting this to pressure parts – and look at the type of shop supervision and the controls, they are not so very different, given that the configuration of the plant is not the same. There may be differences in documents or material specifications, but the approach to consistency and reliability seems essentially the same.

4.7 PUBLIC AND POLITICAL OPINION

We conclude with some comments on public and political opinion. We all know about the nuclear debate, which is much concerned with safety and the environment, and even talks darkly about secrecy and civil liberties. Manufacturers could play a greater part in this than they do. After all, they have the enormous advantage of being impartial, in the sense that they are as happy to supply plant to one market or one set of customers as to another, and have no special reason to prefer the nuclear business. This makes the manufacturer an important independent witness. He can vouch that nuclear contracts are not necessarily more 'lucrative' than others (taking the phrase from one well known critic); in fact over the years and given the vacillations in policy, probably the opposite. He can also testify that no special conditions of secrecy or security attach to his nuclear work, and that staff engaged in it are not desperately protecting their careers because it is normal for them to move on readily to non-nuclear work and back again. In short, he can help expose some of the more absurd statements of the anti-nuclear people.

There is another and more serious level of debate which concerns us, namely, the tremendous reluctance of public and politicians to take the long view. Earlier, it was explained why a manufacturer could not content himself with responding to immediate business, but must look at his market for the longer term. When he does so, the key points are the growing scarcity of oil – and to a lesser extent of gas – as primary fuels, and the need to substitute; and that nuclear power (and coal too) required investment over a period to build up a significant contribution. Finally, he knows that the cheapest way of doing that is through a steady programme.

Leaving aside mere anti-nuclear sentiment, this simple thesis fails to be accepted by public opinion, apparently for two reasons. The first is a suspicion that the manufacturer is simply trying to promote his own business. This arises because on past occasions, when the necessity to place orders with the power plant industry was being advocated, the need to support employment has been the most strongly emphasized of the various arguments: whether the prospective orders were for conventional or nuclear plant. It may be that the jobs argument is the one which commends itself most to politicians and has been over-emphasized for that reason. In reality, all that it justifies is smoothness of work-load, not ordering the unwanted. By the same token, if the nuclear programme really were unnecessary, the manufacturer's best interest would be to leave it and to concentrate on his other work. His difficulty is that he can see the need and can predict that he will be called upon for the plant sooner or later. What he is pleading for is timely decision.

There is a more important factor: the deep-rooted aversion of public and political opinion to thinking ahead more than a few years. This permeates our society but its causes would take us well outside the scope of this chapter. We noticed some symptoms of it earlier: the preoccupation with current consumer behaviour, the emphasis on electricity demand in the present decade, the failure to realize that construction takes a long time, the lack of understanding of the time constants in the manufacturing process. One could have added that when our critics look at the performance of other countries, they often get enthusiastic about the reactor system used there, or the form of industrial organization, or some other feature, but they seldom think to copy the real key to those countries' success, the simple virtues of perseverance and continuity.

Perhaps it is in insisting everywhere on the importance of a consistent and long-term view that manufacturing industry can best influence national policy for nuclear power.

ACKNOWLEDGEMENTS

This paper is based upon one published in *Nuclear Energy*, the Journal of the British Nuclear Energy Society, whose permission to publish here is gratefully acknowledged.

URANIUM PRODUCTION AND DISTRIBUTION

5

R.M. Williams
PROSPECTS AND CONCEPTS

> Immense – and immensely misconstruable – figures for mineral resources fail to impart a clear picture of mineral resource adequacy and long-term mineral supply. Such figures are almost routinely mistaken for material that will be available at acceptable prices when and where needed, as if the world were economically frictionless.
> Jan Zwartendyk (1980) Chief, Resource Evaluation Section Mineral Policy Sector, EMR Canada

5.1 INTRODUCTION

Considerable effort has been expended during the past ten years to develop improved methods for analysing current and future mineral supply. The mineral fuels have received particular attention, prompted by an increasing awareness that the world's supply of petroleum may be severely limited in the very near future. In response to such concerns, analyses of the world's supply of uranium have become increasingly comprehensive, and significant effort has been focused on developing illustrations of future supply possibilities that are both meaningful and easily understood.

A group that has played a key role in this development has been The Joint Working Party on Uranium Resources of the Nuclear Energy Agency (NEA) of OECD and the International Atomic Energy Agency (IAEA). The Working Party was established in 1964 by the NEA, then known as the European Nuclear Energy Agency, for the purpose of preparing assessments of the world's uranium and thorium resources. The scope of the Working Party's studies was broadened in 1967, under the joint auspices of the IAEA, and by December 1979 it had published eight major assessments of world* uranium supply. The assessments, now standardized on a biennial basis, have become essential references for planners and policy makers in the international nuclear community.

There has been a pronounced shift in emphasis in the NEA/IAEA Working Party's analyses since it published the results of its first assessment in August 1965. The first of the OECD 'Red Books', as they are commonly referred to in the nuclear trade, consisted essentially of a tabulation of estimates of the world's uranium (and thorium) resources according to a prescribed resource classification scheme developed for the purpose. Successive Red Books have placed less and less emphasis on the resource

---

* Data was not forthcoming nor included for USSR, Eastern Europe and the Peoples' Republic of China.

estimates themselves and increasing emphasis on projections of production levels that could, under varying assumptions, be supported by the resources that have been identified.

This chapter reviews briefly the evolution in thinking associated with the NEA/IAEA assessments, noting the progress that is being made in the field of uranium supply analysis in several organizations that contribute to the NEA/IAEA exercise. Special reference is made to methodologies that were used to examine the world's uranium supply capabilities in the course of studies carried out in 1977, for the Conservation Commission of the World Energy Conference (WEC), and in 1978 and 1979 for the International Nuclear Fuel Cycle Evaluation (INFCE). Although perceptions of the world's uranium supply capabilities are constantly changing, those developed as part of the recent INFCE exercise reflect the current 'state of the art' of uranium supply analysis and broadly illustrate the current world uranium supply situation.

5.2 RESOURCE CLASSIFICATION PRINCIPLES

Comprehensive resource estimates are the fundamental building blocks of any mineral supply analysis. All resource estimates, whether they be on a

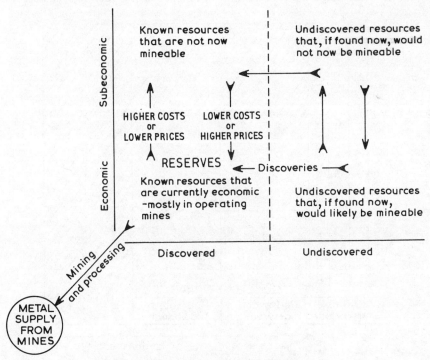

Fig. 5.1 The flow of resources over time.

local, regional or global scale, need to be tied to a recognized system of classification so that the significance of the individual estimates can be viewed in proper perspective. Since much of the debate associated with mineral supply analysis is related to how resource estimates are labelled and categorized, it is important to appreciate the principles behind the classification schemes that are in common use.

Various resource classification schemes have been devised in recent years, the most common and widely accepted schemes being derived from one developed by the United States Geological Survey (USGS) in the early 1970s, often referred to as the McKelvey system, after the then Director of the USGS, Vincent McKelvey. It is a two-dimensional system, the principles of which can be illustrated by Fig. 5.1 [1].

This two-dimensional system provides an opportunity to show resource quantities in gradations of geological assurance (along the horizontal axis) and in gradations of economic attractiveness (along the vertical axis). The first deals with the level of confidence that the estimator has in the geological information that is available to him for making his estimates. The range in the level of confidence of the reported quantities is usually expressed by a series of descriptive terms, such as measured, indicated, inferred, etc., the distinctions between which are not always easy to define. The second dimension calls for judgements about mining and processing methods, capital and

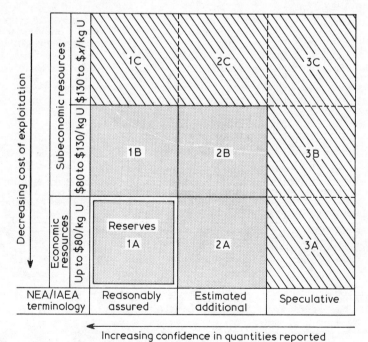

Fig. 5.2 NEA/IAEA uranium resource classification scheme.

operating costs and possible markets, factors that can be equally elusive [2, 3].

The dynamic nature of the mineral supply system is also illustrated in Fig. 5.1. Resources can flow, over time, from one category to another as geological knowledge improves as a result of exploration efforts, and as economic factors vary such that there are changes in costs or prices associated with production of the resource. To the non-technical person, resources are often thought of as being finite and static. Because of this perception, such persons are not always aware of the effects of, for example, changes in taxation and regulatory requirements and improvements in extractive technology on the dynamics of the system [4].

5.3 THE NEA/IAEA CLASSIFICATION SYSTEM

The uranium resource classification system currently employed by the NEA/IAEA, as illustrated in Fig. 5.2, follows the principles described above. (Dollars referred to in Fig. 5.2 are United States dollars; the $80 and $130/kg U levels have been used since 1977.) The current definitions* for the three resource terms are shown below.

> 'Reasonably Assured Resources refers to uranium that occurs in known mineral deposits of such size, grade and configuration that it could be recovered within the given production cost ranges, with currently proven mining and processing technology. Estimates of tonnage and grade are based on specific sample data and measurements of the deposits and on knowledge of deposit characteristics. Reasonably Assured Resources have a high assurance of existence and in the cost category below $80/kg U are considered as reserves for the purpose of this report.'
>
> 'Estimated Additional Resources refers to uranium in addition to Reasonably Assured Resources that is expected to occur, mostly on the basis of direct geological evidence, in extensions of well-explored deposits, little-explored deposits, and undiscovered deposits believed to exist along a well-defined geological trend with known deposits. Such deposits can be identified, delineated and the uranium subsequently recovered, all within the given cost ranges. Estimates of tonnage and grade are based primarily on knowledge of the deposit characteristics as determined in its best-known parts or in similar deposits. Less reliance can be placed on the estimate in this category than for Reasonably Assured Resources.'

Note: In presenting its definitions, the NEA/IAEA emphasizes that 'the distinctions drawn between Reasonably Assured Resources, Estimated Additional Resources and Speculative Resources based on differing degrees of geological evidence, make it essential that each category be regarded as a discrete entity. Therefore, great care should be taken in the use of resources estimates (e.g. not taking the sum of estimates of each of the categories to obtain "total resources").'

'Speculative Resources refers to uranium, in addition to Estimated Additional Resources, that is thought to exist mostly on the basis of indirect evidence and geological extrapolations, in deposits discoverable with existing exploration techniques. The location of deposits envisaged in this category could generally be specified only as being somewhere within a given region or geological trend. As the term implies, the existence and size of such resources are highly speculative.'

The two principal resource terms, 'Reasonably Assured Resources' (RAR) and 'Estimated Additional Resources' (EAR) were first introduced in 1965* and have since gained fairly wide international acceptance [5, 6]. However, there have been three noteworthy modifications to the wording of the definitions, the first two being related to the problem of clarifying the nature of the undiscovered resources that may be included in the EAR category, and the third being related to the method of subdividing the resource estimates as to their economic attractiveness.

In 1970 the definition of EAR was modified to restrict the inclusion in this category of only those undiscovered uranium resources that were associated with 'known uranium districts'. Criteria used to define a known district varied from country to country but generally referred to areas where uranium was (or had been) known to occur in sufficient tonnage that it could be (or had been) profitably produced. Although this modification had little impact for most countries, it considerably narrowed the scope for others (e.g. Canada and the United States) where there were extensive, geologically favourable areas for additional discoveries [7].

This question was again addressed on the occasion of an IAEA Advisory Group Meeting on the evaluation of uranium resources, in Rome in late 1976. It was realized that the 'known district' criterion was not in all cases a valid one for expressing the 'level of assurance' with which a given tonnage is known to exist. Some deposits are known to exist outside 'known districts'; conversely, some tonnages estimated to exist within known districts are not more assured than some tonnages that can be estimated to exist outside known districts.

It evolved that, for the purpose of the NEA/IAEA's December 1977 Red Book, the definition for EAR was modified to describe more explicitly the nature of resources to be included in the EAR category, i.e. uranium resources in (1) extensions of well-explored deposits, (2) little-explored deposits, and (3) undiscovered deposits believed to exist along a well-defined geological trend with known deposits [8, 9]. It also evolved, partly as a development of the Rome meeting, and partly as an element of an NEA/IAEA study of world uranium potential, that the term 'Speculative Resources' was introduced in 1978 to encompass those undiscovered

* The original term 'Possible Additional Resources' was modified in 1967 to 'Estimated Additional Resources' to remove the connotation of maximum.

uranium resources that are more speculative in nature, i.e. those that are thought to exist in deposits, the location of which can be specified only as being somewhere within a favourable region or geological trend [10].

Initially the NEA/IAEA subdivided its resource estimates according to three levels of economic attractiveness bounded by limiting price levels. The lowest category referred to resources of current economic interest, the second to resources of interest in the immediate future, while the highest category referred to resources that might be of interest in the longer term. Because few countries were in a position to assess their very low-grade resources, and because it was unlikely that such resources would be exploited in the near future, the third category was dropped in 1970 [7]. It is anticipated that the third category may be reintroduced in future Red Books.

The price concept was retained until 1975 when, in the wake of the oil crisis of the winter of 1973–74, consumer members of the NEA/IAEA Working Party proposed that the uranium resource estimates be subdivided according to production cost levels, rather than price [11]. With the shift from a buyers' to a sellers' uranium market that occurred at that time, consumers expressed concern that retention of the price concept would allow producing countries to incorporate unnecessarily large profit elements into their resource calculations, thus distorting the overall assessment of uranium supply. Although such fears were unfounded in the view of the major producers, the proposal was accepted on the presumption that it would be a relatively simple task to reach a consensus as to the various cost elements that should be included in the cost categorization procedure. Although this task proved more difficult than expected, considerable progress has been made toward reaching a concensus.

5.4 COMPARABILITY OF CLASSIFICATION SYSTEMS

Variations to the two-dimensional resource classification system have been developed in a number of countries. The approximate correlation of terms used by the major contributors to the NEA/IAEA exercises is shown in Fig. 5.3. The terms are not strictly comparable as the criteria used in the various systems are not identical, particularly as the resources become less assured. Nonetheless, based on the principal criterion of 'geological assurance of existence' the chart presents a reasonable approximation of the comparability of terms [12].

One significant difference amongst the major systems is readily apparent. Australia and South Africa take the position that no undiscovered resources should be included in the NEA/IAEA's EAR category, while Canada and the United States feel that such a view is restrictive and misleading. It is argued that there can be actual physical extensions to known deposits beyond the limits set for the equivalent of Canada's inferred category and

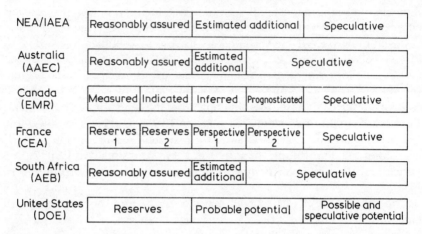

Fig. 5.3 Approximate correlation of terms used in major resource classification systems.

that there can be undiscovered satellite deposits, or deposits of a repetitive nature, existing along the same well-defined geological trend in which a known deposit occurs. More simply, it is believed that there are two types of undiscovered resources, one that is directly associated with known deposits and another that is surmised to occur in virgin uranium areas, or in areas where only uranium occurrences are known. It is submitted that the first type should be included as part of the EAR category.

Although the terms illustrated in Fig. 5.3 are primarily distinguished one from another by criteria related to certainty of existence, other criteria are also sometimes used. For example, United States Department of Energy (USDOE) criteria for distinguishing between resource categories include the relationships of the resource to a productive formation or a productive province. In other words, an element of production feasibility is superimposed on the criteria related to certainty of existence. Most classification systems seek to avoid such mixing of X and Y axis criteria, although judgements about confidence levels are often inadvertently coloured by the realization that exploitation of a resource is not technically feasible.

Although NEA/IAEA uranium resource estimates ostensibly refer to quantities of uranium that are recoverable, recoverability factors are not treated in the same way by all reporting participants. Generally, reported tonnages refer to quantities of uranium contained in mineable ore, conforming to the traditional method of reporting metal resources – mineable ore being the quantity of ore recoverable by mining, with due consideration of ore dilution. While ore processing losses are considered in the choice of cut-off grades to be used for the resource calculation, such losses are not normally applied to the final resource estimate.

In most cases, uranium ore processing losses are no more than 10%, so that the distinction between uranium contained in mineable ore and uranium recoverable from such ore is not overly important. In some cases, however, where processing losses are very high (e.g. Swedish uraniferous shales), or where uranium is recovered as a byproduct (e.g. South African conglomeratic deposits) the application of processing losses to final resource estimates is essential. Indeed, as increasingly lower grade resources are exploited in the future, the distinction between mineable resources and recoverable uranium may become increasingly important. Clearly, there is a lack of consistency in the methods used to account for recovery, a fact that is often not fully appreciated.

Procedures used for subdividing uranium resource estimates according to different levels of economic attractiveness also vary from country to country. Most uranium resource evaluations are based on the 'forward cost' concept, where the cost of exploiting the resource is considered to include not only the direct costs of mining and processing, but also those future costs associated with delineating a deposit and providing and maintaining the required production facility. Various expenditures incurred prior to the date of the resource estimate are generally but not always excluded (e.g. land acquisition, exploration, mine development and mill construction), and provision is not always made for the inclusion of depreciation, income taxes, interest on borrowed capital, the cost of marketing and general overhead, and profit.

Although there has been considerable debate about the inclusion or exclusion of particular costs, there is wide agreement that the objective should be to include as many of the applicable costs as possible. Indeed, the Canadian view has been that, if all costs are considered, including the cost of using money, cost levels for the purpose of resource classification are synonymous with minimum acceptable price levels.

Notwithstanding the perceived shortcomings of the various systems and the realization that resource and cost/price categories are not strictly comparable, the aggregated uranium resource estimates compiled by the NEA/IAEA Working Party on Uranium Resources are believed to represent the best available estimates of the world's uranium resources. Efforts have been made in successive Red Book exercises to provide clear descriptions of resource classification procedures used in the major reporting countries, in the belief that clarity and comparability in resource terminology are more attainable goals than consistency and standardization [13].

A number of positive efforts have been made recently in the field of resource classification and terminology. Particularly noteworthy was a study prepared by an expert group convened by the United Nations' Economic and Social Council to develop a mineral resource classification system to be used in reporting data to the United Nations [14]. The proposed classification is very similar to that used by the NEA/IAEA. Also of importance is the recent publication of modifications to the USGS system [5], and a study currently underway in the United States under the auspices of the American Society for Testing and Materials (ASTM), which is attempting to develop standards in the area of 'uranium resource evaluation'.

5.5 URANIUM DISCOVERY POTENTIAL

Increasing concern about long-term energy supplies has resulted in efforts to expand assessments of the world's nuclear fuel resources. In 1974, the USDOE initiated a National Uranium Resource Evaluation (NURE) programme to develop an authoritative, comprehensive assessment of United States uranium resources. To carry out the NURE programme, several large-scale field investigations were carried out by the USDOE and the USGS and through state organizations, companies and universities. Extensive areas of the United States were covered by aerial radiometric surveys and hydrogeochemical surveys, supplemented in selected areas by geological investigations and drilling. A major status report on the NURE programme was released in October 1980 [16].

To provide a framework for estimates of United States uranium resources that might exist beyond the USDOE categories of 'reserves' and 'probable' potential resources, two further categories of potential* resources were developed (see Fig. 5.3), based on their spatial relationship to defined resources [16]. Estimates for 'possible' and 'speculative' potential resources have been made and published by the USDOE since 1975, based largely on data generated through the NURE programme.

In 1975 the Geological Survey of Canada (GSC) embarked on a Uranium Reconnaissance Programme (URP) modelled on Canada's Federal–Provincial Aeromagnetic Programme. The objective of the programme was to ensure that all areas in Canada that may contain uranium resources were publicly identified and delineated, so as to facilitate the discovery of deposits by industry and assessment of resources of the government. Large areas of Canada were covered with airborne gamma-ray spectrometric surveys and hydrogeochemical surveys before federal funding of URP was suspended in 1979, as part of federal government economy measures. URP data has contributed significantly to national uranium resource assessment activities. A measure of Canada's uranium discovery potential, in the form of estimates of speculative resources, has been published annually since 1978 by Energy, Mines and Resources (EMR), Canada [17].

Programmes similar to the United States NURE and the Canadian URP have also been undertaken in several other countries, although few quantitative estimates of uranium discovery potential have been made on a

---

*All three of the USDOE's categories of 'potential' include both undiscovered and partly defined resources; see comment in Section 5.4 re mixing of X and Y axis criteria.

Prospects and concepts 125

national basis, other than by the United States and Canada. Some countries do not perceive a need to quantify estimates of uranium resources beyond the 'known' categories and others have reservations about doing so because of the fundamental conceptual (and practical) difficulties associated with assessing the undiscovered.

In recognition that the NEA/IAEA assessments of uranium supply did not constitute a complete appraisal of the world's uranium resources, the NEA/IAEA Joint Steering Group on Uranium Resources* undertook an expert study to define better the possible extent and location of the world's undiscovered uranium resources beyond RAR and EAR (i.e. 'speculative resources'). Data was compiled for 185 countries, and for each country a judgement was made of the order of magnitude of this additional potential, based on the geological favourability for the existence of uranium deposits of conventional types (very low grade resources where exploitation costs would exceed \$130/kg U were excluded). To facilitate the process of judging the relative favourability of each country, the Steering Group devised a ranking scheme, to which quantitative tonnage-ranges were subsequently assigned, for the purpose of obtaining an aggregate illustrative measure of the world's total discovery potential, based on then-current geological knowledge [10].

The study constituted Phase 1 of the International Uranium Resources Evaluation Project (IUREP) and was completed in 1978. It represented the most comprehensive international effort of its kind ever undertaken for any mineral commodity. A summary of the Steering Group's report appeared in December 1978 and the full report of Phase 1 was published in 1980 [10, 18]. The concepts developed during the IUREP study, together with the debate about the usefulness of the results that subsequently took place in INFCE fora, have provided significant impetus to international efforts to improve uranium supply analysis techniques. The 'speculative resource' tonnages developed in Phase 1 of the IUREP study are presented in a later section of this chapter.

5.6 RESOURCES AS A MEASURE OF SUPPLY

For many years the adequacy of reserves† of a mineral commodity was measured using the 'life index' principle. The life index of reserves is obtained by dividing a country's total reserves of a mineral commodity by

---

*A nucleus of the members of the NEA/IAEA Joint Working Party on Uranium Resources formed for the purpose of providing guidance and direction to NEA/IAEA uranium resource assessment activities. Current members are Australia, Britain, Canada, France, Italy, Japan, South Africa, United States, West Germany, and the Commission of the European Communities (CEC).

†The term reserves, as used in this paper, refers to the 'low-cost' portion of Reasonably Assured Resources, as defined by NEA/IAEA (see Fig. 5.2).

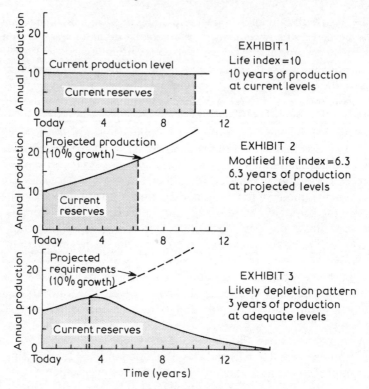

Fig. 5.4 The life index and its limitation (after reference [19]).

the current rate of production. The conclusion is then drawn that the nation's reserves are sufficient to last X years (see Fig. 5.4). The life index gives the false impression that reserves and production have a direct and straightforward relationship to each other. It ignores the fact that there are differences in extraction rates and life expectancies of individual production operations, and can lead to a false sense of security as illustrated by Fig. 5.4. Neither the classic nor the modified version of the life index depicted in the first two exhibits bears much relation to the reality of the likely pattern of reserve depletion shown in the third. Notwithstanding these limitations, a decline in the life index from year to year can be used as a crude 'early-warning' indicator of possible future supply problems. The problem must be verified by other means, however, because a decline in the life index can be caused by a number of factors that may not be cause for alarm [19].

The earliest NEA/IAEA assessments of the world's uranium resources consisted primarily of a tabulation of resources by country. Later assessments analysed aggregate resource levels using the 'forward-reserve' concept, a minor improvement on the life index principle in that it takes into account projections of future requirements. Because of the time required to replace reserves that are being produced, it is recognized that a viable industry must at all times maintain reserves sufficient to meet an appropriate period of forward requirements. Reserve levels equivalent to cumulative requirements for a future 10-year period were considered to be sufficient to maintain a viable industry. For the hypothetical example illustrated in Fig. 5.5, it is shown that current reserves (1.1 million tonnes U) are equivalent to cumulative requirements for the next 15 years, implying that reserve levels are more than adequate.

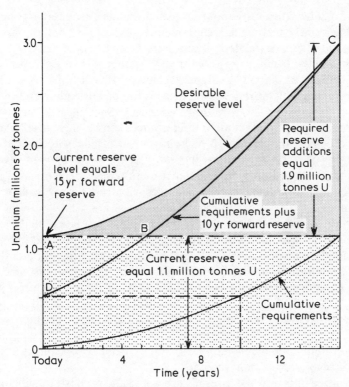

Fig. 5.5 The forward reserve concept as a measure of resource adequacy.

The forward-reserve principle was also used in a very qualitative way to give a rough indication of desirable future levels and the rate of annual additions to reserves that would be required to attain these levels. In Fig. 5.5, the appropriate minimum reserve levels are determined by moving the cumulative requirement curve 10 years forward in time. The new curve, D–C, illustrates that reserve additions of 1.9 million tonnes U will be required during the next 15-year period to maintain a minimum 10-year forward reserve. Beginning today, at a level of 1.1 million tonnes U, the desirable growth in the reserve level can be depicted by the third curve,

A–C. An alternative, but unrealistic, growth pattern would be defined by A–B–C [20].

Comparison of the forward-reserve situation was made from one NEA/IAEA assessment to the next in an effort to give an indication of the adequacy of exploration and development activity. Unfortunately, the value of this technique is somewhat limited because the quantity of reserves that must be identified to meet a given projected requirement can vary considerably depending on the order in which deposits of different geologic types, sizes and grades will be discovered and developed. Moreover, the lack of a clear consensus about projected uranium requirements limits the usefulness of the technique. Consequently, decreasing emphasis was placed on forward-reserve related illustrations in successive NEA/IAEA Red Books, and the technique was omitted altogether in their last report [12].

Any type of supply analysis technique that compares cumulative requirements with current resource levels can be misleading and can result in false conclusions about resource adequacy. Such illustrations make the incorrect assumption that all of the resources can be made available during the projected time-frame. To make matters worse, resources in different categories of reliability and at all cost levels are often summed in order to make the comparison with cumulative requirements. To do so is akin to adding apples, oranges and pineapples and can be misleading in the extreme.

Certain resources, even though in the RAR category, may never be produced because of environmental or other constraints, and optimum or controlled extraction rates can sometimes extend the life of other deposits beyond the time-frame of a supply analysis. In the case of resources in the EAR category, it cannot be assumed that all of the resources will be discovered and, if discovered, that they will be developed and produced in the given time-frame. Unfortunately, comparisons of uranium resource levels and cumulative requirements, without adequate consideration of discoverability and availability factors, have been all too common amongst nuclear planners and policymakers.

Resource data must be carefully presented and properly qualified. Even a simple misuse of resource terminology can lead to false complacency about resource adequacy. The term 'reserves' is routinely used indiscriminately with reference to resource estimates of differing degrees of reliability and of varying degrees of economic interest, thus leading to misconceptions about resource adequacy. Because of this risk, some authorities now avoid using the term altogether [17].

The generation of estimates of undiscovered resources of uranium, especially those assigned to the NEA/IAEA Speculative Resource category, has increased the necessity of presenting resource estimates in proper perspective. In presenting the Speculative Resource tonnages developed in Phase 1 of IUREP for example, the NEA/IAEA Joint Steering

Group took great care to state that the figures 'should not, under any circumstances, be used for nuclear power planning purposes. Rather, they should be viewed as a qualitative measure of the present state of geological knowledge, with all the inherent uncertainties, and looked upon as a guide for establishing priorities for future evaluation efforts' [10]. Unfortunately, such resource estimates are frequently taken out of the context in which they were presented and given an unwarranted sense of reality. This practice highlights the need for nuclear planners to understand the meaning and limitation of resource estimates [21].

Unwarranted emphasis is often placed on the significance of subdividing uranium resource estimates according to different levels of economic attractiveness. The subdivision between the NEA/IAEA $80/kg U and the $80 to $130/kg U categories is not absolutely sharp, nor can it be in view of the continued debate about the scope of the costs to be considered. The principal objective of the subdivision, which is often overlooked, is to distinguish those resources that are of economic interest at the time of the assessment from those that are not. Despite a detailed examination of costs, judgement factors also enter into the categorization process, making the result more qualitative than quantitative.

The appropriate method of accounting for expected ore-processing losses in uranium resource estimates has also been a topic of much discussion. An examination of USDOE uranium resource evaluation practices by the United States General Accounting Office (USGAO) during 1978, highlighted the fact that United States uranium resource estimates (in common with those of Canada, France and many other countries) are expressed in terms of uranium contained in mineable ore, rather than as uranium recoverable from such ore. The USGAO report, and its rather unorthodox recommendation that, in the future, all USDOE demand projections be estimated at the mine, rather than at the mill, became an international issue in INFCE fora. The issue served to emphasize still further the danger of comparing resource estimates directly with projections of cumulative requirements.

Because of these many limitations of resource estimates as measures of supply, considerable effort has been made and continues to be made to develop improved methods of mineral supply analysis. At least one authority has expressed the opinion that resource estimates may be too raw and undigested to be presented as such to the public and politicians, and that they should be used only as working material for developing scenarios of possible supply [3]. While the suggestion may be viewed with scepticism by some, it emphasizes that comprehensive resource estimates are but the first step in supply analysis.

5.7 PROJECTIONS OF PRODUCTION CAPABILITY

The NEA/IAEA Working Party on Uranium Resources compiled its first report on production capabilities in 1969 [22]. Aggregate national estimates were presented for current production and for production capability attainable within three years, based on facilities that were either operating, under construction or capable of being reactivated. Successive reports expanded slightly on this theme by providing capability estimates for existing, planned and projected facilities, as well as a projection of the maximum level of production that could be achieved within five years, supported by known deposits, markets permitting. In most cases, these estimates were fairly crude and judgemental in nature, and influenced to a large degree by industry's announced intentions.

It was not until the NEA/IAEA's 1977 assessment that a projection of production capability was presented on a year-by-year basis; annual figures were compiled for the period 1977 to 1985, with an additional projection for 1990. National projections were prepared with varying degrees of sophistication depending on the number of production centres that were being considered and the nature of the methodology that had been developed by the reporting organizations. The USDOE in particular had accumulated a great deal of experience in this field, dating back to the 1950s. With the build-up of nuclear power capacity beginning in the 1960s, production capability analysis had been carried out for nuclear planning purposes for time-spans of 10 to 15 years, while more recent USDOE studies had covered periods up to 30 years [23].

EMR Canada also had gained some experience in production capability analysis, beginning in 1973, as part of a major study which analysed Canada's requirements for new metal mines and the associated needs for exploration expenditures, capital investment and labour to the year 2000. The study, known as the 'Mineral Area Planning Study' (MAPS), covered seven major metals – nickel, copper, lead, zinc, molybdenum, iron ore and uranium [24]. A computer program was developed which was applicable to all of these commodities and projections of production capability were prepared on a mine-by-mine basis, year-by-year to the end of the century. The MAPS methodology was later modified to better accommodate uranium input data so that the program could be used for more specialized uranium supply analyses [25].

Production capability projections are basically supply scenarios showing the possible availability of a mineral commodity from different categories of resources and production centres. Such scenarios illustrate the quantity of resources that might be exploited in a specified time-frame to meet given requirements, given the lag-times involved in discovery, evaluation and development. Figure 5.6 illustrates how, as one looks into the future, the mixture of supply sources contains increasingly uncertain elements. These scenarios help to demonstrate the reality that considerable exploration and development efforts and related investments of time, money and manpower are required to achieve future production goals [2, 3].

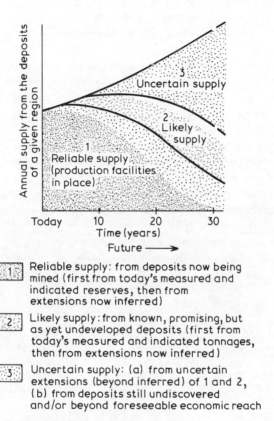

Fig. 5.6 Supply of mineral commodity – mine production from a given region to meet given requirements.

All methodologies for projecting production capability rely on a detailed mine-by-mine analysis, usually utilizing the 'production centre' concept. In the case of the uranium industry, a production centre consists of one or more ore-processing plant(s) and all resources that are likely to be mined and processed in those plants. Production centres may then be divided into classes of varying reliability, each class being analysed separately as illustrated in Fig. 5.6. For each production centre, determinations must be made of such factors as the tonnage of resources that are tributary to the production centre, mining rates and grades, ore-processing plant capacities, recovery factors and expected start-up dates in the case of production from new facilities. The more sophisticated methodologies, like those developed by the USDOE, include varying degrees of economic analysis to determine the viability of the production centres under the economic conditions established for the projection [23]. In the case of methodologies employed by the South African Atomic Energy Board, the influence of the production of gold as a co-product must also be considered [26].

Figure 5.7 presents the output from such a mine-by-mine analysis for the Canadian uranium industry, carried out for the INFCE exercise in early 1979. The projections are divided into four classes of production centres, case A representing centres that were in operation in mid-1979, case B including those centres under construction or committed for construction, and case C including those centres that were anticipated but for which no production decisions had yet been taken. Case D included carefully selected additional production centres that could be envisaged, assuming that a major portion of the inferred and prognosticated resources tributary to the centres could be realized in the projected time-frame. This latter case was judged to be too uncertain to be considered for any long-term planning purposes, and case C was chosen as the 'best-guess' maximum attainable uranium supply scenario [27].

It is pertinent to emphasize at this juncture that there is not wide support for the idea of making projections of production capability based on

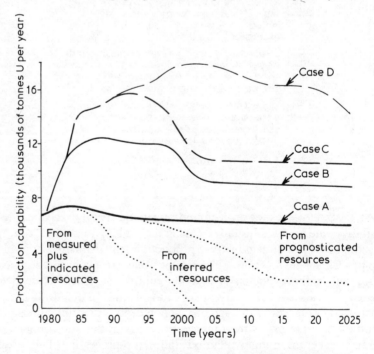

Fig. 5.7 Projected Canadian production capability to 2025.

production centres that are postulated on the realization of undiscovered resources. In the case of currently viable production centres, however, it seems justified when making projections beyond 10 years to consider undiscovered resources in the immediate vicinity (i.e. the undiscovered component of EAR). Normally the realization of such resources will not increase the maximum attainable level of production, but will merely extend the life of the operation.

The first attempt to prepare a world projection of uranium production capability for the period beyond 1990, was made as part of the INFCE exercise during 1978 and 1979. The major producing countries (i.e. Australia, Canada, France, South Africa and the United States) agreed to prepare a projection to the year 2025. The projection incorporated the maximum attainable levels of production from production centres that were operating, under construction or committed for construction, and those that were possible based on known development proposals. The supply scenarios in each case were supported by the RAR and EAR that were tributary to the production centres [12, 28].

For countries unable to prepare their own projections, an aggregate projection was made using a mathematical model developed by the USDOE as part of its internal contribution to the INFCE exercise. The Reserves and Production Projection (RAPP) model was designed to generate annual uranium ore reserve build-up and subsequent production from different geographical settings. While RAPP can be used to generate projections based on RAR and EAR only, it was designed with the objective of facilitating the determination of feasible limits to production from Speculative Resources. For this purpose, the model incorporates a number of additional input variables related to the lead-times and resource development rates for each geological type of uranium deposit for varying country situations, depending on their state of industrial development and past activity in uranium exploration and development [29].

During the INFCE exercise, considerable debate was devoted to a United States proposal to use the RAPP model to examine future supply possibilities from the Speculative Resource tonnages developed as part of the NEA/IAEA's International Uranium Resources Evaluation Project (IUREP). Strong reservations were expressed by a number of INFCE participants, related primarily to the uncertain nature of the Speculative Resource tonnages and to the high degree of judgement that was involved in developing the figures. Notwithstanding these reservations, it was agreed that it was essential to place the Speculative Resource numbers in perspective and that the RAPP model could provide guidance toward this end. Results of these exercises are presented in a later section of this chapter.

Efforts are continuing to refine and improve methodologies for projecting production capability. The NEA/IAEA Steering Group on Uranium Resources held a workshop on the subject in early 1980 to review in more detail procedures currently in use in major producing countries for projecting production capability from RAR and EAR. The USDOE's RAPP model was also reviewed and is being subjected to further studies in an effort to examine its potential usefulness for future NEA/IAEA uranium supply studies. In addition, the NEA/IAEA is hoping to prepare a manual that will provide guidelines for the preparation of production capability projections as an aid to organizations that contribute to its Red Book exercise.

The USDOE is further refining its methodologies to take into account the probability distribution characteristics of its uranium resource estimates which were developed during the NURE programme. Demand-driven projections are also being developed in recognition that, ultimately, supply flows will develop in line with actual requirements. Of particular interest in this regard is the development of a sophisticated Uranium Supply Analysis System (USAS) which will incorporate current USDOE production capability activities into a total system of uranium supply analysis for the United States [23, 30, 31].

5.8 MEASURING DISCOVERY REQUIREMENTS

One of the most difficult aspects of mineral supply analysis is the illustration of discovery requirements. The actual quantity of resources that must be discovered, to support production levels sufficient to meet a given demand scenario will vary considerably depending on the order in which deposits of different sizes and grades will be discovered and developed. For example, if a large number of small-sized, medium to high-grade deposits were discovered and developed on an appropriate time scale, future production requirements could be met with a minimum of forward reserves (see discussion on forward-reserve concept in Section 5.6). More realistically, a mix of different types of deposits will be discovered consisting of different sizes and grades and production capabilities.

By using as 'measuring sticks' the resources and production capabilities of a number of known deposits, the order of magnitude of discoveries required to support the required new production can be put into perspective. The key elements in this kind of analysis are assumptions about 'mine-life' reserves, that is to say the total reserves that would be mined during the life of each of several different mine types, and the average annual production capability of each of these mines, based on their 'mine-life' reserves. EMR Canada has utilized this method quite successfully in illustrating the magnitude of discovery requirements for a number of mineral commodities, including uranium [24, 25].

The method was employed on a world scale as a means of illustrating the amount of effort required to meet illustrative uranium requirement scenarios in a study carried out for the Conservation Commission of the

Prospects and concepts

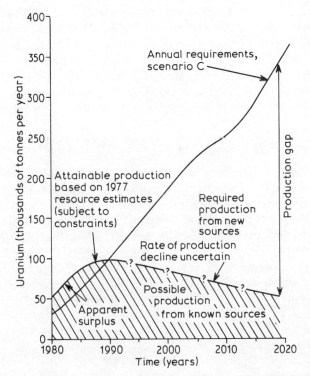

Fig. 5.8 Schematic illustration of world uranium supply problem, 1980 to 2020.

World Energy Conference (WEC) during 1977 [32, 33]. Figure 5.8, which illustrates schematically the world uranium supply problem as perceived in 1977 at the time of the WEC study, served as the framework for the exercise. The figure shows that some 300 000 tonnes of annual uranium production would have to be developed from new sources between 1990 and 2020, to meet the illustrative (base-case) requirement scenario. It was concluded that most of this new production would have to come from new discoveries.

Six mine-types of varying sizes and grades were chosen as illustrative examples of the kind of discoveries that could be forthcoming to provide the required additional production. An average time-lag of five years was assumed between discovery and initial production. It was estimated that some 329 new discoveries, representing a mixture of the six mine-types would be required by the year 2015 to provide the levels of production needed by 2020, under the assumptions of scenario C. Total 'mine-life' reserves contained in these 329 new discoveries were estimated at some 9.0 million tonnes U.

Clearly, quite a different mixture of mine-types could have been chosen and the order and number of discoveries could have been varied. The object of this type of exercise is simply to present an order-of-magnitude illustration in terms that can be readily identified with known producing operations. New mine requirements numbered in hundreds (rather than tens) is much more meaningful and readily understood than resource requirements measured in millions of tonnes U. Moreover, the illustration can be extended with relative ease to provide a measure of required effort in terms of exploration and development expenditures and manpower.

5.9 CURRENT PERCEPTIONS OF URANIUM SUPPLY

Several major studies of world uranium supply have been published since 1977, all related either directly or indirectly to the NEA/IAEA biennial assessment exercise. The comprehensive Canadian study carried out in 1977 for the WEC's Conservation Commission [32], which examined world uranium supply possibilities to 2020, relied heavily on the NEA/IAEA's 1977 assessment [9]. The NEA/IAEA's 1979 assessment [12] was scheduled and carried out as a major contribution to Working Group 1 of INFCE, the results of which were released in early 1980 [28]. Finally, the results of a major survey of the world's total energy resources – including nuclear fuel

Table 5.1 1979 Estimates of world[a] uranium resources. (Source: NEA/IAEA, December 1979)

Country	Reasonably Assured Resources (RAR) (1000 tonnes U)[b] Recoverable at costs[c] up to:		Estimated Additional Resources (EAR) (1000 tonnes U)[b] Recoverable at costs[c] up to:	
	$80/kg U	$130/kg U[d]	$80/kg U	$130/kg U[d]
Australia	290	299	47	53
Brazil	74	74	90	90
Canada	215	235	370	728
France	40	55	26	46
Gabon	37	37	0	0
Namibia	117	133	30	53
Niger	160	160	53	53
South Africa	247	391	54	139
Sweden	0	301	0	3
United States	531	708	773	1158
Others	144	193	37	123
Total (rounded)	1850	2590	1480	2450

[a] Excluding USSR, Eastern Europe and People's Republic of China.
[b] 1 tonne U (1 metric ton of uranium) is equivalent to 1.3 short tons U_3O_8.
[c] Costs expressed in July 1978, US dollars.
[d] Includes resources recoverable at costs up to $80/kg U.

Prospects and concepts

resources – carried out on behalf of the WEC by West Germany's Federal Institute for Geosciences and Natural Resources was released in the fall of 1980 [34]. The WEC's 1980 survey drew heavily on the results of INFCE.

Because of their interrelationships, all of these reports presented similar perceptions of the world uranium supply situation. The 1979 NEA/IAEA estimates of the world's uranium resources are summarized in Table 5.1. The geological nature of these resources, their distribution and importance are well described in the studies referenced above. The results of the NEA/ IAEA's IUREP exercise are presented in Table 5.2. Again, the potential for uranium discovery that is illustrated by these figures is well described in the two published reports on IUREP [10, 18]. The IUREP reports emphasized that the Speculative Resource totals 'are not meant to indicate ultimate resources of uranium, since the perspective of the Group was restricted by current knowledge which is itself severely limited in many areas of the world.'

The long-term projection of production capability developed as part of the INFCE exercise is presented in Table 5.2. It should be re-emphasized that the projection represents the maximum level of production that could be achieved, under optimum conditions, based on the resources presented in Table 5.1. Attainment of these levels would be dependent on an adequate availability of manpower, equipment and materials, appropriate financing

Table 5.2 Speculative uranium resources by continent. (Source: NEA/IAEA, December 1978)

Continent	Number of countries	Speculative resources[a] (million t U)[b]
Africa	51	1.3–4.0
America, North	3	2.1–3.6
America, South and Central	41	0.7–1.9
Asia and Far East[c]	41	0.2–1.0
Australia and Oceania	18	2.0–3.0
Western Europe	22	0.3–1.3
Total	176	6.6–14.8[d]
Eastern Europe, USSR, People's Republic of China	9	3.3–7.3[e]

[a] Resources that would be geologically comparable to resources in known deposits that are judged to be exploitable at costs of $130/kg U or less.
[b] 1 tonne U (1 metric ton of uranium) is equivalent to 1.3 short tons U_3O_8.
[c] Excluding People's Republic of China and the eastern part of USSR.
[d] A small portion of the potential represented by these Speculative Resource tonnages has likely been discovered during the period 1977 to 1980, although such discoveries would not appreciably alter the judgements contained in the table.
[e] The potential shown here is 'Estimated Total Potential' and includes an element for RAR and EAR although those data were not available to the NEA/IAEA.

Table 5.3 Projection of maximum attainable production capabilities from the world's[a] known uranium resources (1980–2025). (Source: NEA/IAEA, December 1979)

Country	1980	1985	1990	1995	2000	2005	2010	2015	2020	2025
				(1000 tonnes U/yr)[b]						
Australia	0.6	12.0	20.0	17.0	10.0	2.0	—	—	—	—
Canada	7.2	14.4	15.5	15.4	12.5	10.8	10.7	10.6	10.5	10.4
France	3.5	4.0	4.4	3.1	1.6	—	—	—	—	—
Namibia	4.1	5.0	5.0	4.6	4.6	—	—	—	—	—
Niger[c]	4.3	10.5	12.0	10.2	5.5	3.5	—	—	—	—
South Africa	6.5	10.6	10.4	10.0	10.0	10.0	10.0	10.0	10.0	10.0
United States	19.9	31.4	40.8	46.7	51.6	46.7	40.7	25.4	12.3	—
Others[d]	3.0	6.6	6.2	10.0	11.0	11.0	9.5	7.0	11.0	—
Subtotal	49.1	94.5	114.3	117.0	106.8	84.0	70.9	53.0	43.8	20.4
Phosphates[e]	1.0	3.5	5.0	6.0	8.0	10.0	12.0	13.0	14.0	16.0
Total	50.1	98.0	119.3	123.0	114.8	94.0	82.9	66.0	57.8	36.4

[a] Excluding USSR, Eastern Europe, and People's Republic of China.
[b] 1 tonne U (1 metric ton of uranium) is equivalent to 1.3 short tons U_3O_8.
[c] Figures for 1995 to 2005 from USDOE estimate.
[d] USDOE estimate.
[e] Byproduct from phosphoric acid production.

arrangements, and a minimum of regulatory delays. Critical to this projected growth, however, would be the availability of markets. Under these conditions, production could peak at levels in the order of 110 000 to 120 000 tonnes U/year during the 1990s. Thereafter, production levels from known resources would decline due to the depletion of some deposits and the mining of lower-grade ore from others. Incremental additions could be made available as a byproduct of phosphoric acid production.

The projected supply scenario from known resources (i.e. RAR plus EAR) is illustrated graphically in Fig. 5.9, together with INFCE's perception of supply possibilities based on IUREP Speculative Resources. The diagram is reproduced here exactly as it appeared in the INFCE Working Group 1 report. Rarely has so much effort gone into the preparation of a single diagram. The format of the illustration, the positioning and length of the arrows and the text for the title, legend and labels were all the subject of much debate.

The INFCE report placed considerable emphasis on the uncertainties associated with the existence of the Speculative Resources, noting that little could be said about their discoverability or availability and that only a small portion of these resources could conceivably be discovered, delineated and made available before 2025. The report's overall conclusion was that 'given the present state of knowledge of uranium resources and in view of the many constraints on exploration and development, it is unlikely that, by the year 2025, uranium could be made available at rates greater than 1.5 to 2.5 times (i.e. some 175 000 to 305 000 tonnes U/year in 2025) those currently believed possible from presently known resources in the mid-1990s'.

These conclusions were not much different than those reached by the Canadian study carried out for the WEC's Conservation Commission in 1977. Estimates of RAR and EAR and the projection of peak production capability that could be supported by these resources were of the same order-of-magnitude. With respect to longer-term supply possibilities, it was concluded that exploration and development momentum could not likely be built up sufficiently to achieve production levels in excess of 350 000 tonnes U/year by the first quarter of the next century. The implication of both sets of conclusions was that high nuclear power growth rates supported largely by conventional light-water reactors were unrealistic and that, if nuclear power was to play a truly major role in the world's future energy supply, fast breeder reactors (FBRs) and advanced fuel cycles would need to be developed and introduced.

INFCE's projection of long-term uranium availability based on known resources is compared in Fig. 5.10 with the results of its deliberations on

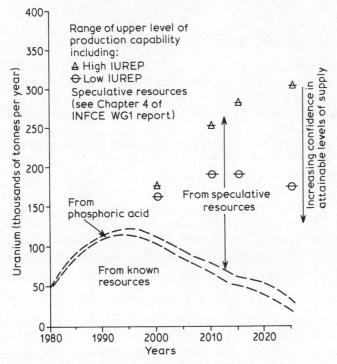

Fig. 5.9 Current perception of annual world uranium availability to 2025. (Source: INFCE Working Group 1.)

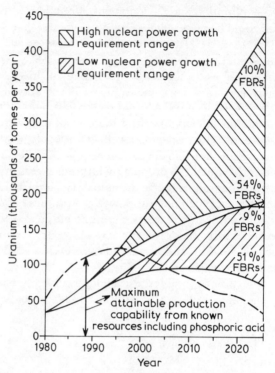

Fig. 5.10 Long-term outlook – world uranium supply and requirements. (Source: INFCE Working Group 1.)

future requirements. Over two dozen illustrative fuel requirement scenarios were developed in the INFCE exercise, for both high and low nuclear power growth situations, to produce about 50 projections of uranium demand. Those summarized in Fig. 5.10 were considered the most plausible for comparison with projections of supply. Clearly, with a range of 90 000 to 430 000 tonnes U/year in 2025 – depending on the choice of reactor strategy and the rate of nuclear power growth – there was considerable uncertainty about the role that nuclear power will be expected to play over the long-term. It is pertinent to note that this uncertainty imposes severe constraints on many aspects of uranium supply analysis. For example, it is extremely difficult, if not impossible, to judge whether exploration and development efforts are adequate, without a clear idea of the production goals that must be achieved.

Comparison of the INFCE supply and requirement projections indicates an apparent surplus of uranium supply capability lasting well into the 1990s. Moreover, a closer look at the short-term situation (Fig. 5.11) indicates that there is likely to be considerable competition in the uranium market during this period. The imbalance between supply and demand that was apparent from the INFCE analysis has been augmented since then by a continued decline in nuclear power expectations in most countries. These events have contributed to a build up of uranium inventories, to a reduction of short-term market opportunities, and to considerable downward pressure on uranium prices. Many companies, particularly in the United States, have reduced their level of production, and there have been some reports of mine closures and deferrals of uranium development projects.

In the absence of major new sources of production, however, it is unlikely that the maximum level of production that can be supported by known resources has changed materially since INFCE. Because of the cut-backs and deferrals noted above, the productive life of some operations will be extended and the shape and timing of the supply curve shown in Figs 5.10 and 5.11 will require some minor modification. Despite these modifications, it can be expected that, without major new discoveries, the overall nature of the supply curve will remain much the same. All of the post-INFCE supply developments will be considered and reflected in an updated projection of uranium production capability which is currently being prepared for the NEA/IAEA's December 1981 Red Book.

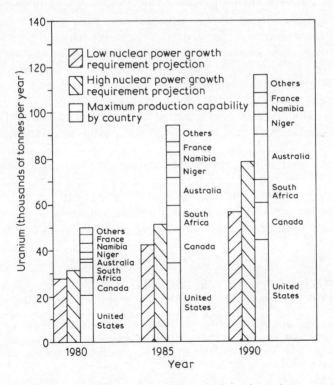

Fig. 5.11 Short-term outlook – world uranium supply and requirements. (Source: INFCE Working Group 1.)

5.10 CONCLUSIONS

In response to increasing concern about the future availability of the world's energy supplies, considerable effort has been expended in recent years to develop improved methods for uranium supply analysis. Biennial assessments of world uranium supply carried out since 1965 by the NEA/IAEA Joint Working Party on Uranium Resources have both contributed to and reflected changing concepts associated with such analyses. The NEA/IAEA uranium resource classification system has evolved through continued refinement to provide an internationally recognized framework for describing uranium resource estimates, and successive NEA/IAEA assessments have resulted in expanded and more comprehensive analyses of world uranium supply possibilities.

There has been a growing awareness that uranium resource estimates, in isolation, cannot provide a truly meaningful illustration of uranium availability. Indeed, resource estimates taken out of context can lead to false conclusions about resource adequacy. Consequently, there has been a gradual shift in emphasis away from uranium resource estimates as a measure of supply to projections of production capability that show the possible availability of uranium from different categories of resources and production centres over specified time-frames. It is believed that such supply scenarios provide a much more meaningful illustration of uranium availability for long-term planning purposes, and considerable effort is underway to improve and refine methodologies for developing these types of supply portrayals.

Recent studies carried out as part of the International Nuclear Fuel Cycle Evaluation (INFCE) provided considerable impetus to international efforts to improve uranium supply analysis techniques. INFCE's assessment of world uranium supply, which included the first long-term (i.e. to 2025) projection of uranium production capability on a world basis, was one of the most comprehensive such studies to date. The results reflect the current 'state of the art' of uranium supply analysis and continue to be broadly illustrative of current perceptions of future uranium supply possibilities.

REFERENCES

[1] USGS (1976) *Principles of the Mineral Resource Classification System of the US Bureau of Mines and the US Geological Survey,* USGS Bulletin 1450-A.
[2] Drolet, J.-P. (1980) *Future Prospects for Resources in the Face of Demand,* paper presented at the 26th International Geological Congress, Paris, July.
[3] Zwartendyk J. (1980) *Economic Issues in Mineral Resource Adequacy and in the Long-Term Supply of Minerals,* paper presented to Society of Economic Geologists, Las Vegas, February.
[4] Shank, R.J. (1980) *Dynamics of a National Resource Assessment Programme,* paper presented at Conference on National and International Management of Mineral Resources, Institution of Mining and Metallurgy, London, May.

[5] ENEA (1965) *World Uranium and Thorium Resources*, OECD, Paris, August.
[6] ENEA/IAEA (1967) *Uranium Resources, Revised Estimates*, OECD, Paris, December.
[7] ENEA/IAEA (1970) *Uranium Resources, Production and Demand*, OECD, Paris, September.
[8] IAEA (1976, 1979) *Evaluation of Uranium Resources*, Proceedings of an Advisory Group Meeting, Rome, 1976, IAEA, Vienna, 1979.
[9] NEA/IAEA (1977) *Uranium Resources, Production and Demand*, OECD, Paris, December.
[10] NEA/IAEA (1978) *World Uranium Potential, An International Evaluation*, OECD, Paris, December.
[11] NEA/IAEA (1975) *Uranium Resources, Production and Demand*, OECD, Paris, December.
[12] NEA/IAEA (1979) *Uranium Resources, Production and Demand*, OECD, Paris, December.
[13] Zwartendyk, J. (1976, 1979) Resource Classification and Terminology, in *Evaluation of Uranium Resources*, Proceedings of an Advisory Group Meeting, Rome, 1976; IAEA, Vienna, 1979.
[14] Schanz, J.J. (1979, 1980) The United Nations' Endeavour to Standardize Mineral Resource Classification, in *Uranium Evaluation and Mining Techniques*, Proceedings of an NEA/IAEA Symposium, Buenos Aires, 1979, IAEA, Vienna, 1980.
[15] USGS/USBM (1980) *Principles of a Resource/Reserve Classification for Minerals*, USGS Geological Survey Circular 831.
[16] USDOE (1980) *An Assessment Report on Uranium in the United States of America*, USDOE, GJO-111(80), October.
[17] EMR Canada (1980) *Uranium in Canada, 1979 Assessment of Supply and Requirements*, EMR Canada, Report EP 80-3, September.
[18] NEA/IAEA (1980) *World Uranium – Geology and Resource Potential*, Report on Phase I of the International Uranium Resources Evaluation Project (IUREP); published on behalf of OECD by Miller Freeman Publications, Inc., San Francisco.
[19] Zwartendyk, J. (1974) *The Life Index of Mineral Reserves – A Statistical Mirage*, The Canadian Mining and Metallurgical Bulletin, October.
[20] Williams, R.M. (1976) *Uranium Supply to 2000, Canada and the World*, EMR Canada, Report MR 168.
[21] Robertson, D.S. (1979) *The Interpretation of Uranium Resource Data*, paper presented to the Uranium Institute's Fourth Annual Symposium, September.
[22] NEA/IAEA (1969) *Uranium Production and Short Term Demand*, OECD, Paris, January.
[23] Kleminic, J. (1980) *Uranium Production Capability in the United States*; a paper presented to an NEA/IAEA Workshop on Methodologies for Forecasting Uranium Availability, Grand Junction, Colorado, February.
[24] Martin, H.L., Cranstone, D.A. and Zwartendyk, J. (1976) *Metal Mining in Canada, 1976–2000*, EMR Canada, Report MR 167.
[25] Martin, H.L., Azis, A. and Williams, R.M. (1976, 1979) Estimating Long-Term Uranium Resource Availability and Discovery Requirements – A Canadian Case Study, in *Evaluation of Uranium Resources*, Proceedings of an Advisory Group Meeting held in Rome, 1976; IAEA, Vienna, 1979.
[26] Camisani-Calzolari, F.A.G.M. and Toens, P.D. (1980) *South African Uranium Resource and Production Capability Estimates*, South African Atomic Energy Board, Report PER-51, September.

[27] Martin, H.L. (1980) *Canadian Methodology for Projecting Uranium Production Capability*, a paper presented to an NEA/IAEA Workshop on Methodoligies for Forecasting Uranium Availability, Grand Junction, Colorado, February.
[28] INFCE (1980) *Fuel and Heavy Water Availability,* Report of International Nuclear Fuel Cycle Evaluation Working Group 1, IAEA, Vienna.
[29] De Vergie, P.C. (1980) *Projection of Long-Term Uranium Production Capability Using A Mathematical Model*, a paper presented to an NEA/IAEA Workshop on Methodologies for Forecasting Uranium Availability, Grand Junction, Colorado, February.
[30] De Vergie, P.C., Anderson, J.R., Miley, J.W. and Lu, F.C.J. (1980) *Production Capability of the US Uranium Industry*, USDOE Uranium Seminar, Grand Junction, Colorado, October.
[31] Lu, F.C.J. and Welborn, L.E. (1980) *Systems Approach to US Uranium Production Capability Analysis*, a paper presented to an NEA/IAEA Workshop on Methodologies for Forecasting Uranium Availability, Grand Junction, Colorado, February.
[32] Duret, M.F., Williams, R.M. *et al.* (1978) The Contribution of Nuclear Power to World Energy Supply, 1975 to 2020 (Volume 2 – *Nuclear Resources*), in *World Energy Resources 1985–2020*, IPC Science and Technology Press, London, July.
[33] Williams, R.M. (1978) *World Uranium Requirements in Perspective,* EMR Canada, Report ER 78-4.
[34] WEC (1980) *Survey of Energy Resources 1980,* prepared for the World Energy Conference by the Federal Institute for Geosciences and Natural Resources, Hanover, September.

6

T. Price
THE BALANCE OF SUPPLY AND DEMAND

6.1 INTRODUCTION

Uranium is a commodity, and like any other is subject to the laws of supply and demand. In a number of respects, however, it is unique. First, it is a very young commodity: prior to World War 2 it had virtually no industrial uses, if we exclude minor applications like ceramic decoration. When the military demand was created, a whole new mining industry had to be brought into existence. Later the civil nuclear power programme started, which meant that a rapidly rising demand had to be matched by a similar increase in uranium availability – despite lead-times on both sides of the industry of around 10 years. The major shifts in energy policy in the 1970s were inevitably reflected in correspondingly severe fluctuations in the market conditions for uranium.

Secondly, for all practical purposes uranium has only two uses, one civil, one military. In its civil application there is no alternative primary fuel, so the many possibilities of substitution, which for other commodities do something to cushion market fluctuations, do not exist for uranium.

Thirdly, the uranium market is strongly price-inelastic: the operator of a nuclear power station, costing around one thousand million pounds, has no choice but to buy the uranium he needs. In such circumstances the fact that uranium accounts for only 6% of the cost of power sent out (this figure is for German PWRs in 1981) is a reflection of the reality of international competition in the mining industry, not of what the market could bear in its absence.

Fourthly, its military applications make uranium a highly political commodity. The fear of nuclear proliferation, and the determination of nations supplying uranium and nuclear hardware to do nothing to assist would-be proliferators, have led to a regulatory regime, which in the period 1977 to 1981, created some operating problems for both miners and uranium consumers.

Fifthly, of all the industrially developed sources of energy currently available, uranium is potentially the largest. This would not be true if we could only rely on the thermal reactors (PWR, AGR, Magnox, CANDU) which are the sources of nuclear electricity today. But the development of the fast breeder reactor opens up the possibility of increasing the energy obtainable from uranium by a factor of 50 times. The price which has to be paid for this enormous technical advance is a good deal of recycling of the fuel, which in turn requires chemical reprocessing. During this reprocessing the fissionable byproduct, plutonium, may be separated in pure form. The military applications of plutonium have become a symbol for opposition to nuclear fuel reprocessing, to the concept of the fast reactor and, by extension, to nuclear power generally. The nuclear opponents reject the dazzling prospect of an energy source which can provide a massive underpinning of industrial civilization, as being not worth the increased risk of nuclear proliferation – while in their private lives voting in effect for additional electricity supplies every time they buy a dishwasher, power drill, or electric fire. The irresistable pressures of economic demand are the surest reason for believing that the fast reactor will eventually come into its own; but, as we shall see later, the pace at which that happens will turn on the continued availability of uranium in the quantities needed for thermal reactors.

These unique attributes of uranium, superimposed on the normal supply–demand behaviour of a mineral commodity, have moulded its geopolitical context, and given rise to the policy issues we shall be discussing in this chapter.

6.2 HISTORY

Throughout its 40 years of history as an energy commodity, the uranium market has never been in a state remotely approaching equilibrium – this remains true today. The first period of growth began as a result of the military requirement. The first American needs were met from mines in Canada and Zaire, no indigenous sources being available. After World War 2 the continued military requirement led to the creation of the Combined Development Agency, a body set up jointly by the United States, Canadian, and United Kingdom governments, which assisted Canada and South Africa to develop their uranium production, including byproduct uranium from gold mining. The first Australian mine at Rum Jungle was also started with CDA support. As seen in Fig. 6.1, the industry developed rapidly until 1959; but by then progress had began to saturate the US military market, which accounted for 90% of purchases. Some contracts were not renewed; but even so world* production continued to climb, reaching 33 000 tonnes in 1959 (a figure which was not reached again until 1978). By then the indigenous US industry was coming into operation. Options on Canadian production were not taken up; an effective moratorium on imports into the US began. This was given legal status in 1964 as one of the provisions in the *Private Ownership of Special Nuclear Materials Act*. The embargo remained

---

*In the context of this paper 'world' means the world outside the Centrally-Planned Area (WOCA). Very little information is available regarding uranium production in that area.

The balance of supply and demand

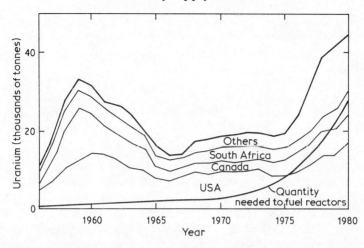

Fig. 6.1 Historical uranium production of the Western World; Source [8].

fully in force until 1977, when it began to be lifted, phased over a period of six years.

Unfortunately for the uranium miners, this collapse of the military market was not mitigated by any immediate upsurge in production for civil purposes. The inevitable result was large-scale closures. In the Elliot Lake area of Ontario, nine of the eleven integrated mines and mills shut down in the early 1960s, and production was assured only by stretching out the balance of deliveries still to be made under government contracts. In 1963 and 1965 the Canadian government provided a measure of support through a government stockpiling programme, which by 1970 had led to an accumulation of 7000 tonnes of uranium. South Africa fared slightly better, since uranium was mainly a byproduct from gold mining, which was able to continue. Even so, of the 26 uranium-producing mines and seventeen plants which supplied about 6000 tonnes to the CDA in 1959, nineteen producers had stopped and ten plants had been shut down by 1965 – production had fallen to 2650 tonnes. Only the United States industry, with the protection of the import embargo, was able to remain in a reasonably healthy state.

Confidence in the future of nuclear power gradually strengthened in the second half of the 1960s. A number of new mines were opened, and the Agnew Lake province in Ontario was discovered. Exploration expanded to a new peak (Fig. 6.2) and prices rallied from $4 to $6 per pound of U_3O_8. Hopes proved short-lived however, and construction delays and steadily more stringent safety requirements held back the nuclear programme. By 1971 uranium prices were once again down to $5. Australia went out of production altogether, and stayed out of the market for almost the whole decade. Exploration funding once again fell, uranium in this respect behaving exactly like any other mineral commodity in a falling market. Those

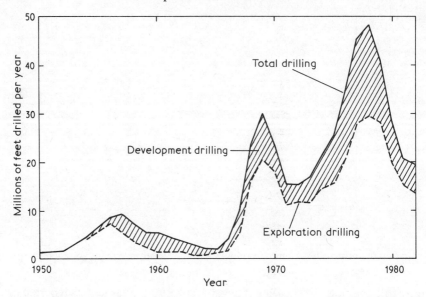

Fig. 6.2 Exploration and development drilling in US, 1950–1982; Source [9].

mines that remained in production had to operate wastefully, using only the most accessible and highest-grade ore to keep costs down, and in some cases permanently bypassing lower-grade areas.

A slow recovery in the nuclear power industry began in the early years of the seventies, but did not amount to much until the events of 1973 dramatically altered the entire energy scene. The immediate generally held expectation was that nuclear power had at last come into its own. There was a scramble for the relatively small amounts of uranium that were not committed on long-term contracts, and the price for immediate delivery rose very sharply.

Three other factors conduced to a rapid price rise. One was world-wide inflation, which in the years immediately following the 1973 war accounted for more than a doubling of mining costs generally. The second was a coincidental alteration in the terms for uranium enrichment contracts in the US – which at the time had a virtual monopoly of uranium enrichment for reactor purposes in the Western World. The changes were designed to assist the smooth working of the enrichment plants by requiring utilities to commit themselves to firm ten-year programmes, with severe penalties for late delivery of feed material; and the flexibility which had previously existed in the phasing of deliveries was removed. The result was a rush to cover the requirements of the new contracts. A third jolt came from the discovery that a major operator in the uranium market had gone short of uranium on the massive scale of 1.25 times the world annual production, and that these contractual obligations were not covered by orders on the mines.

The balance of supply and demand

Since all these factors operated simultaneously it is impossible to disentangle their effects. However, the result was that the price of uranium in the small marginal or 'spot' market rose very sharply in 1974 and 1975. The spot price went for a time to over $40, (Fig. 6.3) though meanwhile the long-term contract prices on which the bulk of uranium continued to be traded remained much lower: $16.10 on average in the US in 1976, and $21 in Canada (both official figures). This disparity between the price that the short-term market would appear to bear, and the long-term prices that had been negotiated prior to the world-wide inflation of 1973–1975, led to a period of contract readjustment – since it was not in the nuclear industry's interests to force the mining industry into liquidation. It also led to a policy on the part of the Canadian government, followed later by Australia, of setting floor prices for contracts. A spokesman for Canada announced in March 1977 that all new export contracts would be based on an escalating floor price, or an annually negotiated price based on world market value, which ever was the higher.

For a few years uranium again experienced boom conditions. But by about 1978 it was realized that the long-term effects of the oil crisis would slow down economic growth so much that demand for *additional* electric generating capacity would be much less than had previously been expected. Meanwhile, nuclear power was losing some of its economic attraction as a candidate for the reduced number of generating stations which were being ordered. The oil crisis had led to high interest rates, which bore particularly harshly on nuclear stations, with their relatively high capital cost. In addition, regulatory authorities scrutinized nuclear developments more closely after the Three Mile Island incident in March 1979, and the associated delays added to carrying charges. And supporters of strong non-proliferation

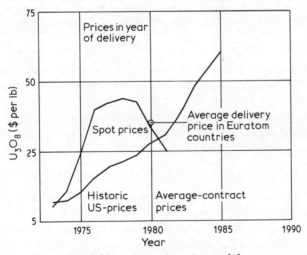

Fig. 6.3 Uranium prices; Source [8].

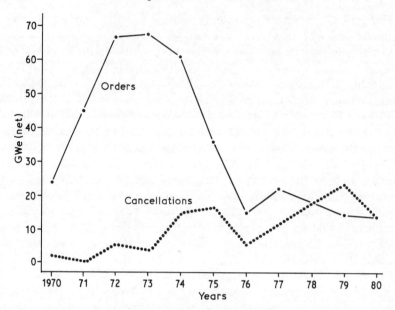

Fig. 6.4 Nuclear reactor orders and cancellations woca – annual – year end; Source [10].

policies increased their general opposition to nuclear power, as part of a worldwide political campaign. The inevitable result of all this was a string of nuclear cancellations (Fig. 6.4), and a marked fall in the price of uranium from the high levels reached in the mid-seventies. Moreover, forward provisions made for reactors which had later been cancelled or deferred created a massive stockpile, which by the autumn of 1981 was estimated to have reached the equivalent of 3 to 7 years supply (the exact figure depending on the extent to which government stocks are included). This will inevitably have some effect on the short-term development of the industry.

Despite all these problems the nuclear power industry continues to make progress on its economic merits. Reactor demand for uranium will by the mid-1990s reach around 70000 tonnes U, which will be equivalent to a doubling in 15 years. This is a rate of development which most industries would regard as entirely satisfactory.

6.3 URANIUM MINING AS A BUSINESS

Mining is an activity with long lead-times, both for exploration and construction. Exploration is always a chancy business, and as the deposits with surface expressions and radioactive signatures are identified and exploited, ways have to be found of exploring successfully at greater depths. A battery of complementary techniques is now available, but the uncertainty about a successful outcome in a given location remains high; so the future of the industry depends essentially on maintaining an exploration portfolio on an appropriate scale. Unfortunately, as we have seen, exploration programmes are not always easy to sustain in a down-turn (Fig. 6.2).

Once exploration has identified a promising ore-body, approval must still be obtained from government regulatory agencies before work can start. This in some cases can take years, as we saw with the huge deposits in the Northern Territory of Australia, which were brought progressively into production only after a two-year enquiry under Mr Justice Fox, and several years of further negotiation. The whole process proved to be a highly political affair, concerned not only with uranium markets, but also with environmental quality, safety, non-proliferation, and the territorial rights of the Aborigines.

Finally, once all the approvals have been obtained, the mine development may take five years or more. All this means that the actual uranium mining capacity of a country in a given year is something which it is difficult to predict even as little as 10 years in advance.

The commercial problem for the uranium industry, therefore, is how to keep the production rate in step with a nuclear demand which, in spite of recent reactor cancellations and deferrals, will be rising steeply over the next two decades (Fig. 5.10, p. 140). One way is via long-term contracts, let when a new reactor is ordered – a process made possible by the fact that the lead-time for mine development of a known ore-body need be no longer than the time taken to build a reactor. Such contracts provide a stabilizing force; but they have proved unable to insulate the uranium industry from the normal rigours of free markets, whose nature is hinted at in the dry statistic that in the two years after January 1980 there were cuts and delays of previously announced production of some 6000 tonnes and 5500 tonnes respectively, against a world total annual production of 44 000 tonnes (for 1980).

The country most affected by the recent changes has been the United States, where the announced cuts represented a 30% fall between 1980 and 1982. And throughout the world the industry is undergoing major restructuring. The outcome, which will affect the total amount of ore which is economically recoverable, will depend [1] on a number of factors

1. The nature of producers' supply contracts, with particular regard to quantity and price. Those with firm long-term contracts containing cost escalation clauses will have a distinct advantage.
2. The magnitude and structure of costs incurred by each producer. Those able to vary output without experiencing large fluctuations in cost, and those able to bear the cost of interim stockpiling, will have greater flexibility of choice.
3. The producer's market objectives and strategies. Large, diversified

companies may be better able to absorb high production costs during a period of low prices.
4. Political considerations, e.g. stockpiling policies, government subsidies, and the imposition of import embargoes.
5. Consumer strategy, e.g. stockpiling and diversification policies.
6. Environmental difficulties. In some countries, notably the US, prohibitive costs may be incurred in meeting the stringent environmental requirements relating to the permanent closure of mines and mills, and to the re-opening of those which have been temporarily closed.
7. Byproduct relationships.

6.4 INFLUENCE OF NATIONAL POLICIES ON THE URANIUM INDUSTRY

In many countries, the regulation of mining activities by government has continued to increase. Nowhere is this more true than in the uranium industry where, in many areas, the net effect has been to decrease flexibility, increase lead-times and, arguably, raise capital and operating costs.

Controversy over the environmental impact of uranium mining and milling is greatest in the United States, where steadily increasing pressure from various groups within and outside government has resulted in a plethora of regulations governing the exposure of both industry workers and the general public to radiation from uranium mines and their associated facilities. The regulatory bodies have wide powers of discretion, and the cost of meeting their stringent requirements, coupled with the long delays and uncertainties associated with licensing procedures, are a substantial burden for existing or potential producers – even for an industry which is conscious of the importance of maintaining high standards of pollution control.

In the United States, Canada and Australia there has been conflict over land tenure and mineral rights. The interests of local population groups vary greatly, from restriction or prevention of uranium exploration and mining to their encouragement. At one end of the spectrum there is concern that uranium development could interfere with traditional ways of life and economic pursuits; whilst at the other there is the wish to ensure that local economic benefits are obtained. Such issues are, perhaps, currently of greatest importance in Australia, where negotation of land rights with Aboriginal groups has been a precondition for the approval of several major projects.

Ownership and foreign participation is another issue. Foreign participation in energy projects was discouraged by the Australian Labour government in 1974–1975. The succeeding coalition government brought in a more open policy, which permitted foreign equity participation in uranium mining ventures to the extent of 25%. In Canada any individual or company may explore for uranium; but when a mine is developed to the production stage, foreign participation must normally be reduced to no more than one-third.

The number of countries which can base their nuclear programme in whole or part on indigenous uranium is small: the United States, Canada, South Africa, France and possibly India in the non-Communist world. A high proportion of the world's uranium, therefore, is destined to be exported for use elsewhere. The conditions under which uranium can be licensed for export consequently have a direct impact on the industry. These conditions can have three functions: to conserve sufficient indigenous uranium to ensure that a country's own reactor programmes will not be hampered by exports of uranium (Canada); to ensure that the exporting country obtains a sufficient return for its resources (Canada and Australia); and to ensure that exported material is not diverted to military uses. The effects of this third factor are sufficiently pervasive to warrant a section of their own.

6.5 NUCLEAR NON-PROLIFERATION POLICY AND ITS EFFECTS ON URANIUM

Since World War 2 it has been a prime political objective for the overwhelming majority of states to establish some kind of political fire-break between the civil and military uses of uranium. Weapon production requires its own special facilities, some of which have little to do with civil applications. But there are inevitably points where the two technologies touch. Once it had become clear that nuclear power would become a major resource, at least for developed nations, very strong political pressures developed to link the trade in nuclear fuels and equipment, and the provision of technical advice and assistance, to an obligation on the part of the recipient or purchaser not to undertake weapons production.

The first major political step towards keeping the two separate was the negotiation of the *Nuclear Non-Proliferation Treaty*, which was signed in 1968 and came into force on 5 March 1970. The Treaty was in effect a bargain between the nuclear weapons states and the rest. The understanding was that the nuclear weapons states would restrict the 'vertical' proliferation of their own nuclear weapons, while making nuclear technology for civil purposes more freely available. By February 1981 one hundred and fourteen states had ratified their adherence to the *NPT,* although a few nations, unfortunately including some which are regarded as possibly significant in the context of proliferation, still remain outside.

The satisfaction created by the signing of the *NPT* was shaken by the explosion by the Indians of a nuclear device in May 1974. Although not technically an infringement of the letter of the Treaty – the device was described as a 'peaceful nuclear explosion', and in any case the Indians had not accepted the *NPT* – it was clearly felt to be contrary to the mood of world political opinion. This was particularly true of Canada, which had supplied the research reactor hardware, and of the United States, which provided the heavy water – an essential part of the reactor from which the plutonium for the explosion was produced. The Canadian government resolved to use the leverage available from being a substantial exporter of uranium to secure binding agreements from recipient countries, requiring that all subsequent usage (e.g. reprocessing) should be subject to Canadian approval. On 20 December 1974, only 7 months after the Indian explosion, the Canadian Minister of Energy, Mines and Resources announced that in future Canada would not supply nuclear material, equipment or technology without binding safeguard assurances that they would not be used to produce a nuclear explosive device, whether the development was stated to be for peaceful purposes or not. In December 1976 the policy was strengthened by making export of uranium and reactors conditional upon ratification of the *Non-Proliferation Treaty* by recipient countries, or acceptance of international safeguards on their entire nuclear programme. This policy, which inspired a similar approach on the part of the Australian government, demonstrates the political leverage possessed by the supplier of such a valuable commodity as uranium.

Although Canadian policy objectives were fully supported by her European and Japanese customers for uranium, they had an unforeseen side-effect which led to a temporary cessation of Canadian exports to West Germany for several months. Germany had, of course, signed the *Euratom Treaty*. Under the provision of that Treaty the signatories have the right to move nuclear fuel freely within the Euratom Community. Indeed, such freedom of movement is a fundamental requirement of the *Treaty of Rome*, and one of the duties of the Euratom Supply Agency is to monitor contracts to ensure that this right of freedom of movement is not infringed by supply contract provisions. However, two Euratom countries – France and the United Kingdom – possess nuclear weapons; and this was regarded as opening up the possibility, in principle, that Canadian uranium might find its way into weapons.

Although West Germany was not herself a nuclear weapons power, and had declared by unanimous vote in the Bundestag that she had no intention of becoming one, the Canadian government was unwilling to create a precedent which might have hampered her determination to press forward vigorously with non-proliferation measures. The result was an embargo, imposed in January 1977, and lasting throughout that year, on the supply of Canadian uranium to Germany. The problem was eventually resolved by an interim agreement. The United Kingdom was also affected. The subsequent Australian agreement with the United Kingdom of July 1979 built on this experience; it made specific reference to the need to establish a further agreement directly with Euratom, and permitted retransfer of nuclear material within the European Community.

The Canadian determination to ensure that non-proliferation controls

The balance of supply and demand

were effective was matched by similar attitudes in the United States. The strictly worded and immensely complex *Nuclear Non-Proliferation Act* (*NNPA*) which became law in March 1978, during President Carter's administration, was in fact an essentially bipartisan measure, commanding the support of both Republicans and Democrats. It contained a number of provisions which *inter alia* restricted trade in nuclear fuels, for reasons which aroused considerable opposition for technical reasons on the part of the United States' friends and allies.

The problems sprang from differing views on the best future course for nuclear power. Before taking office, President Carter had been advised by a group which, in the Ford–Mitre report [2] of 1977, argued strongly that uranium supplies would be adequate for a long time to come to satisfy the world's needs for nuclear power, on the basis of generation in thermal reactor systems, without any need for the fast breeder and all the reprocessing which goes with it. From this it was a short step to a policy which sought to restrict United States' uranium and enrichment services to nations which did not intend to reprocess their spent fuel. Deliveries of enriched uranium were held up for a time while the United States reviewed the position.

Other nations took a different view. While they agreed that the fast reactor would not be needed on a commercial scale for at least another two decades, they argued that it would take that time to develop it fully. On a somewhat shorter time-scale they felt that reprocessing – which allows uranium and plutonium to be recycled in useful quantities – was worth pursuing, at least on a pilot-project scale, again because the learning period is long. Moreover, in some countries, such as Sweden, it is necessary as a matter of law to provide the regulatory authorities with a statement of how the spent fuel is to be treated before a reactor can be licensed; so reprocessing could well be relevant to the future of reactor programmes. And finally, there was the paradox that the self-same interruptions in fuel supply which had been created by producer governments in pursuit of non-proliferation policies would themselves come to make the fast reactor and reprocessing seem more attractive, since they would be essential components of an independent fortress-type energy strategy, albeit one only attainable in the long term.

In an attempt to win the world round to the American viewpoint President Carter launched, in 1977, an International Nuclear Fuel Cycle Evaluation, INFCE. This grew into a major exercise, lasting 2½ years, with over fifty nations taking part in eight working groups. It was a useful contribution to clear thinking on this complicated subject, because it brought out for the first time that non-proliferation policy has strong effects on the nuclear industries –including the uranium mines and fuel services – which can seriously interfere with planning.

The industry seized the opportunity afforded by INFCE to remind those taking part that they were discussing the future of an industry which was already significant, and which by the end of the century, for some countries, would be approaching the importance of oil. In France, nuclear power electricity is expected to account for about one-third of primary energy requirements by the year 2000, about the same as oil; and a number of European countries will be making half their electricity by nuclear means by 1985. The regulations governing such an industry must take into account the need for operability in the practical everyday circumstances of a fuel market which involves as much prompt decision-making as any other. Particular exception was taken to the so-called 'case-by-case determinations' which were enshrined, for instance, in the United States' *NNPA*. Such determinations, for various bureaucratic or political reasons, could last as long as 12 months even for allies such as Germany and Japan. For other countries years could go by without decisions being made.

The Uranium Institute, in a series of publications [3, 4] during and immediately after INFCE, argued for non-proliferation policies which were more stable, and which, to the maximum extent possible, streamlined the procedures by making the requirements of the various exporting countries clear. If possible (the Institute argued) it would be best if one universally applicable set of rules could be adopted. Although the ideal of a single world-wide non-proliferation framework remains elusive, some convergence of national requirements of the producer nations is a goal worth working for.

At the time of writing (1982) there seems to be much closer accord between the regulators of nuclear fuel trade and the regulated. Early in the Reagan administration, in July 1981, it was made clear that the industry's need for greater predictability was well recognized; moreover, it was explicitly stated that the United States would take into account the political record of consuming nations – so avoiding the anomaly of treating its closest friends on exactly the same footing as other nations whose dedication to non-proliferation was less clearly established. A somewhat similar change of mood had occurred a little earlier, at the end of 1980, in both Canada and Australia. There are thus hopeful signs that a reasonable balance will come to exist between the need to police non-proliferation agreements (e.g. via the inspection teams of the International Atomic Energy Agency) and the need for the industry to conduct its affairs in a business-like fashion.

6.6 URANIUM AND THE FAST REACTOR

A major policy issue, whose impact spills over into national energy policies, is for how long uranium supplies will be adequate to fuel the world's thermal reactor stations. In other words, at what stage should a shift begin, on more than merely a demonstration scale, towards the fast breeder reactor, with its dramatically improved uranium utilization? The resource figures given in Table 5.2 (p. 137) if taken at their face value – and depending somewhat on

The balance of supply and demand

the course of the world economy over the next thirty years – would imply a date around the end of the first quarter of the 21st century. This is further away than the design life-time of thermal reactors currently being brought on-stream; so provided the cost of uranium is not likely to rise too far from present levels during their operating lifetime the question can be regarded as one to be settled at some time in the future.

There are, however, large current research and development commitments to the fast reactor and its associated reprocessing facilities. Should these continue at their present level, be cut back, or expanded? It is impossible to give a firm answer at this stage; but there are reasons for believing that there is unlikely to be any great urgency, unless considerations of national self-sufficiency in energy supervene.

The key unknown is how far the cost of power sent out from a breeder reactor can match that sent out by existing and improved types of thermal reactor. Several reasons suggest that the fast reactor will not find it easy to match thermal reactor economy. First, the fast reactor will have to compete against thermal reactors whose performance had already been optimized over a period of five or six decades. Then it is likely to have high carrying charges for its concentrated fossil fuel inventory. Furthermore, to gain the theoretically possible fifty-fold increase in uranium utilization, a great deal of reprocessing will be needed. These are grounds for believing that the breeder will be unable to overhaul thermal reactor economics for at least as long as natural uranium prices remain in the present range. At that level they amount to only about 6% of the cost of electricity sent out from German PWRs. So if the overall generating cost of the breeder is, for the sake of argument, 30% more than that of the PWR, the latter could afford to pay up to 5 times as much for its uranium as at present (in real terms) and still break even. Given the certainty that any such price increase would stimulate a high level of exploration, the question then becomes one for geologists to answer. The cost of a mineral being, to a first approximation, inversely proportional to the ore-grade, the issue is what total tonnage may be expected to exist in uranium deposits with average grades one-fifth of those at present used commercially.

Table 5.1 (p. 136) indicates that a useful improvement in resource availability occurs, by about 10% for each 10% increase in real costs; but such figures do not take us very far. A somewhat higher price elasticity of availability has been predicted in a study by Deffeyes and MacGregor [5]. Their theoretical model shows a log–log relationship between cumulative tonnage and ore-grade of uranium, which implies that halving the workable ore-grade (i.e. doubling the cost or price) would result in an increase in the economically accessible material by a factor of around five. While the discrepancy between the two elasticity estimates indicates that more work needs to be done, the message is clear enough: substantial changes in the price payable for uranium will also result in substantial changes in the quantity that can be mined at a profit.

This leads on to the question of whether there are any natural lower limits to the grades that could be commercially worked. Three possible kinds of limit can be envisaged. The first is when the energy consumed in extracting the uranium becomes comparable with what can be derived from using the uranium in thermal reactors (we are here only concerned with the thermal reactor context: the problem would not arise if fast reactors, with their far better uranium utilization, were the main electricity producers). According to an estimate made by Mortimer [6] this energy limit comes for practical purposes at an ore-grade of about 50 ppm. This is so far below the grades now being worked – which are rarely below 1000 ppm, and can be much higher – that the exact figure is not an immediate concern. A much more pressing question is whether exploration can find such additional resources in the first place.

A second limit of a kind would arise from the possibility of extracting uranium from sea-water, which beyond a certain uranium price would make it unnecessary to take the laborious route of mining and breaking up low-grade rock. No-one yet seems to be in a position to give reliable figures for the cost of such a process, and the environmentalists concerned with possible physical damage to marine life have yet to be heard. Unless substantial improvements to present technology can be achieved, it seems likely that sea-water uranium could well use up a large part, if not all, of the cost advantage thermal reactors may have even over prototype fast reactors; and there would seem little point in using sea-water extraction permanently as a basis for fast reactors, because the limited fuel supplies that would then be necessary could quite certainly be acquired from more conventional sources, uranium being a relatively abundant and widely-dispersed material.

A third limit to low-grade ore extraction could be environmental. To extract the 100 000 tonnes of uranium which will be needed annually in the early years of the 21st century, all from low-grade ores of say 300 ppm average grade, would require the mining and subsequent disposal of 300 million tonnes of tailings per year. Environmental considerations would then certainly be important, but there is no reason to believe that they would be insuperable.

In the present state of knowledge it is premature to assume that appreciably lower grade ores cannot be worked; or, to put the point more positively, if there is sufficient economic incentive it is likely that this is what will happen. Assuming that the Deffeyes and MacGregor analysis is broadly correct this would then provide large additional resources, which could be brought into play during an extended period of economic competition between fast and thermal reactors. Utilities would then be likely to continue with their programmes of thermal reactors for as long as they could be certain that the uranium supply would be reasonably secure for a further 20 to 30 years; and in the circumstances we are assuming this assurance would seem likely to be available at least until the year 2010.

The balance of supply and demand

Furthermore, thermal reactor technology would not necessarily stand still. In the early years of reactors a key design factor was 'neutron economy' – the conversion of ^{238}U to ^{239}Pu – since this greatly affected the attainable burn-up, metallurgical considerations apart. It was also the thermal reactor designer's natural response to the same problem of fuel conservation which has led to the fast reactor. It was one of several reasons for starting work on the High Temperature Gas Cooled Reactor – which, in the late fifties, was seen as holding promise of a high conversion factor, coupled with robust fuel elements. In the intervening years this consideration has been lost sight of, so far as thermal reactors are concerned, partly because engineering robust reactors has proved difficult without using considerable amounts of neutron-absorbing material in the core structure, and partly because uranium supplies have shown no signs of coming under immediate strain. But nuclear power is now fully established and uranium consumption is increasing, so the issue of neutron economy once more needs to be given proper attention. The INFCE figures for what might be attainable from design improvements are given in Table 6.1. They show potential gains of up to 2.25 times over conventional PWRs without recycle and without resorting to thorium.

Table 6.1 Dependence of life-time requirements on reactor technology [7]

Type	Lifetime uranium requirements* (tonnes U)
LWR	
Current technology	4300
Improved technology	3100–3700
HWR	
Current technology	3600
Improved technology	2500
With Pu recycle	1800
With Th recycle	1200
HTR (Thorium cycle)	1600–2400
FBR	
Post 2000 technology	50

*1 GW for 30 years at 70% capacity factor 0.2% enrichment tails assay

Putting all this together, we can see our way to closing some of the enormous gap, a factor of 50 in uranium utilization, that previously appeared to exist between the 'breeder' and the thermal reactor. We have now identified a factor of perhaps 5 from supply elasticities (Deffeyes would put the figure rather higher) and a further 2 or more from better reactor design, making a foreseeable improvement about a factor of 10 in all. If uranium supply is indeed the key factor determining the timing of the commercial fast breeder era, this shows that it is premature to rule out the possibility of a longish period of competition between thermal and fast reactors – even without bringing in the supplementary thermal reactor fuel, thorium.

6.7 CONCLUSION

The purpose of this chapter has been to draw the reader's attention to the special geopolitical identity which uranium has acquired. To understand its role in the nuclear scene it is not enough to look only at the statistics of medium-term resource availability which were the subject of the previous chapter. We also need to bear in mind the political sensitivity of uranium; the politics of environmental considerations, which can affect mine approvals; and its significant role in nuclear non-proliferation policy – which in turn affects the conditions of trade. If these conditions were to impede the smooth flow of uranium without good reason we could then find that the understandable desire of several countries for energy independence would give a boost to the fast breeder and reprocessing. If on the other hand a workable regime of non-proliferation controls is maintained, then there will be a different outcome, with the fast reactor taking over only gradually. It need occasion no surprise that the future of the most advanced of reactor systems will turn, in the end, on the ready availability of the basic fuel.

REFERENCES

[1] *The Uranium Equation* (1981) Mining Journal Books, London.
[2] *Nuclear Power Issues and Choices* (1977) Ballinger Publishing Company, Cambridge, USA.
[3] *Government Influence on International Trade in Uranium* (1978) Uranium Institute, London.
[4] *Bilateral Agreements and the Evaluation of the International Safeguards System* (1981) Uranium Institute, London.
[5] Deffeyes, K.S. and MacGregor, I.D. (1980) World Uranium Resources. *Scientific American*, January, p. 5.
[6] Mortimer, N.D. (1980) in: *Uranium and Nuclear Energy,* Mining Journal Books, London, p. 79.
[7] *International Nuclear Fuel Cycle Evaluation** (1980) Summary INFCE/PC/2/9, International Atomic Energy Agency, Vienna.
[8] Stephany, M. (1982) in: *Uranium and Nuclear Energy: 1981,* Butterworth Scientific, London, p. 105.
[9] *Uranium Exploration Expenditures in 1980* (1981) Report GJO-103 (81), US Department of Energy.
[10] Darmayan, P. (1981) in: *Uranium and Nuclear Energy: 1980,* Westbury House, Guildford, UK, p. 13.

* International Fuel Cycle Evaluation, an exercise in which many countries took part in order to assess the relative merits of reprocessing or not reprocessing irradiated nuclear fuel before storage or disposal.

NUCLEAR FUEL CYCLE SERVICES

7

A.J. Hyett
THE STRUCTURE AND ECONOMICS OF THE NUCLEAR FUEL CYCLE SERVICE INDUSTRY

7.1 INTRODUCTION

The UK, USA and USSR have been using nuclear power to produce electricity commercially since the 1950s (then with prototype reactors); the first full-sized reactors to come on stream in the UK were the Magnox reactors developed and built in the late 1950s and early 1960s; similar reactors were built in France, while developments in Canada and in the USA produced alternative commercial designs. The 1950s was also the decade in which the American 'atoms for peace' programme initiated the worldwide dissemination of information on nuclear power technology. By the end of 1978 there were some 220 reactors operating worldwide with a total capacity of some 120 GW [1] an energy equivalent of 3 million barrels of oil per day, or roughly 2% of world energy consumption. For the USA nuclear power was then providing some 13% of electricity; in the UK some 14%; Belgium 20% [2]. France has the most ambitious target of all of meeting 20% of her energy needs by 1985, requiring nuclear power to provide some 50% of total electricity production [3, 30]. The 1981 Report of the International Atomic Energy Agency showed that by end 1980 the number of reactors worldwide had risen to 253, generating some 8% of world electricity. In addition a further 230 reactors were then under construction and 118 were at the planning/ordering stage.

Because the energy available from uranium is so great in comparison to that from fossil fuels, a substantially more complex fuel cycle and more elaborate capital infrastructure are made economically feasible. Although the output of electricity is by no means the only role suggested for nuclear reactors, it is their only current commercial use; a recent proposition for Magnox reactors has been to use the high-temperature steam (350°C at 2500 psi) they can provide to recover heavy oil or bitumen at depths of 1500 to 3000 feet. There is a claimed conversion efficiency of 300% (measured as the calorific value of heavy oil produced to the energy value of steam production and injection) (*Atom*, June 1981 and *Financial Times*, 23 September 1981). See [1] and [4] for a discussion of the use of nuclear reactors in providing process heat. The Swedish company ASEA-Atom have devised a reactor project named SECURE (the Safe and Environmentally Clean Urban Reactor) to produce hot water for district heating. This aims to provide 400 MW of heat at an outlet temperature of 110°C to cater for a community of 50 000 (*Financial Times*, 18 July 1980). The USSR fast reactor Bal 350 at Shevchenko on the Caspian Sea is designed to purify 120 000 tons of water a day as well as generating 150 MWe of electricity (*Atom*, November 1979).

Thus, electricity output is one element in an entire cycle of operations concerning nuclear fuel, and resulting in a worldwide industry. Conventionally the nuclear fuel cycle is divided into front-end and back-end activities. The front end of the cycle relates to the mining and milling of uranium ore to produce 'yellowcake' (80% pure uranium oxide (U_3O_8)), the purification of the yellowcake and conversion to uranium hexafluoride gas; the enrichment of this gas, production of uranium dioxide and fuel fabrication. The back end of the fuel cycle concerns the methods of dealing with the irradiated fuel rods once they leave the reactor. These involve spent fuel storage, reprocessing and waste solidification and disposal. The possibility of reprocessing the fuel makes it necessary to divide the back end of the cycle further, into either an open or a closed cycle. The open or 'once-through' fuel cycle eliminates the reprocessing option and is concerned with the processes necessary to store, safely and indefinitely, the spent fuel and accumulated waste products. The closed cycle still involves the storage of waste, but includes the reprocessing of the spent fuel to extract energy-rich components which it still contains and which may then be recycled to the reactors.

7.2 MINING AND MILLING OF URANIUM ORE

Mining of uranium-bearing rock initiates the nuclear fuel cycle. Uranium is a mineral which occurs naturally in ore in varying levels of concentration; the generally higher-grade ores currently mined have uranium concentrations in the region of 0.1% to 0.5% by weight. In order to obtain (after refining) 150 tonnes of natural uranium, the requirement would involve mining, at the most, some 300 000 tonnes of ore [5]. Such a quantity of natural uranium burnt in a reactor for one year would generate as much electricity as would a coal-fired station burning over two million tonnes of coal. The nuclear plant would not require the mining of an equivalent tonnage of ore until the average uranium oxide content of the ore fell to a range of 0.006% to 0.007%, which is some twenty times lower than the current average [6]. This comparison of relative mining tonnages also leads to comparison of land areas devoted to nuclear and coal fired plant. A 1000 MWe nuclear plant would represent annually mining of some 20 to 50 acres of land (at an ore concentration of 0.2%) plus a further 30 to 70 acres to accommodate milling plant and storage of residue (tailings). The equivalent output coal plant would require 100 to 400 acres of mining land plus space to dispose of ash and sludge (although some of this land can be reclaimed when mining and waste dumping is completed) [6].

Naturally, as mining continues, the more accessible and richer ore reserves will be depleted, leading to increased mining costs as greater volumes of rock will have to be processed to achieve given outputs of uranium oxide. (Where open-cast mines are in operation the possibilities for very large scale activities are offered, but the more intensive mining could impose significant and expensive environmental problems.) Once mined the ore yields its uranium content in the milling processes. The rock must be crushed and ground, and from the slurry that is formed a purified uranium salt known as 'yellowcake' is separated. This powder will contain some 80% uranium oxide (U_3O_8).

The OECD has published a series of reports on world uranium potential compiled in a joint study by OECD, NEA and IAEA [7]. These estimates of resource availability are necessarily dependent on the cost allowed for exploitation. Should the price of a mineral rise then reserves previously unexploited for economic reasons become desirable. Thus 'Reasonably Assured Resources' refers to the expectation of that which will be recoverable from known deposits, using existing technology at current prices. 'Estimated Additional Resources' refers to predictions that deposits will be discoverable and exploitable at current prices within as yet unexplored areas either of known deposits or of probable ore-bearing rock. A discussion of this approach to analysis of uranium availability, together with discussion of the possibilities of exploiting the vast reserves of very low grades of ores where 'the problems of recovery are at least as forbidding as the quantity is attractive' is given in [4].

Despite their limitations as measures of an absolute availability, these estimates serve to give some guide as to the tightness, or otherwise, of supply over time, and thus to the need for continued exploration to meet the industry's desired position of having exploration precede production by a decade. The rate of exploitation itself will be directly related to the rate of change of the market price net of extraction cost. Should this be rising at a rate less than the available interest rate it will pay mine operators to deplete reserves as fast as possible; similarly a net price whose rise is at a rate greater than the interest rate will encourage a decrease in exploitation of a resource which now offers a better return than the money market [8].

A recent claim holds that, on current evidence, not much more than 100 000 tonnes of uranium could be retrieved annually from known reserves, and while active exploration could increase this there remains a lead-time of ten to fifteen years between prospecting and bringing a mine on stream, the fall in the spot market price from $115/kg in 1979 to $66/kg in 1981 would not be an encouragement to such investment (*Atom*, July 1981, and [9]).

Kroch [10] suggests that a typical mine processing some 1000 tonnes of uranium ore per day would represent capital costs in the range $10 million to $20 million with operating costs of some $1.10 per kilo of U_3O_8 ($ 1974).

Given the evident dependence of the largely private-enterprise mining and milling industry on the final demand for uranium in electric power stations, it is clear that any uncertainty in orders for, and development of, these stations can result in substantial problems for the industry which will lead to a reduction in the flexibility of output. Uranium exploration is more costly than is the case for oil; it is expensive to close and reopen mines, and the time lag between exploration and mine operation may be up to eight or ten years. Similarly the electric power generating industry suffers greatly in periods of uncertainty; planners must operate on the construction time for a reactor being at least ten years and the plant having an expected lifetime of over thirty years. Since, in general, electricity producers are required to meet demands placed on them, stability of fuel supplies becomes a highly significant element in the decision process.

By 1980 a uranium glut was becoming apparent, showing itself in falling prices, mine closures and cancellations of future mine projects. Since it can cost more to reopen a closed mine than to start afresh, a prolonged weak market for uranium would reduce the chances of mines being reopened and further limit the responsiveness of output.

What little elasticity of supply there is in the short term at this point in the cycle is provided either by stocks or by increasing the recovery rate of U_3O_8 in the mill by leaching more oxide from what would otherwise be the waste; the latter will inevitably lead to increased costs. Should the nuclear fuel cycle be closed, however, the reprocessing of irradiated fuel rods will generate more fuel for recycle to reactors, thus alleviating to some extent the pressure on the mining and milling sector; furthermore the introduction, on a commercial scale, of breeder reactors burning plutonium would expand the timescale of usefulness of existing stocks of uranium dramatically. Lellouche [3, 30], quoting André Giraud, then head of the Commissariat à l'Énergie Atomique: 'France owns estimated natural uranium reserves of 100000 tonnes . . . consumed in Light Water Reactors they represent 800 Mtoe or one-third of the North Sea Oil reserves. Through the use of breeder reactors this uranium can produce 50000 Mtoe, the equivalent of all the Middle East oil reserves.'

Thus, IIASA [4] argues that 2030 would be the exhaustion date of even the most optimistic resource estimates (except very dilute uranium sources) if consumed in a once-through cycle, so that uranium resource estimates serve to define the period of useful life of existing reactor systems and to indicate the proximity of a move toward commercial breeder reactors. 'The "once through cycle" . . . a sensible alternative for a country rich in uranium but not for countries lacking such resources' [11].

Unusually for such a commodity, there is no international exchange nor any 'marker' price for uranium, although private organizations do attempt to publish up-to-date contract prices. In 1972 the Uranium Marketing Research Organization (UMRO) was formed amongst companies and governments outside the USA concerned with uranium mining. This organization existed to exert pressure to raise prices – an event which, by 1975, had occurred to an extent sufficient for UMRO to cease to exist (although prices have since fallen again, and both USA and Canada initiated legal action against UMRO).

Uranium mining and milling is an industry comprising a comparatively small number of companies exploiting reserves which are (currently) fairly highly-concentrated geographically. It is by no means easy for new entrants to gain access to such an industry, whose output is a commodity without direct substitute. Given these conditions a cartel might seem a highly probable outcome; nevertheless, Buckley et al. [12] argue that any further moves toward a cartel are unlikely for two reasons: (1) producers would be unwilling to be the cause of similar economic problems to those resulting from the OPEC cartel (although given the degree of difference between the uses served by oil and the sheer volumes of output this argument seems rather weak), and (2) any oligopolistic behaviour would be counteracted by vertical integration from the large users in the private sector, whose unique dependence on uranium would otherwise leave them facing ever higher prices. In France such vertical integration already exists, as CEA, through its subsidiary COGEMA*, has gained control with a legal monopoly over the entire commercial services of the fuel cycle from uranium mining to enrichment and reprocessing [13]. In 1978 the first uranium was produced from the mine owned by Cominak (La Compagne Minière d'Akouta) in Niger, with a target full-capacity of 2000 tonnes per year by 1980. CEA holds shares in Cominak, together with ENUSA (Spain), OURD (Japan) and the Niger government (*Atom*, November 1979).

7.3 THE NUCLEAR ENERGY PROCESS

The next stage in the front end of the nuclear fuel cycle involves the purification of yellowcake and subsequent manufacture into fuel for the reactors whose types vary throughout the world. The most widely used are the light water reactors (LWR) which include both pressurized water reactors (PWR) and boiling water reactors (BWR) [1], developed in the USA in a manufacturing industry dominated by Westinghouse and General Electric. Both the UK and France developed gas-cooled graphite reactors (the UK Magnox reactor) but these have been superceded in the UK by the indigenous Advanced Gas Cooled Reactor (together with current govern-

---

*COGEMA, La Compagne Générale des Matières Nucléaires. Created 19 January 1976, it is 100% owned by CEA and exists to operate all industrial and commercial activities concerned in the nuclear fuel cycle, and so is separate from the public service duties of CEA.

ment plans [14] for a third phase of reactors based on the Westinghouse PWR design), and in France by the PWR whose monopoly supplier is Framatome, under licence from the American company, Westinghouse. In the mid-1970s CEA became a shareholder in Framatome and Westinghouse agreed to a gradual decrease in its own participation [3, 30]. In Germany the largely privately-owned industry is based on a reactor type derived from the basic Westinghouse model; by the end of 1978 there were 25 nuclear power plants in operation or under construction in Germany. Twenty-one of these were the responsibility of one company: Kraftwerk Union; which had also built the nine reactors exported from Germany with orders from Spain, Switzerland, The Netherlands and Austria [15, 30].

The Canadian development of nuclear power has led to a unique design of heavy water reactor known as CANDU which has found some export markets in Argentina, India, South Korea and Pakistan [16]. At the end of 1978, twenty-two countries were operating a total of 223 power reactors with an average size of over 800 MWe [4].

The fundamental process on which all of these rely is the chain reaction of fission in nuclei of uranium. Natural uranium consists of two isotopes; 99.3% of it is made up of ^{238}U, the remaining 0.7% is ^{235}U (the numbers 238 and 235 refer to the mass of the atomic nucleus; thus the two isotopes differ in atomic mass but not in their chemical properties). The isotope ^{235}U is said to be unstable, as spontaneous fission may occur. The nucleus of the atom breaks up and this will cause excess neutrons (which form part of the nucleus) to be emitted at high speed. If one of these neutrons should be absorbed by the nucleus of another atom of ^{235}U, the second nucleus becomes 'excited' and so unstable that it too will undergo fission, again throwing off excess neutrons, and releasing energy in the form of heat. If it were possible to arrange for this process to be continued, then we would have a chain reaction with an output of heat; if, however, the excess neutron collides instead with a nucleus of ^{238}U then it is simply absorbed, there is no new fission, and thus no chain reaction occurs.

Evidently the engineering objective is to design a nuclear reactor such that a continuous chain reaction is possible. The conditions of this chain reaction should be such that one of the neutrons emitted in each fission should go on to cause a further fission and allow the chain reaction to repeat. Should more than one neutron cause additional fissions, the chain reaction would grow exponentially; and similarly it could decline.

However, natural uranium contains a relatively low concentration (0.7%) of the fissile ^{235}U, and this necessitates the introduction of a mechanism to slow down, or 'moderate', the emitted neutrons. By so-doing it increases the probability that a neutron will be absorbed by a nucleus of ^{235}U which will undergo fission and so allow the continuation of the chain reaction. (Natural and low-enriched uranium contain a low concentration of the fissile material. The effect of the moderator in slowing the neutrons is to increase the probability of a chain reaction by increasing the fission cross section or apparent size of the nucleus as seen by the incident neutron.) Light water reactors (PWR and BWR) are moderated by ordinary (light) water; the British Magnox and AGR use graphite and the CANDU uses heavy water (deuterium oxide). Once the reaction is made possible (by bringing together a sufficient quantity of uranium fuel), control over start-up and shut-down is exerted by withdrawal and insertion of control rods. These are made of a substance which has the property of absorbing neutrons; when the rods are inserted into the core of fuel in the reactor they will act to prevent emitted neutrons from causing further fission; as they are removed so the chain reaction may begin again. Boron is an ideal substance for manufacture into control rods. (Further safety and control devices are part of the engineering design.)

Finally, a means must be found to draw off the heat generated and drive the turbines. The PWR uses light water under high pressure to act as both the moderator and the coolant; the Magnox and AGR use carbon dioxide gas to cool the fuel and transfer the heat. Each of these reactors uses a secondary water circuit heated by the water or gas to generate steam for the turbines. (The BWR uses the steam produced in the process of cooling the reactor, and does not require a secondary circuit.)

Thus, these 'variations on a theme' (collectively grouped as 'thermal' reactors, since slow neutrons are also known as thermal neutrons) rely on the instability of the ^{235}U isotope, with neutrons slowed by a moderator. And in the cases of the British Magnox and Canadian CANDU reactors the fuel input is pure natural uranium. However, the other reactors mentioned so far operate on fuel in which the naturally occurring ratio of ^{235}U has been increased – fuel which is *enriched* so that its concentration of ^{235}U is between 2.0% and 3.0%. (Even this concentration is still low enough to require a moderator to slow the neutrons to ensure the chain reaction). The major benefit of enriching the fuel input to this level is that the reactors can operate with a substantially lower fuel load and thus smaller core size than is required by CANDU and Magnox. Fuel enrichment means that both higher rating (measured as MW/tonne) and higher burnup (MWdays/tonne) are achieved; the smaller core size is made possible as each fuel element is contributing a greater energy output. The cost lies in the economics of the enrichment process.

7.4 ENRICHMENT

To produce fuel for reactors operating on enriched uranium, the purified natural uranium is converted to uranium hexafluoride gas (UF_6, known as 'Hex') which is a suitable form for handling. The Hex is then transported to the enrichment plant. There are enrichment plants in the USA, UK, France, West Germany, the Netherlands and South Africa, operating one of three methods: gaseous diffusion, gas centrifuge, or the nozzle process.

The work performed in the enrichment plant is measured in Separative Work Units (SWU) and defined in kilograms, the capacity of the entire plant being defined in SWU per year. Since the object of the process is to produce enriched uranium, a byproduct is uranium depleted in ^{235}U, known as the 'tails'. The 'tails assay' refers to the ^{235}U which is allowed to remain in the stream of depleted tailings. Thus, operating the plant with a tails assay of 0.3% means that there will be 0.3% ^{235}U and 99.7% ^{238}U in the residue remaining to be stored. This high concentration of ^{238}U renders the tails stream useless as a fuel within the framework of the nuclear fuel cycle presented so far; but should the fast reactor become a commercial proposition, this inventory of depleted uranium will take on enormous importance.

The differing possible combinations of the various elements of the enrichment process may now be seen. A given quantity of enriched uranium may be obtained by simultaneously lowering both the tails assay and the original input of natural uranium feed; the reduced tails assay will imply more SWU per kilogram of product, but this increase in cost may be traded off against the smaller expenditure on uranium. Buckley et al. [12] point out that reducing the tails assay from 0.3% to 0.2% reduces the natural uranium required by 18%, and Gordon and Baughman [17] show comparisons for the case of the typical American LWR: using 'fuel enriched to 3%, the requirement would be 3.425 SWU/kg of product with a tails assay of 0.3% or 4.306 SWU/kg of product if the tails assay is 0.2%'. Similarly, reducing the tails assay with a given input of natural uranium feed will raise the enriched output quantity. There will thus be some optimum tails assay which will vary with the price of natural uranium and with the price of energy used in the enrichment plant.

The original process (whose origins lie in nuclear weapon development and which is still to some extent secret), is gaseous diffusion. US Department of Energy plants operate at Oak Ridge Tennessee, Paduca Kentucky, Portsmouth Ohio; the USSR operates this process, as does France through the Eurodif plant at Tricastin which began production in early 1979, with a planned maximum output of 10.8 million SWU by the end of 1981 (*Atom*, November 1979). In the gaseous diffusion process Hex is pumped through a porous membrane more than one thousand times to obtain product enriched to 3%. The ^{238}U atoms are marginally denser than the ^{235}U atoms, and so are just slower to move through the membrane. Thus as it is pumped through the cascade barriers, the Hex becomes very slightly enriched at each. The energy requirements for this process are substantial as there must be many repetitions.

The second major enrichment process is the gas centrifuge method. In March 1970 Britain, the Netherlands and West Germany signed the *Almelo Treaty*, which was an agreement to pool their research and development programmes in uranium enrichment. The joint sales organization is URENCO and the two gas centrifuge enrichment plants now operating are at Almelo in the Netherlands and Capenhurst, UK, operated by British Nuclear Fuels Ltd (BNFL) who also operate a diffusion plant at Capenhurst. This produces enriched uranium for the AGRs and recycles uranium recovered from the Magnox reactors, enriching it to natural ^{235}U concentration. The current plan is to phase out the diffusion plant by the end of 1982 [18]. There is a further enrichment plant planned for operation in 1984 in Gronau in West Germany to accommodate the West German electricity companies' requirements for an independent source of enrichment capacity (*Financial Times*, 19 June 1979).

The gas centrifuge spins the Hex and causes the heavier ^{238}U atoms to move outwards faster than the ^{235}U, thus achieving the desired separation. Enrichment to 3% takes some twelve repetitions and only about 10% of the power requirement of the gaseous diffusion method [6], although reference [5] quotes figures of 2500 KWh of electricity per SWU for a gaseous diffusion plant compared to 100 KWh for a gas centrifuge. A further benefit of the gas centrifuge is that small additions to capacity can be made economically, a desirable property when there is no firm commitment to permit expenditure on a substantial development; it appears that the maximum size required to achieve all available economies of scale is approximately one-third of that needed for a gaseous diffusion plant. Kroch [10] quotes the optimum scale of a diffusion plant to be 9 million SWU, while a gas centrifuge plant has exhausted economies of scale by 2 or 3 million SWU. Against this, however, must be set the drawback that the great stresses on the centrifuges make them very expensive both to manufacture and to replace.

URENCO claims (*Financial Times*, 19 June 1979) that, alone with Eurodif, it offers a commercial enrichment facility. Their argument is that the USA, with more than 60% of world enrichment capacity, quotes prices which barely cover energy costs and that the USSR (with 7% of world capacity) simply sets its prices 5% below those of the USA. The Soviet Union has been supplying enriched uranium to West European countries (particularly West Germany) since 1974; in 1980 $43.8 million worth of Soviet enriched uranium was imported into the USA, some to be fabricated into fuel for re-export to West Germany, some for use in US power stations (*Financial Times*, 18 August 1981). The capacity shares of Eurodif and URENCO are 25% and 5% respectively. By the middle of 1980 a succession of world-wide cancellations and cuts in orders for nuclear power stations had led to a position of substantial over-capacity in the enrichment industry.

The 1976 US charge for a fixed amount of enrichment was $61 per SWU [17] and [16]. The Gordon and Baughman model [17] estimates a government charge of $75 per SWU ($1975) until 1990. This matches Kroch's [10] estimate of ($1974) costs for a diffusion plant. Prior [19], however, puts enrichment costs at about $110 to $120 per SWU, arguing that

Structure and economics

costs have risen as the process has become increasingly commercial (and correspondingly less militarily based), thus explaining in large part the CEGB evidence that fuel for nuclear plant increased by more than 22% annually from 1968 to 1978. In the UK context Prior gives the breakdown of costs as: mining and milling 33%; conversion 1%; enrichment 30%; fuel fabrication 9%; reprocessing 25%; reconversion 1%; high level waste storage 1%; this is for a mid-1978 fuel cost of some £0.35 per GJ. This is in contrast with Greenwood et al. [20], who state that the overall significance of the costs of enrichment as part of a nuclear power programme is relatively small, at about 5% of the total cost of generating power.

The IIASA study [4] estimates that an enrichment plant with the capacity to provide for forty LWRs can be built in the same time as a LWR, at four or five times the LWR construction cost. Their world-wide total for enrichment capacity (outside centrally-planned economies) at the end of 1978 is just under 24 million SWU per year. A 1000 MWe LWR at a 65% load factor would need approximately 100000 SWU per year of enrichment service [20].

The third enrichment process currently operating commercially is the aerodynamic technique developed by UCOR in South Africa. This is of a type similar to that invented in Germany by Becker, and involves mixing Hex with either helium or hydrogen, compressing this and pumping it over a curved surface where it is separated by a knife edge [20]. The power requirements for this process are substantially greater than for the gaseous diffusion method (some 50% per SWU more), although it is a simpler technology. It has proved acceptable enough for Brazil to have purchased it as part of its nuclear development programme, since there are substantial hydroelectric resources at distances from population centres too great to make it worthwhile transmitting electricity from them.

The IIASA [4] end of 1978 estimate for aerodynamic capacity was of the order of 6000 SWU per year in pilot or laboratory facilities. The possibility for South African expansion to 5 million SWU per year is stated in reference [6].

The fourth alternative is laser enrichment, which has been demonstrated in laboratory experiment, and is predicted in reference [6] to be a probable commercial technique in the 1980s. The expected benefits of the technique (which involves exciting, by laser, only ^{235}U atoms so that they may be separated from ^{238}U) are that the power consumption should be comparatively low and the degree of separation in a single stage very high. Not only might virtually total separation be feasible, but the tails from previous enrichment processes would become suitable inputs to laser enrichment, which would thus represent a means of stretching uranium supplies.

Following enrichment, the enriched Hex is transported to the fuel fabrication plant; the particular manufacturing process depends on the type of reactor to be fuelled. At BNFL's plant at Springfield, the Hex is converted to small ceramic pellets of uranium dioxide (UO_2) and sealed into stainless steel AGR fuel rods. By the end of the 1970s the Springfield factory was producing some 2000 tonnes of nuclear fuel a year, with a target of 5000 tonnes by 1990 (*Financial Times*, 27 July 1979). The plant is also to use Westinghouse technology to fabricate the fuel assemblies should the CEGB's PWR programme [14] go ahead (*Atom*, September 1980). The Springfield plant also exports Hex, powdered pellets and finished fuel, particularly to France, Italy and Japan [18]. In the process for fuel manufacture for LWRs the pellets of UO_2 are loaded into Zircaloy fuel rods. Kroch [10] estimates an operating cost of $70 per kilogram of uranium; Gordon and Baughman [17] cite a range of estimated fabrication costs from $70 to $150 ($1975).

7.5 BURNUP

The fuel rods are now ready to be inserted into the reactor where they are 'burned' in the core over three to five years. The chain reaction of fission is now allowed to take place to generate power; in addition fission products will appear and begin to build up in the fuel. These products have neutron-absorbing properties and will thus inhibit the carefully balanced chain reaction until a point is reached where that reaction can no longer be sustained by the fuel which must then be replaced. The actual procedure for reloading the reactor varies according to type – some need to be shut down completely others, such as the AGR, are designed so that they may be reloaded while still on stream.

The period of core life and average core burnup (measured in megawatt days per tonne uranium) at the end of the life of the fuel rods will vary with reactor and with the position of the fuel rod within the core. Thus, the BWR has a lower burnup per tonne uranium than the PWR. A representative PWR burnup would be 33 thermal megawatt days per kilogram at 33% thermal efficiency [6]. The AGR has rating and burning which are both lower than those of the LWR, but its higher thermal efficiency (or Carnot efficiency), demonstrated by the high temperature heat it provides, is compensation for these deficiencies. Although there is little to choose in the amounts of natural uranium consumed by thermal reactors, the Magnox reactors using pure uranium have comparatively low burnup, due to their relative ^{235}U poverty and graphite moderator. The use of heavy water as moderator and coolant in the CANDU allows a substantially higher fuel utilization (as heavy water does not absorb as many neutrons as does light water), although the cost of heavy water detracts from this benefit. Similarly, fuel pins placed in core positions with higher neutron flux and power densities will need to be replaced earlier than others.

Prior to their insertion into the reactor the fuel rods do not present any radiation risk and they may be safely handled without the use of any shielding. Having undergone burnup in the reactor, however, they emerge emitting intense radioactivity and heat. Shielding becomes vital and the used rods are transferred immediately to cooling ponds. These are large metal-lined concrete pools of water which will allow the heat to dissipate from the rods, and will present a satisfactory storage area for the period of more than one year necessary to allow the initially very intense radiation to decay to a level where the rods may undergo further processing.

Although no longer useful in the reactor core, the rods contain 97% uranium; this is almost entirely ^{238}U but there will still be some ^{235}U (the actual quantities of the constituents of the spent fuel rod will depend on the reactor being used). In addition, some of the atoms of ^{238}U will have absorbed neutrons; this produces ^{239}U which decays and becomes plutonium (^{239}Pu).

Thus, all reactors produce plutonium, in varying quantities, and as it is a fissile material it contributes to the power output by sustaining the reaction. The yield of plutonium in tonnes per year from a 1000 MWE station operating continuously would be (*Hansard*, 3 March 1980):

AGR	0.25
PWR	0.33
CANDU	0.51
Magnox	0.75

7.6 REPROCESSING

The uranium and plutonium represent an immense source of still untapped energy. Whether or not this energy is to be used depends on the decision concerning the nuclear fuel cycle. If the cycle is to be left open, then the contents of the fuel rods must be dealt with to allow indefinite safe storage. Alternatively the cycle may be closed and, in this case, the fuel rods will be reprocessed to separate out the energy rich materials so that they may be recycled. The decision revolves around the balance to be struck between the extra costs involved in reprocessing (against the comparatively small cost of indefinite storage) relative to the benefits of the availability of depleted uranium and plutonium which will fuel a fast reactor. The value of the uranium is determined by the mining, conversion and enrichment costs; the value of plutonium must reflect these as well as additional costs caused by increased safeguards that are needed.

If spent fuel is reprocessed then the result will be stocks of ^{238}U, ^{235}U, plutonium and fission products, and actinides, which are waste and must be dealt with as such (actinides are heavy elements with atomic weights above actinium and include thorium, plutonium, uranium, americium and curium; they are created when uranium or another heavy element absorbs neutrons). For example, irradiated Magnox fuel contains more than 99% depleted uranium, 0.27% plutonium and 0.4% of other actinides and fission products (*Hansard*, 3 March 1980) while the figures for AGR and LWR fuel are 97% uranium (which is still slightly enriched in ^{235}U) up to 1% plutonium and 2 to 3% fission products and actinides [21]. For every tonne of uranium that enters the LWR (enriched to 3.3% ^{235}U), 24 kg of ^{238}U and 25 kg of ^{235}U are consumed. This is converted to 35 kg of fission products, 8.9 kg of isotopes of plutonium, 4.6 kg of uranium-236, 0.5 kg of neptunium-237, 0.12 kg of americium-243 and 0.04 kg of curium-244. The remainder is unburned uranium containing 0.8% ^{235}U [22]. The figures for fission products and transuranic elements (those elements with atomic weight above uranium) will be increased by longer burnup. On exit from the reactor this tonne of material will be emitting 300 million curies of activity and 3300 kilowatts of heat (the latter reducing to 1 kW after ten years) [6]. The UK has been reprocessing fuel from Magnox reactors for some years at Windscale (Sellafield) and the result of the 1977–1978 Windscale Inquiry was to accept the BNFL plan to construct the thermal oxide reprocessing plant (THORP) to reprocess the oxide fuel from the AGR.

The French reprocessing capability is based on Cogéma's plant at La Hague and is planned to be 650 tonnes of LWR fuel by 1984, and eventually 800 tonnes, which is the initial capacity intended for the new plant. 1984 is also the planned date for the Traitement Oxydes Rapide plant to be built at Marcoule to reprocess the spent fuel from the Phenix reactor. (*Atom*, November 1979). Cogéma has since 1977/78 entered into reprocessing contracts with Germany, Japan, Austria, Belgium, the Netherlands, Sweden and Switzerland [3, 30].

The German reprocessing developments were based on the formation of the 'Deutsche Gesellschaft für die Wiederaufarbeitung von Kernbrennstoffen' (DWK) which built a small scale prototype facility at Karlsruhe. The full size plant planned for Gorleben was to take a throughput of 1400 tonnes of spent fuel giving an output of 14 tonnes of plutonium per year. The objective of this development was to create an integrated waste management centre, as it was also intended to use the site for the construction of plant to manufacture mixed oxide fuel elements, and to deposit high level waste permanently in salt rock caves deep underground. The whole project is now in abeyance waiting for approval from Lower Saxony [15].

United States reprocessing at West Valley and Hanford has been halted since April 1977 when President Carter called for an investigation into the significance of the 'plutonium economy', with the initiation of the International Fuel Cycle Evaluation [9] (the civil plant at West Valley has reprocessed 630 tonnes of spent fuel between 1966 and 1972 (*Atom*, March 1979)).

It is argued [6] that a reprocessing plant only achieves economies of scale once it is capable of a throughput in excess of 1000 tonnes/year, which represents the output from a very large programme – perhaps reactors generating over 50 000 MWe. If so the export of reprocessing facilities becomes a significant part of the economics of such a plant. In the mid 1970s France exported small reprocessing plants at relatively low prices to Pakistan and South Korea, in contracts whose value was estimated at $200m and $10m, while reprocessing contracts at La Hague have earned some 12 billion francs ($2.5 billion) [3, 30] and have avoided weakening the French market power in this field.

The optimum size for a reprocessing plant is given in reference [20] as 1500 tonnes per year with an estimated cost of between $500 and $800 million; operating costs will vary with throughput since there are economies of scale available. A range between $180 and $250 per kilogram of spent fuel is cited as an estimate for the costs of the back end of the fuel cycle, covering initial storage, transport, reprocessing, conversion and waste management. Since they also quote a range of $50 to $150 per kg as long term fuel storage costs, there is little to choose on this evidence alone between an open or a closed cycle. (A cost of $100 per kg is also quoted in reference [6] as the total cost of waste disposal in 1976). They conclude that the back end of the fuel cycle would represent up to 10% of the total fuel cycle costs (and therefore only a few per cent of total electricity costs), again in contrast to Prior [19].

The Windscale Inquiry Report [21], notes that any estimate for long term spent fuel storage costs must be speculative as no detailed work had been done on the alternatives. Using BNFL estimates of £225 000 per tonne for dry storage (in inert gas) and £150 000 per tonne for wet storage (in cooling ponds) Parker concludes that reprocessing is the desirable solution. The BNFL estimates for reprocessing are £260 000 per tonne (or £200 000 after allowance is made for the recovered uranium, valued in the report at $30 per pound) which makes it immediately preferable to dry storage and, in fact, to wet storage, since there are probably corrosion problems associated with spent AGR fuel. The size of reprocessing plant desired by BNFL does, at 1200 tonnes, conform to the above estimates of the optimum.

Having completed their term in the cooling ponds the fuel rods are transported to the reprocessing plants in containers designed to shield the radioactivity, to avoid excessive heat build-up and to remain intact in case of accident; in the UK the quality of the containers is established by the Department of Transport, based on overall safety considerations laid down by the Health and Safety Executive. After further cooling the fuel rods pass through a robot operated 'decanning' plant which chops them up and strips away the outer cladding leaving the used fuel to be dissolved in nitric acid prior to further chemical treatment whose result is the separation and extraction of the uranium, plutonium and waste. Some of the uranium will be suitable for transport back to the enrichment plant as its ^{238}U content is sufficiently high to make economies in that process. The 1980 BNFL annual report stated that their Capenhurst diffusion plant was enriching to natural concentration the uranium recovered at Sellafield from spent Magnox fuel, with a consequent reduction in the UK's requirement for uranium imports of several hundred tonnes per year. BNFL has undertaken an investment programme of more than £16 million for the 1980s for the refurbishment and development of the Magnox fuel reprocessing plant, which is currently working to deal with fuel from UK reactors as it is discharged. With additional Japanese and Italian Magnox reactor fuel BNFL is anticipating a throughput of 900 to 1000 tonnes a year in the 1980s; the investment programme should give them a capacity of 1400 tonnes of fuel a year by 1983. The construction of the reprocessing plant (THORP) is due to begin in 1982, and large scale reprocessing of spent oxide fuel from both British and foreign reactors is planned for the late 1980s.

Remaining uranium will add to existing stocks of depleted uranium held at enrichment plants. (Now it no longer represents the handling problems of intense radioactivity that was the case prior to reprocessing.)

A further point concerning the effect of reprocessing is that it reduces substantially the quantity of long-lived radioactivity in the remaining waste (by removing almost all the plutonium which has a half-life of 25 000 years). Other long-lived isotopes remain and do represent management and handling difficulties, although the volume and heat output of the waste resulting from the reprocessing operation are probably little different from those in the original spent fuel [6].

United Kingdom estimates of the amount of depleted uranium separated during the reprocessing of irradiated fuel elements are:

1977–1978	850
1978–1979	690
1979–1980	750–800 tonnes uranium (UO_3)

plus some 2500 tonnes of uranium as Magnox fuel and 100 tonnes uranium as oxide fuel in storage facilities either at the power stations or Sellafield (*Hansard*, 3 March 1980).

The recovered plutonium could now be used together with recycled uranium as mixed oxide fuel in LWRs. Marshall (*Atom*, April 1980) gives the overall fuel savings from this as: 'recycle of uranium alone economizes on the supply of uranium fuel by 23% and the recycle of plutonium saves up to an additional 16%'. This is a comparatively small saving whose value will increase with any growth in the price of uranium ore. Similarly the Ford/Mitre study [6] argues that plutonium recycle in LWRs represents only a small economic benefit by lowering the fuel cycle cost by 10% and electricity costs by 2%.

7.7 FAST REACTORS

The alternative recycling solution is to develop fast reactors whose fissile material is ^{239}Pu, or perhaps a mixture of plutonium and ^{235}U, surrounded by ^{238}U. The relative concentration of plutonium to the uranium is 20:80; at this concentration the desired chain reaction in the plutonium will occur. In this case the speed of the neutrons is not moderated in any way (hence the name *fast* reactor); in addition the number of neutrons produced when an atom of plutonium splits is greater than was the case for ^{235}U fission in a thermal reactor. When these are absorbed by ^{238}U the result is the formation of more plutonium. Thus the operation of the fast reactor involves fission of plutonium generating the heat to drive the turbines, plus the conversion of some of the blanket of ^{238}U surrounding the core to plutonium. Because it behaves in this manner in the reactor, the ^{238}U is referred to as the fertile material.

Since the reactor not only 'burns' plutonium, but also generates more, it is also known as a breeder reactor. Thus the fast breeder reactor saves uranium over its lifetime by generating its own fissile material. The neutron flux in a fast reactor is much higher than in a thermal reactor and is sufficient both to provide heat and to convert ^{238}U simultaneously. The predicted uranium saving is substantial: 'over the lifetime of each 1 GWe of fast reactor capacity installed there will be a reduction of some 4000 tonnes of uranium compared with using a similar thermal reactor capacity' (*Atom*, November 1979). The thermal reactor also makes use of the plutonium it produces, but it can only operate with ^{239}Pu and the longer the fuel remains in the reactor the greater are the amounts of ^{240}Pu, ^{241}Pu and so on that are formed. These, however, represent fuel to the fast reactor which can consume them perfectly satisfactorily. The objective of the fast reactor design can be such that after a given period there will have been produced a surplus of plutonium, over and above the operating requirements of the reactor itself, sufficient to provide for another reactor; this period is known as the linear doubling time. The core of the first reactor is compact yet produces great amounts of heat; this is no longer transferred by carbon dioxide, but by liquid sodium which has a number of desirable properties. It is very efficient in transferring heat without the prerequisite of being at high pressure and it does not interfere with the action of the fast neutrons since it is not a good moderator. The liquid sodium serving to cool the reactor core becomes radioactive; it transfers its heat to a secondary sodium circuit which is not radioactive and which finally heats water to steam for power. The major significance of the fast reactor, however, is not that it breeds plutonium (all thermal reactors do that). Current designs of LMFBR have a fuel doubling time of 25 to 30 years (*Atom*, July 1981). This implies that the breeding gain (that part of the total fissile find produced in the reactor, over and above the quantity of primary fuel destroyed which is thus available for use in new plant) is not particularly high. (The breeding gain will depend on (among other things) the losses in reprocessing, fuel design factors, rating and burnup.) Thus provision of plutonium inventories for fast reactors is helped by a continued thermal programme based on reactors having a high uranium to plutonium conversion ratio.

The fast reactor also offers an immensely greater ability to extract energy from its fuel input; while the thermal reactor can utilize some 1% of the energy in its fuel input, the fast reactor is able to make some 60% of the energy in the original uranium available. The significance of this is shown in

Table 7.1 Sources [5, 29]

	Specific energy content (therms per tonne)	Annual fuel required for 1 GWe power station, 30% efficiency, 70% load factor
Coal	230–300	2.3m tonnes
Oil	420–440	1.5m tonnes
Gas	500	
Uranium in thermal reactor	4800–8000	26 tonnes enriched uranium (derived from 150 tonnes natural uranium)
Uranium in fast reactor	480 000	

Table 7.1. The economic significance of this is further highlighted by the assertion that 1 tonne of depleted uranium from the reprocessing of Magnox fuel will, if fissioned in a fast reactor, produce the same quantity of electricity as would be produced by burning 2.1m tonnes of power station coal or 8m barrels of oil (*Hansard*, 3 March 1980). Existing stockpiles of depleted uranium in the UK (end of 1979) were some 20 000 tonnes, which if used in a commercial fast reactor programme would have an energy equivalent of 40×10^9 tonnes coal (equal to some 400 years' supply at current UK coal extraction rates or the whole of the estimated coal reserves of the UK [1]) (*Atom*, March 1980). In France the domestic dependence on imported fossil fuel jumped from 36% to 77% between 1955 and 1976; the Messmer plan of February 1975 called for the construction of 40 nuclear power plants to produce 45 000 MWe by 1985 and so provide some 55% of total electricity production and 25% of total energy needs [3, 30]. While this target has been subsequently reduced to 20%, as the power station programme fell behind schedule, by May 1981 France had 18 000 MWe installed and in operation, 30 000 MWe under construction and 15 000 MWe on order (*Atom*, July 1981). In addition, the 1200 MWe Super Phénix fast reactor at Crays-Melville was planned for commission at the end of 1983; the nuclear programme would require by 1995 to 2000 some 1000 tonnes of uranium per

year, which would then exceed existing French national and foreign resources, but would also have generated the accumulation of a 250 000 tonne stockpile of depleted uranium by 2000 (*Atom*, March 1980). The Windscale Inquiry Report [21] also noted (paragraphs 8.34 and 8.35) that should nuclear power be used for a substantial part of electricity supplies 'it is in the public interest that we should, unless the price of doing so is too great, minimize reliance on imported fuel', which implies, at least, leaving the reprocessing option open. The United States' position, in evidence to INFCE, was that reprocessing neither reduced dependence on foreign energy sources nor was a necessary prerequisite for final waste disposal. Its value lay only in providing fuel for fast reactors, and they were justified only where the electrical grid had attained a certain minimum capacity. Conversely, the German argument saw reprocessing as an indispensable component of safe waste disposal [6, 11].

United Kingdom fast reactor research and development has been based on the 60 MWth (15 MWe) Dounreay Fast Reactor, between 1959 and 1977, superceded from 1975 by the 600 MWth (250 MWe) Prototype Fast Reactor, also at Dounreay. The latter has a core fuel charge of some 4 tonnes, involving 1 tonne of plutonium plus 2 tonnes in the fuel cycle outside the reactor. Developing this to a commercial demonstration fast reactor would increase the core fuel charge to 20 tonnes, of which 6 tonnes would be plutonium (*Atom*, January 1981). The fast reactor fuel fabrication plant at Sellafield has the capacity to take plutonium nitrate in solution (prepared on site at Dounreay) from which powder is derived to form granules to be made into fuel pellets. The alternative fabrication route loads fuel directly from granules formed by gel precipitation from plutonium–uranium nitrate solution; a pilot plant for this is under construction at Sellafield. The stock of plutonium available for civil use in the UK in early 1981 was approximately 12 tonnes (*Hansard*, 6 April 1981), together with plutonium contained in irradiated fuel awaiting reprocessing; by then 6 tonnes had been used in the fast reactor research programme. The major American research reactor is the Clinch River Breeder Reactor Plant, using mixed oxide fuel; a project which began in 1969. The US breeder programme has an almost exclusive emphasis on plutonium cycle LMFBR [6], but has experienced substantial setbacks since the Carter Administration's ban on reprocessing.

France is the only Western nation to have developed fast reactor technology to a commercial scale with the Super Phénix plant now under construction. To encourage the economic viability of this project, and so avoid being left with a purely scientific monopoly, France entered the Paris Agreement in 1977 with Germany (whose subsidiary partners were Belgium and the Netherlands) [3, 30]. The Agreement was based on the exchange of technical information and the coordination of research and development, together with the creation of a joint marketing and licensing company SERENA (Société Européene pour la Promotion des Systèmes de

Réacteurs Rapides à Sodium). The French participants are CEA and Novatome (a joint CEA–Creusot–Loire Subsidiary), the German company is Kenntnisverwertungs Gesellschaft Schnelle Brutreaktoren (KVG) composed of Interatom 51% (a KWR–Siemens Subsidiary), Gesellschaft für Kernforschung Karlsruhe 19% (GFK) plus Belgonucleaire (15%) and Neratoom (15%).

The Phénix reactor – forerunner of Super Phénix – has reached a burnup of up to 9.2% of the fuel, while the Dounreay PFR has achieved 7 to 8%. Fast reactor fuel will be more expensive per tonne than is the case for thermal reactors, but the costs per unit of electricity produced should be lower because of the high rating and long burnup the designers expect to achieve. Similarly, the capital costs of fast reactors will outweigh those of thermals but if uranium prices rise then the expected lifetime fuel input costs for a thermal reactor will be sufficient to justify the introduction of the fast reactor.

The LMFBR is not the only cycle which is possible: there are also gas-cooled fast reactors, and converter reactors operating on a Thorium/^{233}U cycle have also reached the research stage. ^{233}U is an alternative fissile isotope of uranium, not found naturally but produced from Thorium (^{232}Th). Reserves of thorium are known to exist in large quantities in India, Norway, the United States, Canada, Brazil and Australia, with their total extent being perhaps as great as that of uranium [20, 4]. The research reactors originally operating on a thorium–uranium fuel cycle were high-temperature gas reactors, helium cooled and graphite moderated. These converter reactors can utilize some 4% to 5% of the total energy available in thorium, and they operate with a high thermal efficiency.

7.8 WASTE DISPOSAL

The final stage of the nuclear fuel cycle is the disposal of waste. In the UK the total level of nuclear waste holdings at the end of 1979 was given as:

Concentrated high level waste in liquid form:	1 000 m³
Fuel cladding sludges and miscellaneous waste from earlier processes:	19 000 m³
Plutonium contaminated waste:	3 500 m³
Wastes stored at power stations:	20 000 m³

(*Hansard*, March 1980)

Waste treatment in the UK is carried out at BNFL's Sellafield plant. In the US at the end of 1975, some 1200 tonnes of spent fuel were being held at reactor sites and reprocessing facilities [6]; there were plants at West Valley Hanford, Idaho Falls (the National Reactor Testing Station) and Richland (operated by the Batelle Memorial Institute). Waste treatment in France is carried out at the Atelier de Vitrification à Marcoule (AVM), whose process was to have been developed in Germany by DWK at Karlsruhe.

If the fuel cycle is closed, ^{238}U and plutonium are removed from spent fuel and stored for later use, leaving a comparatively small bulk of waste of high, medium and low level. The second is essentially solid waste which may be buried in shallow trenches or encased in concrete within metal drums and taken out for dumping into the sea; the last has been treated so that the activity associated with it is at a level low enough for discharge into the environment. The remaining high level wastes are currently held as liquids in acid solution in water-cooled stainless steel tanks. (By 1980, ten tanks at Sellafield, UK, held some 770 m^3 of high level waste, which represented almost the entire accumulation from 25 years of nuclear research, development and power output).

The high level waste (associated with the fuel elements), has a relatively low volume but has both high heat output and, obviously, high radioactivity. Medium level waste (where the heat ouput is far less but the volume to be handled – including fuel-rod cladding, waste from processing operations and laboratory equipment – is substantially greater), can be quite satisfactorily dealt with by enclosing it in a sufficient quantity of concrete, sealed in drums and then dropping it into the ocean.

The standard sequence of events in dealing with high level waste begins with storage of liquid-form waste for a period of years in water cooled tanks, which allows the dissipation of some heat, and a decline in radioactivity. The most advanced technology for further stages has been developed in France at AVM, with the PIVER process of conversion of the liquid waste to glass cylinders. By April 1980, AVM had vitrified 230 m^3 of fission products with a total activity of some 25 m curies, and had produced 313 cannisters comprising 108 tonnes of active glass (*Atom*, December 1980). France is also planning second and third vitrification plants at Cap de la Hague for 1986 and 1987.

Atomic Energy of Canada Ltd has continued research on vitrification, as has BNFL of the UK, who developed the HARVEST process. However the first Sellafield vitrification plant will be based on the AVM process, because of the latter's successful commercial-sized operation since 1978 (*Atom*, August 1981).

In Australia, research by A.E. Ringwood has concentrated on the possibilities of fixing waste in synthetic rock (SYNROC). Swedish research work on waste disposal has been based on the law passed in April 1977, which prevented the commissioning of any new nuclear power station unless it could be shown that a completely safe way of disposing of all waste products has been found. This is comparable to the position taken in the UK in the Flowers Report [23], which argued that there should be no commitment to a large programme of reactors 'until it has been demonstrated beyond reasonable doubt that a method exists to ensure the safe containment of long-lived highly radioactive waste for the indefinite future'. The Swedish power industry formed the Nuclear Fuel Safety project (Karn Bransle Sakerhet) in 1976, and research has followed the route of containment of waste in cannisters (possibly synthetic corundum which, it is estimated, should resist the leaching effects of groundwater for some hundreds of thousands of years), to be buried deep in stable geological structures.

The significance of vitrification lies both in the heat tolerance of glass and its ability to withstand groundwater and in the reduction in volume that is made possible. In the UK, the Magnox reactors generate about 1000 tonnes of spent fuel a year, which is reprocessed very soon after arrival at Sellafield (to avoid corrosion of the cladding). This annual quantity of fuel would produce some 60 m^3 of high level fission product waste liquor for storage; this in turn would produce a bulk of 15 m^3 per year if vitrified. Similarly the AGR when fully operational should generate about 300 tonnes of uranium per year as spent fuel to be reprocessed at THORP; this would amount to 30 m^3 of high level fission product waste which would reduce to 15 m^3 if vitrified (*Hansard*, March 1980). Similarly, an estimate for a 20 GW nuclear programme in the UK is that, once vitrified, the total high level waste would amount to less than 500 m^3 [24]. It is these very small quantities which justify the development of highly sophisticated, capital intensive, waste management techniques.

Once converted to glass the cylinders are enclosed within steel cladding; these cylinders are then to be subject to further artificial cooling (either air or water) in stores which will allow inspection for a further period of possibly up to three decades. The length of time the waste is stored in each repository will determine how far it cools as will the quantity and type of waste (how much is fission product) formed into each glass block. After this cooling period the waste would be finally sealed and deposited in their ultimate permanent repository (possibly involving further titanium and lead shielding to guarantee even greater security).

These canisters will remain hot over centuries, but the temperature will gradually decline to the original level of the surround: 'The overall level of radioactivity reduces to a very low figure within a few hundred years. Indeed, in a few thousand years it reduces to the order of that in the original ore from which the fissile material in the fuel elements is fabricated' [24]. It is the fission products which are responsible for the radioactivity and heat in the first few hundred years of life of the waste: but the total activity of these fission products (e.g. strontium-90, cesium-137, iodine-129, krypton-85) has reduced by a factor of about ten million 700 years after production [6]. What remains is the α-activity of the actinides which, while a substantial cancer risk in very small quantities, is easily contained. The significance of this for the area to be devoted to waste disposal is shown in the comparison given in reference [25]. There, it is stated that at the time of solidification each cubic foot of waste is some 10^5 times more radioactive than a cubic foot of natural uranium; the decline in radioactivity will be such that after 1000 years the activity in the waste will be 150 times that of the original uranium which produced the waste. This would possibly allow the use of space between or slightly above old canisters for burial of new wastes. No decision has been reached on just where the final disposal of waste will take place, nor is there any pressing need yet for such a decision, as a period of fifty years is perfectly feasible between vitrification and final storage [24]. One hundred years of cooling would have dissipated the bulk of the decay heat and increased by 30 times the number of canisters able to be buried in any one area [25]. Cohen [22] discusses the area which would have to be devoted to waste disposal arguing that a 1 GW nuclear plant would create waste which would fill ten canisters each year of its operation. If each canister occupies 100 m^2 then the year's waste will need 1000 m^2. He then estimates that an all nuclear US electric power system (of four hundred 1000 MW plants) would produce, in total, waste canisters occupying less than 0.5 km^2 each year. But, as argued above, delaying burial would allow the heat output of each canister to fall substantially, so reducing the space required for heat dissipation. Three major possibilities are: on or under the ocean floor or burial on land. The ocean itself evidently represents an ideal means of heat transfer and dilution of activity, although with blocks designed to resist corrosion or decay for more than 500 years, the latter should not be relevant. Burial beneath the ocean floor loses the benefit of heat removal but the sediment might be expected to absorb any leakage.

One area proposed for land burial is deep in deposits of hard rock. Once there, the only routes for radioactivity to reach man would involve leaching into groundwater from a corroded or broken container, or a natural (or man-induced) catastrophe. While deep in the rock structure – chosen for its geological stability and predictability – the heat transfer would be slow which might imply extra stress on the containers. Analysis in Sweden shows that failure of a single container would cause a change in the level of radiation within the local variations occurring naturally in the areas likely to be used. Catastrophic changes in rock structure are predictable by geologists within the time period when the waste containers would represent the greatest risk, and deep burial would minimize the possibility. The significant point here is that environmental conditions in rock formations 600 m below the surface are not comparable with those on the surface; at that depth the 'characteristic time intervals required for any substantial change are of the order of millions of years' [22]. Areas chosen because of their stability and freedom from groundwater are likely to remain in that state over a few hundred years – in geological terms a short, reasonably predictable period. A further possibility for waste storage is within a salt deposit, as has been shown in a disused salt mine at Asse in West Germany. Since 1967 drums of low level waste have been dumped into the chambers left by the mine-working, and medium level waste has been stored in batches of drums shielded with several tonnes of concrete (*Financial Times*, 24 November 1980). The advantages of salt as a burial medium are that it can conduct heat satisfactorily; its presence indicates the absence of any flowing groundwater, and it moves gradually over time to cover totally whatever has been left in it, so that eventually the drums will be completely encapsulated in salt with no access at all. Cost estimates for waste disposal will vary with the nuclear fuel cycle under consideration. Meckoni, Catlin and Bennett [26] estimate that '70% of the total capital cost of waste management is attributable to the solidification plant for high-level liquid waste and the cost of disposal in a geological formation'. Gordon and Baughmann [17] use a waste disposal price of $100 per kg of spent fuel ($ 1975) for their once-through model of the cycle, and $300 per kg to cover reprocessing activities and subsequent waste disposal (which must be considered in the light of the value of uranium and plutonium recovered for re-use). Thus the Windscale Inquiry [21], in estimating the operating cost of THORP, quotes a reprocessing plus vitrification cost of £260 000 per tonne, from which £60 000 are substracted as credit for recovered uranium; (implying a net reprocessing plus vitrification cost of $384 per kg in $ 1978). Kroch's estimate [10] for waste disposal costs range from $61 per kg of spent fuel from a CANDU reactor to $85 per kg for a LMFBR ($ 1974).

7.9 INTERNATIONAL ASPECTS OF THE NUCLEAR FUEL CYCLE

International trade in all aspects of the nuclear fuel cycle has been common since the 1950s, with transfers of research technology and materials from Canada, the United Kingdom and the United States ('by mid 1976, 14 supplier countries had over 100 agreements in force with other countries and some Third World countries were also beginning to give nuclear assistance to fellow developing countries') [27], although increasingly subject to debate because of the possibility of its becoming either the cause or the means for the proliferation of weapons. The International Nuclear Fuel Cycle Evaluation [9] was set up in October 1977 as a technical study of how elements of the nuclear fuel cycle might be abused for purposes of weapons production. It is evidently not necessarily the case that a nuclear electricity programme is the forerunner of a nuclear weapons programme (Canada, West Germany, Sweden). 'The route to a nuclear weapon through the commercial fuel cycle has not been chosen by any of today's weapon states. Water reactors produce an inferior material for weapons' [2]. Diversion of materials from an electricity programme would increase the complexity and risks of weapons design as the materials would not be suitable and it is not obvious that the development of nuclear weapons has been more possible anywhere because of a previously existing power programme: 'a state that had no nuclear power industry would not be expected to acquire nuclear power generation facilities for the sole purpose of obtaining weapons material, unless it wanted to conceal its intentions . . . it is by no means clear that concealment would be easier than constructing special facilities . . .'

[20]. However, the strategic and prestige effects of the possession of weapons capabilities means that risk of diversion exists and is increased by international trade in the nuclear fuel cycle.

In May 1981 it was announced that since 1971 the UK had exported 1280 kg of plutonium produced in the UK to Belgium, France, West Germany, Switzerland, Japan and the USA, in addition 1930 kg of plutonium derived from irradiated fuel imported and reprocessed by BNFL under contract had been exported to the customer or a country nominated by the customer (Belgium, Canada, France, West Germany, Italy, Japan, USA). This plutonium was for civil use in research and development in fast reactors or recycle in thermal reactors. The figures given were, (*Hansard*, 14 May 1981):

Plutonium exports: 3210 kg
Plutonium imports: 560 kg
Highly enriched ($> 40\%$ U_{235}) uranium exports: 660 kg
Highly enriched ($> 40\%$ U_{235}) uranium imports: 640 kg

In the 1950s and 1960s, France exported a research reactor to Israel, a graphite/gas reactor of 497 MW to Spain, a PWR of 870 MW to Belgium and a reprocessing plant to Japan (Tokai Mura) to enter operation in 1977. In the mid-1970s there were exports of PWRs to Iran and South Africa, of reprocessing plant to Pakistan and South Korea and of large research reactors to Iraq and Iran [3, 30].

In 1974, India exploded a nuclear device said to have been built using fissile material from a Canadian-supplied heavy water reactor (an efficient producer of plutonium) and heavy water supplied by the USA [27, 28].

A remarkably comprehensive nuclear contract was negotiated between West Germany and Brazil in 1975, under which Brazil agreed to buy a complete nuclear fuel cycle, from prospecting through production of uranium compounds, enrichment, construction of power stations and reprocessing [15, 27]. The provision of enrichment and reprocessing facilities has caused concern, as each could be diverted from civil use to production of weapons grade material (highly enriched uranium or separated plutonium), although Brazil has pledged that this will not be so.

A reaction to the dangers of uncontrolled nuclear fuel cycle trade was the formation of the 'London Suppliers Club' in 1975/76 which began with meetings of representatives of industrial nations exporting facilities, materials and services related to the cycle. Initial participants were Canada, France, West Germany, Japan, UK, USA and the USSR, and this membership gradually broadened to fifteen member nations (with the inclusion of Belgium, Czechoslovakia, East Germany, Italy, the Netherlands, Poland, Sweden and Switzerland). The London Club established a common code of conduct in the form of a set of guidelines requiring recipient nations to subject the use of 'sensitive' imported materials to IAEA safeguards and to guarantee not to use these items in the manufacture of nuclear weapons.

The guidelines also required restraint in international trade in sensitive nuclear technology (particularly enrichment and reprocessing) but since they are part of an agreement and not of a treaty they have not prevented the export of such technology [3, 27].

The second major element in control over the undesirable possibilities surrounding international fuel cycle trade is the 1968 *Treaty on the Non-Proliferation of Nuclear Weapons* (*NPT*) (see, for example, [16] and [27]), which was an attempt to freeze the number of nuclear weapons states (then five: USA, USSR, UK, France and China, of which France and China have not signed the *NPT*). By May 1980 there were 113 parties, but the force of the *NPT* in achieving its objectives lies in encouraging non member actual or possible nuclear weapons states (such as India, Pakistan, Israel, South Africa) to accept its articles and in expanding and improving the powers of the IAEA inspectorate in its detection of any diversion of fissile material from civil power programmes. The USA *Nuclear Non Proliferation Act* (1975) operates the safeguard of banning nuclear exports unless it can be shown that such exports can satisfy non proliferation commitments by not representing a possible weapons danger. Enrichment technology (and the export of significant amounts of highly enriched uranium) and reprocessing technology (and plutonium export) are embargoed under the Act. It has been argued that this restrictive export policy serves to amplify the inequalities between nuclear weapons states and non-nuclear weapons states. Where proliferation in the *NPT* meant the acquisition of nuclear weapons, under the *Non-Proliferation Act* it is redefined as the 'capability of acquiring nuclear weapons' – a definition which would have rendered the negotiation of the *NPT* impossible had it been used there [*Hansard*].

The Carter Administration also stopped domestic reprocessing of spent fuel prior to the investigations of INFCE. One justification for this ban lies in the resource estimates. Each 1 GWe LWR requires some 5500 tonnes of uranium in its expected 30-year life. The Carter Administration's argument was that with current resource estimates, gradual increases in reactor orders and a reactor construction time of some 10 years there would be adequate fuel to last well into the next century, without any need for reprocessing. Those studies which showed that recycling of uranium and plutonium in LWR was only marginally attractive at best and probably not economic, reinforced this conclusion [*Hansard*].

Thus, one view of the nuclear fuel cycle is that proliferation risks would be greatly reduced if states had no access to further enrichment or any reprocessing plant. This view would support the adoption of the once-through fuel cycle leaving plutonium unseparated in spent fuel; such a position is adopted in reference [27] where it is claimed: 'the fact is that reprocessing at present has little civilian utility, but could have serious military implications'.

There seem to be two major alternative solutions to this *NPT*-based constraint on reprocessing. The first, proposed by Marshall (*Atom*, April and September 1980), considers the possibility of using the fast reactor (which would obviously be prohibited by the non-existence of plutonium a ban on reprocessing would imply) as a proliferation and terrorist-proof incinerator. The spent fuel rods newly extracted from a reactor represent a radiation hazard so extreme that it is safe from any threat or theft. Yet once stored for some years as waste, the level of radioactivity declines, eventually to a point where handling would be possible for a period long enough for theft to take place (ignoring other safeguards). The fast reactor, if constructed commercially, would operate as a form of incinerator for plutonium in that the plutonium actually enters as a fuel to be consumed. The simultaneous creation of plutonium within the reactor is presumably not directly a terrorist risk since it has the same drawback mentioned above of being in fuel elements containing fission products so intensely radioactive that they would be unapproachable. If, in the process of recycling, the plutonium bearing fuel rods were 'spiked' with powerful gamma emitters, it would remain unapproachable, but would not be of any less value as a fuel. If combined with IAEA monitoring of movements of fissile material, this would act to prevent diversion.

The second alternative is the internationalization of sensitive aspects of the nuclear fuel cycle (see, for example, [11, 25 and 26]). Thus, Kaiser [11] writes 'today it is generally agreed that the multinational approach could improve the existing control system in several fields since multinational control of installations is, in principle, considered to be more proliferation-proof than purely national control, particularly in the sensitive fields'.

The IAEA study project on regional nuclear fuel cycle centres (RFCC) 'envisage several countries joining together to plan, build, and operate facilities necessary to service the back end of the nuclear fuel cycle' [26]. There would be other benefits to be gained from adoption of this approach: economies of scale in reprocessing mean that such plant operates optimally on a throughput of about 1000 tonnes a year of spent fuel, a quantity which would be available on an international basis. 'The unit total cost of reprocessing and recycling operations using a 1500 tonne/year reprocessing plant is about 40% lower than with a 500 tonne/year plant' [26]. International regionalization of reprocessing would simultaneously achieve the economies of scale and limit any incentive for the construction of small reprocessing plants, so limiting the number of purely national facilities. Large economies of scale are also claimed for RFCC in waste management: cost reductions per tonne of fuel processed can be lower by a factor of 4 to 6 in large plant; any increases in transport costs would be negligible. Similarly, Kaiser [11] argues in favour of multinational fuel banks to increase the security of supply and points out that, as reprocessing spreads and more countries have the capacity to produce or process plutonium, the creation of an international regime for the storage of plutonium will become increasingly important.

Georgescu-Roegen has characterized the exploitation of natural resources in his phrase the 'hour glass of the universe' – once energy resources (for example) are used, an available resource slips from the top of the hour glass to become waste in the bottom. Uranium has no major use beyond production of power. If burnt in thermal reactors, removed as spent fuel and disposed of as waste in a once-through cycle, the world's uranium resources represent a briefly lasting stockpile of low entropy in the upper half of the hour glass. If reprocessed and recycled in fast reactors the available energy – suitable especially, but not merely, for the generation of electricity – is multiplied dramatically. The use of economic analysis will lead to optimization of exploration and exploitation rates for uranium, the introduction date for fast reactors on a commercial scale and the size of plant to operate the 'super-hard' technology involved; it may also point to international regionalization as a source of economies of scale and limitation on diversion of civil nuclear material.

REFERENCES

[1] Greenhalgh, G. (1980) *The Necessity for Nuclear Power*, Graham & Trotman Ltd.
[2] Camp, P.W. (1980) Prospects for Nuclear Energy Supplies, in *World Energy Issues and Policies*, Oxford University Press, Oxford.
[3] Lellouche, P. (1980) French Nuclear Policy, in *Nuclear Policy in Europe*, Forschungsinstitut fur Deutschen Gesellschaft fur Auswartige Politik.
[4] Energy Systems Program Group of the International Institute for Applied Systems Analysis – Wolf Hafele, Program Leader (1981) *Energy in a Finite World, A Global Systems Analysis*, Ballinger.
[5] Eden, R., Posner, M., Bending, R., Crouch, E. and Stanislaw, J. (1981) *Energy Economics, Growth Resources and Policies*, Cambridge University Press, Cambridge.
[6] Nuclear Energy Policy Study Group (1977) *Nuclear Power Issues and Choices*. Ballinger.
[7] OECD-NEA/IAEA (1978) *1978 World Uranium Potential*, OECD, Paris.
[8] Solow, R. (1974) *The Economics of Resources or the Resources of Economics*, (AEA Papers and Proceedings, 1974).
[9] *International Nuclear Fuel Cycle Evaluation, Summary Volume* (1980) International Atomic Energy Agency, Vienna.
[10] Kroch, E. (1980) The Nuclear Fuel Cycle and the Demand for Uranium, in *Advances in the Economics of Energy and Resources, Vol. 3*, JAI Press Inc.
[11] Kaiser, K. (1980) Nuclear Energy and Nonproliferation in the 1980s, in *Nuclear Policy in Europe*, Forschungsinstitut für Deutschen Gesellschaft für Auswartige Politik.
[12] Buckley, C.M., Mackerron, G.S. and Surrey, A.J. *The International Uranium Market*, Energy Policy, June 1980.
[13] Commissariat à L'Energie Atomique (1978) *Annual Report*.
[14] Nuclear Power (1980/81) The Government's Response to the Select Committee on Energy's Report on the Nuclear Power Programme, Session 1980–81. House of Commons Paper 114-1, Cmnd 8317, HMSO, London.

[15] Hackel, E. (1980) The Domestic and International Context of West Germany's Nuclear Energy Policy, in *Nuclear Policy in Europe*, Forschungsinstitut für Deutschen Gesellschaft für Auswartige Politik.
[16] Treverton, G. (ed.) (1980) *Energy and Security*, Gower and Allenheld, Osmun.
[17] Gordon, J.B. and Baughman, M.L. (1979) Economics of the Throwaway Nuclear Fuel Cycle, in *Advances in the Economics of Energy and Resources, Vol. 1*, JAI Press Inc.
[18] British Nuclear Fuels Ltd (1980/81) *Annual Report*.
[19] Prior, M. (Winter 1979) Myths and Mysteries of Electricity from Coal and Nuclear Power, *Coal and Energy Quarterly* No. 23. (National Coal Board, London.)
[20] Greenwood, T., Ratjhens, G.W. and Ruina, J. (1980) Nuclear Power and Weapons Proliferation, in *Energy and Security*, Gower and Allenheld, Osmun.
[21] The Hon. Mr Justice Parker (1978) *The Windscale Inquiry, Vol. 1*, HMSO, London.
[22] Cohen, B.L. (1977) The Disposal of Radioactive Wastes from Fission Reactors, *Scientific American*, **236**, No. 6.
[23] *Nuclear Power and the Environment* (1976) Sixth Report of the Royal Commission on Environmental Pollution, HMSO Cmnd 6618.
[24] Second Annual Report of the Radioactive Waste Management Advisory Committee, (May 1981), HMSO, London.
[25] Institute for Energy Analysis, Oak Ridge Associated Universities (1979) *Economic and Environmental Impacts of a US Nuclear Moratorium, 1985–2010*, MIT Press.
[26] Meckoni, V., Catlin, R.J. and Bennett, L.L. (1977) Regional Nuclear Fuel Cycle Centres, IAEA Study Project, (Energy Policy December).
[27] Stockholm International Peace Research Institute (1980) *The NPT*, Taylor and Francis Ltd.
[28] Maddox, J. (1980) Prospects for Nuclear Proliferation, in *Energy and Security*, Gower and Allenheld, Osmun.
[29] Hunt, H. and Betteridge, G. (1978) The Economics of Nuclear Power, *Atom*, **266**, December.
[30] Hackel, E. (ed.) (1980) *Nuclear Policy in Europe*, Forschungsinstitut für Deutschen Gesellschaft für Auswartige Politik.

THE ECONOMICS OF
ADVANCED CONVERTERS
AND BREEDERS

8

A.A. Farmer
ASSESSING THE ECONOMICS OF THE LIQUID METAL FAST BREEDER REACTOR

8.1 INTRODUCTION

From the very beginning of nuclear power it was realized that proven reserves of natural uranium were limited and that a large industrial nuclear programme could not be sustained over a long period of time unless a way could be found for the efficient use of the available uranium by using spare neutrons to convert fertile uranium-238 into fissile plutonium-239. In typical thermal reactors the spare neutrons available for capture in uranium-238 equal about half the number absorbed in fissile nuclides and it is possible to utilize only about 1% of the uranium. Using the once-through fuel cycle, i.e. with no reprocessing of spent irradiated fuel to recover and recycle the uranium and plutonium, the actual utilization of uranium is nearer 0.5% in the current light water moderated and cooled thermal reactors. In thermal reactors of the CANDU type the spare neutrons available equal around 80% the number absorbed in fissile nuclides and a utilization approaching around 2% could be achieved if the spent fuel is reprocessed to recover and recycle its contained plutonium. As a consequence of this realization, work on developing the fast reactor, which can have spare neutrons equal to over 100% (typically 115 to 130%) the number absorbed in fissile nuclides and so can produce more fissile material than they consume, has been carried out since the earliest days of nuclear power. In fact the first electricity ever produced from nuclear fission was generated by a fast reactor at Idaho Falls in the early 1950s. In Britain, work on the development and design of fast reactors was started at Harwell and Risley in 1951, and an on-going programme was endorsed by the energy policy paper (Command 9389) published by the government in 1955.

The main purpose of this paper is to examine the economics of fast reactors but, before doing so, it describes briefly some of their characteristics and states their main attraction, namely to utilize to the maximum the available low-cost uranium resources. This particularly makes fast reactors desirable for nations without large indigenous uranium reserves. Turning to economics, the components that go to make up the cost in a fast reactor, such as capital, fuel fabrication, reprocessing, etc. are considered first. The chapter then deals with the costs of generating electricity from stations taken in isolation (i.e. single station generating costs) and identifies those factors which can help to reduce them to a minimum. Finally, the expenditure of a whole system of thermal and fast reactors is considered over an extended period, where it will be shown that an optimum fast reactor design based on system costs may differ from one based on single station generating costs.

8.2 PLUTONIUM AND FAST REACTORS

Some plutonium will be produced during the operation of all reactors fuelled with uranium, by the conversion of the isotope ^{238}U into ^{239}Pu. There are four major plutonium isotopes, ranging in atomic weight from 239 to 242. Neutron capture in ^{238}U leads to ^{239}Pu and further neutron captures produce the other plutonium isotopes.

Nuclides in the trans-actinium group, i.e. with atomic number greater than or equal to that of actinium (89) may, in general, be divided into three groups: the fissile group, the fertile group and the group which captures neutrons without resulting in fission or the production of fissile material. All the nuclides are fissionable if the neutron energy is high enough, such as in a fast reactor. The first group, however, contains nuclides like ^{235}U and ^{239}Pu for which fission is the most probable result of a neutron absorption over a wide energy range of the incident neutron including low energy, thermal neutrons. The second group contains nuclides for which a neutron capture leads to a fissile nuclide. Examples are ^{238}U and ^{240}Pu which are converted to ^{239}Pu (eventually) and ^{241}Pu respectively. The third group contains nuclides like ^{236}U and ^{242}Pu whose main effect in a reactor is that they capture neutrons in an unproductive manner. If the various neutron absorption processes and nuclide conversion and fission events are followed through, then it can be shown that the use of plutonium in a fast reactor is more attactive than use in a thermal reactor because it leads to a greater fraction of spare neutrons.

In a thermal reactor the ability of a fissile nuclide to absorb a neutron, followed by fission, is extremely high compared to the ability of a nuclide to merely capture a neutron. This arises from the fact that at low (thermal) neutron energies the fission cross-sections of the fissile nuclides are two orders of magnitude higher than the capture cross-sections of the fertile nuclides. As a consequence thermal reactors need only a relatively small ratio of fissile nuclides to fertile nuclides in order to operate (i.e. to become critical). Indeed some reactors can operate with fissile concentrations equal to only the natural concentration of ^{235}U in fresh uranium. In fast reactors however, with their high neutron energies, fission cross-sections are relatively small and of the same order as capture cross-sections. A high concentration of fissile material is therefore required in the case of a fast reactor for criticality to be achieved. Even an infinitely large core, i.e. one from which no neutrons can escape, would need an enrichment of about 7% ^{235}U or 5% ^{239}Pu in ^{238}U to become critical. In practice, fissile material is in short supply and the amount required to operate a power reactor can be kept to an acceptably low value by using enrichments in the range 15 to 30%. Such a concentrated highly fissile fuel mixture with no moderator separating the fuel elements becomes critical at a much smaller size than that of a thermal reactor. The power is normally extracted from a relatively small volume resulting in a specific power (MW/tonne of heavy atoms) an order of magnitude higher than in thermal reactors such as AGR and PWR. This high-power density in fast reactors in turn leads to a finely divided core to provide a large heat transfer area and to the use of non-moderating coolant with good heat transfer properties such as liquid metals. These considerations, together with the solution of problems associated with the endurance of materials in high fast neutron fluxes, has led to the development of ceramic fuels such as mixed uranium/plutonium oxide formed into fuel pins several millimetres in diameter canned in stainless steel, the latter being compatible at high temperatures with both the fuel and liquid sodium coolant. Typical core parameters for an early 1250 MWe commercial fast reactor with oxide fuel and a possible later reactor with carbide fuel are indicated in Tables 8.1 and 8.2.

Table 8.1 shows that fast reactors can produce more plutonium than they consume. This conversion of uranium to plutonium at a faster rate than plutonium is used has also been called 'breeding' – hence the name 'fast

Table 8.1 Fast reactor parameters – common

Parameter		
Station net output	1250	MWe
Reactor heat output	3222	MW
Overall thermal efficiency	38.8	%
Fuel can material	S/S*	
Plutonium out-of-pile hold-up time	9	Months
Plutonium throughput held in process plant residues	0.5	%
Fuel density	80	% theoretical
Core height	1	m
Axial breeder height	2×0.4	m
Radial breeder thickness	0.4	m
Core coolant	Sodium	
Core coolant outlet temperature	537	°C
Core coolant inlet temperature	370	°C
Core coolant pressure drop	7.25	Bar
Peak random can temperature	645	°C
Core maximum sub-assembly outlet temperature	600	°C

*stainless steel

Table 8.2 Fast reactor parameters – variable

Parameter	Units	1	2	3	4
Fuel		UO_2/PuO_2	UO_2/PuO_2	UC/PuC	UO_2[a]
Fuel can inside diameter	mm	5.08	6.0	10	6
Maximum fuel burnup in core	%ha	10	10	7.5	10
Peak linear rating	W/mm	42	50	137	50
Peak core mass rating	MW/tha	261	229	170	229
Fuel cycle	Batch	6	3	3	3
Core fuel inventory	tha/GWe	14.73	16.89	22.2	17.07
Core diameter	m	3.11	3.16	3.06	3.18
Total sodium area/total core area		0.391	0.348	0.336	0.348
Mean fuel feed enrichment	%PuE239	0.16	0.15	0.113	0.19[b]
Initial core inventory	tPuE239/GWe	2.37	2.57	2.51	3.36[c]
In-pile inventory at equilibrium	tPuE239/GWe	2.79	3.02	3.16	0.89[d]
Out-of-pile inventory at 75% load factor	tPuE239/GWe	1.20	1.20	1.32	0.39[c]
Total inventory at 75% load factor	tPuE239/GWe	3.99	4.22	4.48	1.28[d]
Breeding gain					
core		−0.218	−0.178	0.032	2.62[c]
axial		0.209	0.198	0.219	—
radial		0.214	0.195	0.188	—
total		0.205	0.214	0.439	—
Net plutonium production	tPuE239/GWey				0.91[c]
Before reprocessing					
core		−0.211	−0.173	0.031	0.449[e]
axial		0.202	0.192	0.211	0.166
radial		0.207	0.188	0.182	0.164
total		0.198	0.207	0.424	0.779
After reprocessing: total		0.187	0.197	0.412	0.775
Equilibrium linear doubling time (at 75% load factor, 0.5% Pu held in process plant residues)	years	28.4	28.6	14.5	3.53[c]

[a] UO_2 initial and feed fuel
[b] % ^{235}U
[c] t ^{235}U/GWe
[d] tPuE239/GWe
[e] Positive Pu production in core; ^{235}U consumption equals 0.883 t ^{235}U/GWe y

breeder reactor' or 'FBR' (although it is today perhaps unfortunate that the word 'neutron' was omitted from the name, as the word 'fast' refers to the neutron energy and not the rate of breeding). It follows that after recovery during reprocessing the unused depleted uranium together with sufficient plutonium needed for on-going operation can be recycled. The only make-up required is depleted uranium to replace both the material fissioned and the excess plutonium not recycled which can be stored for use in future fast reactors. In this way most of the uranium can be fissioned, the exact percentage utilized depending upon the quantity of uranium and plutonium that is diverted from the fuel cycle, and held in waste residues during the reprocessing and fabrication operations. Furthermore, once operational, the fast reactor requires only a small supply of depleted uranium, so the use of fast reactors in the place of thermal reactors reduces overall natural uranium requirements. Studies indicate that adopting such a fast reactor strategy world-wide could result in uranium requirements for thermal reactors being supplied from low-cost, high-grade sources and avoid the need to exploit high-cost sources such as shale or sea water for several hundred years. In this way, fast reactors will reduce uranium price rises below what they would otherwise have been and help to stabilize the costs of thermal reactor generation in addition to providing their own direct savings.

8.3 URANIUM UTILIZATION

Table 8.3 shows the lifetime (say 25 years) fuel requirements for single 1000 MWe thermal and fast reactors, assumed to operate throughout with a 70% load factor. The fast reactor is as given in column 1 of Table 8.2 for an early oxide-fuelled LMFBR. PWR and CANDU have been chosen to represent the thermal reactors; the latter is generally considered to utilize uranium the most efficiently out of the currently-available designs. It is seen clearly that 1000 MWe of fast reactor capacity consumes 20 to 40 tonnes of waste depleted uranium during its lifetime, compared to the 3000 to 4000 tonnes of fresh natural uranium required for PWR and CANDU on a once-through fuel cycle.

If spent fuel from uranium-fed reactors is reprocessed to recover the contained plutonium it can be recycled in dedicated plutonium enriched thermal reactors. For CANDU the requirement for fresh uranium is then reduced to 1200 tonnes in the plutonium enriched reactors, although the fact that uranium-fed reactors must be operated in support must be taken into account. An alternative mode of operation is to recycle plutonium in the thermal reactor in which it was produced. This is termed self-generation recycle. In this case a reactor would start up with only uranium fuel and, with recycle, the proportion of mixed plutonium/uranium fuel would gradually increase to 30 to 40%. For LWR, the 25-year lifetime uranium requirements would be reduced from around 3800 to 2300 tonnes, whilst for CANDU, the

Table 8.3 Lifetime (25 years) fuel requirements of 1000 MWe reactors at 70% load factor

Reactor type	Fast reactor[a]			Enriched-U-PWR[b]	Nat-U-CANDU[b,c]	Pu-recycle-CANDU[c]
Station efficiency	39			33	29	29
Fuel type	UO$_2$/PuO$_2$			UO$_2$	UO$_2$	UO$_2$/PuO$_2$
Fuel cycle	6 batch off-load recycle			3 batch off-load once-through	Continuous on-load once-through	Continuous on-load recycle
Mean fuel irradiation MWD/t ha at discharge	69 000 (core) 6 000 (breeder)			30 400	7300	~20 000
Initial load						
t natural U	—			303	134	133–132
t depleted U	12 core 44 breeders			—	—	—
t Pu(T)[d]	2.7			—	—	1–2
Fuel loss from cycle to residues in processing (%)	1	2	4	—	—	Not known, say 1 to 2
Replacement fuel						
t nat U pa	—			139	120	44
t depl U pa	0.9	1.1	1.5	—	—	—
Total net lifetime t nat U	—			3780[e]	3130	1230[f]
Requirement: t depl U	22	28	38	—	—	—
Utilization of uranium %	45	60	75	0.5[e]	0.7	1.9[f]

[a] Table 8.2 column 1
[b] Reference [2]
[c] Reference [1]
[d] Nat U fed to thermal reactors to produce initial plutonium not included as assumed same amount plutonium available in final cores. This is pessimistic for FR and optimistic for CANDU. Also, once initiated a fast reactor system can sustain a growth in demanded capacity with no further input of plutonium. This does not apply to thermal recycle of plutonium.
[e] For self-generated Pu recycling figures become 2300 and 0.9% [3]
[f] For self-generated Pu recycling figures become 1600 and 1.4% [3]

figures are 3100 and 1600 tonnes respectively. In the long-term the operation of a system of thermal reactors, with some on uranium feed only, and some as dedicated plutonium enriched, is equivalent in total uranium requirements to operating all the thermal reactors in the self-generation recycle mode.

Table 8.3 also indicates that both the fast reactor and CANDU on dedicated plutonium recycle require about equal quantities of plutonium from thermal reactors at the start of life. Increasing the fuel burnup in fast reactors will tend to increase the utilization of uranium as less fuel will be lost from the fuel cycle to waste residues during processing. The utilization of uranium in reactors using ^{235}U enriched fuel, such as the once-through PWR, is seen from Table 8.3 to be less than that shown for the once-through CANDU. This is because unused ^{235}U remains in both the uranium rejected from the enrichment process and in spent fuel. The adoption of reprocessing, and the recycle of uranium and plutonium, together with decreasing the enrichment of the depleted uranium rejected from the ^{235}U enrichment plants, will tend to increase uranium utilization in enriched thermal reactors towards the figures shown for CANDU. The last line in Table 8.3 shows quite clearly that the utilization of uranium can be increased from around the 1% level to around 50 to 80% by using the depleted uranium and plutonium obtained from thermal reactor programmes in fast reactors. It is figures such as these that lead to the often quoted factor of 50 to 100 times greater utilization of uranium by fast reactors compared with thermal reactors.

8.4 COMPONENTS OF FAST REACTOR COSTS

8.4.1 Methods of assessment

There are three principal ways of looking at the economics of any power station.

1. It may be looked at in isolation and compared with an alternative design used to serve the same purpose, the 'single station generating cost' method.
2. It may be assessed in terms of its effect on the total generating cost of the whole generating system in which it is to be used, taking into account the fact that because of different running costs, the stations to be compared will operate at different average load factors over their lives and will affect differently other stations operating in the system. This is the 'system net effective capital cost' method used when investment decisions are being taken on the choice of the next power station to be incorporated in a generating system.
3. The third method is still more comprehensive and is required to guide decisions on whether to introduce a whole series of stations of a given type

The liquid metal fast breeder reactor

into an existing generating system. This method calls for the stimulation of the operation of the total electricity generating system over a period of time, long enough to cover the installation of a series of stations and their lifetimes, so that a complete economic assessment of the various alternatives can be carried out to determine the minimum total cost of meeting future electricity demands. This third method should allow for the different generation station types available (i.e. nuclear, fossil and other) and their associated capital and operating costs, for the availability of various fuels and their likely future price levels and for the likely time-scales and costs for achieving development objectives.

For any given power station type, the importance of achieving certain performance characteristics should be assessed. In the case of fast reactor generating stations, it would be necessary to judge the sensitivity of the entire system to various reactor parameters such as fuel geometry, rating and burnup. Similarly, the performance characteristics of supporting services must be included and here the hold-up times of fast reactor fuel in fuel processing plants and the fraction of fuel remaining in wastes from such processing plants, are cited as examples. Some aspects investigated by such technical analysis will be seen to have the potential of improving the overall economics of a fast reactor power programme and warrant consideration for R&D, others will not.

The first and third methods will be considered further but before this is possible, some discussion on the components that go to make up fast reactor costs is necessary. The principal items to be covered are:

1. The cost of the initial investment which corresponds with the power station construction cost (provision of a fund to cover decommissioning costs can also be included here).
2. The cost of the fuel used to produce the energy, together with any associated costs for recycling that fuel (fuel costs will include the capital cost of the fuel fabrication and reprocessing plants – cf. cost of coal mines to produce coal).
3. The cost corresponding to the operation of the station such as the personnel and maintenance.

8.4.2 Capital costs

Any electricity generating station using a steam turbine as its prime mover will comprise a steam supply unit together with a 'balance of plant' comprising the turbo-alternator and associated switch-gear, water feed treatment and heating plant and cooling water heat rejection system. In a nuclear power station, the source of heat is of course the reactor core and, as stated before, fast reactors have the highest volumetric power. A comparison of the relative core volumes required to produce a given power output from

200 The economics of advanced converters and breeders

different reactor types is given in Fig. 8.1, and an immediate reaction is that fast reactors should perhaps have the lowest capital costs. However, the extraction of this high specific power necessitates the use of an efficient coolant and sodium has been chosen. A consequence of the use of sodium is that safety considerations lead to additional costs for a secondary sodium circuit which collects the reactor heat from the primary active sodium circuit in intermediate heat exchangers. The secondary sodium then passes to steam raising units and, in this way, the primary active sodium is isolated from the steam/water circuit.

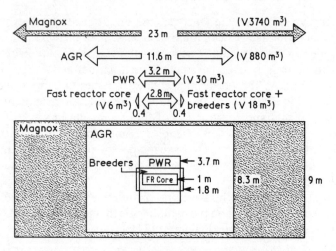

Fig. 8.1 Core dimensions for 1000 MW(e) reactors.

Another consequence of the use of sodium is that high working temperatures (above 500°C) are possible and as can be seen from Table 8.1, fast reactor power stations will operate with good efficiency comparable to the most modern fossil fuel fired and AGR power stations. Hence, it can be expected that the balance of plant will be similar to that used in modern fossil and AGR stations and any major differences in capital costs will be vested in the nuclear steam supply unit.

It is not the purpose of this chapter to give a full description of a liquid metal-fuelled breeder reactor (LMFBR), but a simplified cross-section of a large pool-type reactor is shown in Fig. 8.2. The reactor generates 3200 MW of heat which is sufficient to power turbo-generators producing 1320 MW of electricity of which about 70 MW will be used within the power station itself. The reactor is of the 'pool' type in which all the primary circuit components are immersed in a steel tank about 20 m diameter containing several thousand tonnes of sodium. The alternative 'loop' design of LMFBR has the primary pumps and intermediate heat exchangers connected by large

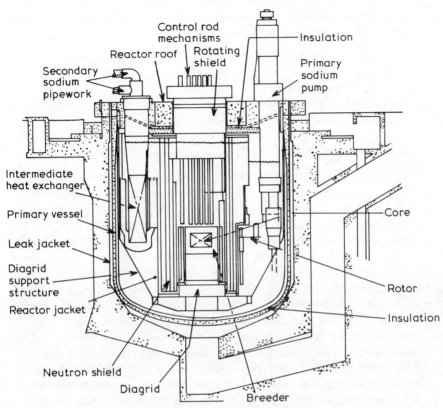

Fig. 8.2 CFR cross-section.

diameter steel pipes containing the primary sodium in an overall arrangement similar to that used for PWR thermal reactors.

Early fast reactor power stations could have a total specific capital cost (i.e. £/kW sent out) as much as 100% greater than contemporary thermal reactor power stations, but with further development based on manufacturing and operating experience, the construction cost of later fast reactor stations should be reduced to less than 40% higher than the lowest thermal reactor stations as presently conceived.

Apart from the addition of a secondary sodium circuit as mentioned above, other reasons why the capital cost of fast reactors is expected to be higher than that of thermal reactors are:

1. Amongst the steels used, the amount of stainless steel is much greater in fast reactors.
2. Trace heating systems have to be used on all sodium pipework and vessels to maintain the sodium molten at all times.
3. Extensive fire fighting systems must be provided to guard against sodium fires in the event of sodium leaks.

Amongst the reasons why the capital cost of developed fast reactors is expected to reduce can be mentioned:

1. The replication effect which is clearly connected with the construction of identical units, thus allowing the manufacturing facilities to be amortized over a larger production.
2. The reduction of the number of secondary sodium circuits with larger intermediate heat exchangers and steam raising units in each circuit.
3. The adoption of a single large turbo-generator rather than 2×660 MW units as proposed for early UK fast reactor generating stations.

Finally, it must be stressed that it is not possible to specify a ratio of fast and thermal reactor capital costs absolutely. Both types of reactor must satisfy the same basic safety requirements of the Nuclear Installations Inspectorate. It is possible that some safety standards will change as the nuclear power programme is developed. Because of inherent differences in the characteristics of thermal and fast reactors, the latter are in some respects easier to bring to an acceptable standard of safety than the former and in other respects more difficult. It follows that a change in design requirement in one type of reactor to meet revised safety standards could be unnecessary in the other, and hence capital cost ratios could alter.

Two examples of features where fast reactors meet acceptable safety standards easily can be quoted. Firstly, tests carried out on existing operating fast reactors indicate that power reactors can be designed so that in the event of a complete loss of both electricity and cooling water supplies (which would initiate a reactor shutdown) the reactor would remain in a safe condition indefinitely. This is achieved by natural convective circulation of the sodium coolant removing the decay heat from the core and rejecting it to atmosphere via natural draft air coolers. Secondly, operating experience with 'pool' designs of fast reactor indicates that the integrated radiation dose to operating personnel is about an order of magnitude down on that experienced in existing thermal reactors which, in any event, meet ICRP recommendations. On the other hand, an example of where it is more difficult for fast reactors to meet safety standards is the requirement to periodically inspect plant components within the reactor primary circuit. In water and gas-cooled thermal reactors, such a requirement is facilitated by the use of transparent coolants. In fast reactors, the use of opaque sodium as coolant renders visual inspection impossible and the problem is avoided by the provision of increased redundancy of plant components and structures.

8.4.3 Fuel fabrication

As mentioned before, a fast reactor utilizes depleted uranium and plutonium as its basic fuel material. The depleted uranium arises out of the thermal reactor programme and hence has already been purchased. The initial plutonium required to launch a fast reactor is taken from either plutonium extracted from spent thermal reactor fuel or from excess plutonium produced by other fast reactors. Thus the plutonium has, in general, been produced in fuel already owned by a nation or electricity utility and purchase of the raw material is unnecessary. But it is difficult to quantify a value for plutonium. It is a byproduct of reprocessing thermal reactor fuel and if this is deemed necessary for safety or environmental reasons separated plutonium is available at no cost. However, it may have a market value if sale to other users is not prohibited on proliferation grounds. Also it may be seen to have value as a substitute for ^{235}U for use in thermal reactors. For simplicity, a zero value will be adopted. This is consistent with the situation of a nation or group of nations comprising a closed system in which the debiting and crediting of reactors with plutonium values will be internal transfer payments which will cancel. The cost of fast reactor fuel can therefore be taken to comprise the non-fuel components required to manufacture the fuel elements together with the cost of its fabrication and later its reprocessing in order to extract the unused fuel from irradiated fuel elements for recycle.

The core of a fast reactor consists of hexagonal sub-assemblies ('SA') each about 4 m long and 15 cm across the flats and weighing some 275 kg in all. In early fast reactors each SA contains some 300 stainless steel clad fuel pins about 2.5 m in length, of which a central section approximately 1 m in length is filled with mixed uranium and plutonium oxide core fuel, formed into pellets about 5 mm diameter. This central one-metre long section of the fuel makes up the core of the reactor. Above and below the core section of the pin is a depleted uranium oxide reflector or 'axial breeder', each some 40 cm in length. The bottom part of the pin is left empty to accommodate the fission gas generated during irradiation. Around the core sub-assemblies are a number of rows (usually three) of blanket sub-assemblies. These are similar to the core SA but have larger-diameter pins filled with depleted uranium oxide.

The incorporation of plutonium as a fissile material in depleted uranium at concentrations up to 30% in fast reactor fuel brings with it all the special requirements of plutonium handling due to its chemical and radiological toxicity. It also poses some problems of criticality control. The economic effect of these special requirements is reflected in the fabrication cost of plutonium-bearing fast reactor fuels by a high capital-cost component. The plant involves the use of glove boxes, special ventilation plant and absolute filters, increased radiation monitoring requirements and the maintenance of plutonium contaminated equipment in fully isolated enclosures. The use of shielded caves, remote manipulators and closed circuit television viewing is also a possibility. In addition, criticality control would be achieved by using small sized plant with small throughputs. Such procedures will be expensive as the throughput per operator is likely to be greatly reduced compared to the fabrication of thermal reactor fuel whilst, at the same time, supervision requirements will be greatly increased as will the training necessary for new operators.

These factors, together with the greater complexity of design of fast reactor sub-assemblies compared with light water reactor fuel assemblies, and the greater number of assemblies per tonne of contained fuel, lead to much higher fuel-fabrication prices expected to be a factor 8 to 10 higher per unit weight of core heavy metal atoms than enriched uranium oxide thermal reactor fuel. These higher fabrication prices do not necessarily lead to higher electricity generating costs, as the throughput of core fuel required per unit of electricity generated is lower by a factor of around 3 in fast reactors because of

1. The higher mean burnup achieved – 70000 to 100000 MWD/tonne compared with around 30000 MWD/tonne in LWR.
2. The higher station thermal efficiency – around 40% in FR compared with about 30% in LWR.

A further factor that reduces fast reactor fuel cost per unit of electricity is the higher fuel rating, which leads to a much smaller initial inventory of core fuel.

It should be noted that although the factor of 8 to 10 on fabrication cost refers to the cost per unit weight of core heavy atoms, it includes also the cost of fabricating the associated axial breeders contained in core sub-assemblies. As radial breeder sub-assemblies in fast reactors contain only depleted uranium oxide, their fabrication cost should be affected only by the greater complexity of design compared with uranium oxide fuel for thermal reactors and a factor of about 1.5 times is expected.

Overall then, the total fuel fabrication costs in fast reactors will be some 2 to 4 times those in thermal reactors per unit of installed capacity.

8.4.4 Fuel reprocessing

At the time of writing it is not general practice to reprocess all spent irradiated fuel discharged from thermal reactors. Policy in this respect varies between countries and with reactor type. For example, the spent fuel discharged from the gas-cooled Magnox reactors is generally stored in water and due to corrosion problems with cladding it is normally reprocessed 6 months to one year after discharge, although some Magnox fuel has been successfully stored in water for considerably longer periods The recovered plutonium and depleted uranium is stored for future use. In these circumstances it could be argued that reprocessing is carried out for environmental and safety reasons to facilitate the storage of spent fuel. The cost of reprocessing can therefore be included in the operating costs of the Magnox reactors and hence it contributes to the cost of the electricity generated.

Furthermore, it could be argued that the plutonium is available for use in other reactors at zero extraction cost.

On the other hand, light water thermal reactors are at present being operated in the once-through mode, i.e. spent fuel is stored after discharge. It is possible that PWR fuel could be stored for extended periods (perhaps several decades) and this could give rise to the option of not reprocessing that fuel until the contained uranium and plutonium is required for recycle. For recycle of both uranium and plutonium in thermal reactors, the time at which reprocessing becomes worthwhile will depend upon the relative economics of enriched uranium price and the cost of reprocessing, refabrication, waste storage and long-term storage of spent fuel. If the uranium only is to be recycled in thermal reactors, whilst plutonium is stored for future use (or disposal), the time at which reprocessing becomes economic could be different from the case of recycling both uranium and plutonium depending upon the relative costs of fabricating plutonium enriched fuels and plutonium storage.

It follows that if reprocessing of spent fuel to obtain the uranium for recycle in thermal reactors is not economic, the cost of reprocessing spent thermal reactor fuel to obtain plutonium for use in fast reactors should be charged to the fast reactor less any credit for saving in spent fuel storage charges and the value of recovered uranium.

Although the merits of reprocessing thermal reactor spent fuel may be doubtful when expressed in terms of the economics of present uranium prices and process costs, it can be stated unambiguously that reprocessing must eventually be carried out for one or both of the following reasons.

1. Reprocessing of thermal reactor spent fuel is necessary for safety of storage (i.e. environmental) reasons.
2. Reprocessing of thermal reactor spent fuel and the recycle of uranium and plutonium in thermal or fast (or both) reactors becomes economic as uranium prices rise.

Furthermore, as already explained, fast reactors, once launched, become virtually self-sufficient in fuel requirements, needing only a small supply of depleted uranium to make up fuel fissioned or diverted to waste residues during recycling. The very nature of fast reactor operation requires the reprocessing of its fuel to realize its purpose.

Immediately on discharge from the reactor core, fast reactor fuel sub-assemblies have a high heat output due to the radioactivity of the contained fission products. They will be stored for some time in liquid sodium to allow the shorter lived fission products to decay and the sub-assembly heat output to fall to a few kilowatts (probably 10 kW or less). It is likely that individual sub-assemblies will be sealed in sodium-filled stainless steel canisters which will then be placed in holes in massive solid steel flasks designed to meet IAEA regulations for the safe transport of radioactive materials. On arrival at the processing plant, the residual sodium could be removed from the fuel sub-assemblies by vacuum distillation with inert gas purge before cutting up the fuel pins and dissolving the fuel prior to the chemical extraction stage. The reprocessing is likely to use a solvent extraction method similar to that developed for thermal reactor fuel to separate uranium, plutonium and fission products, but the plant will be designed to cope with the special requirements of the high plutonium content. Geometrically limited equipment or possibly fixed or soluble neutron absorbers as 'poisons' will be included to prevent criticality. Also, a separate process will be included to remove noble metal fission product alloys which are formed in high burnup fuel, and which are much less soluble than the other fission products and the fuel. Initially it is anticipated that fast reactor fuel will be reprocessed after a long cooling period but as experience is gained, it is anticipated that the necessary radioactive decay cooling will be reduced so that the amount of plutonium held up outside the fast reactor cores is one-third to one-half of that held within the reactors.

Taking into account the more dificult problems associated with handling highly irradiated fast reactor fuel at short cooling times (say 6 months to one year compared with up to 5 years for oxide thermal reactor fuel) and the need for sodium cleaning and disassembly of the fuel prior to reprocessing together with the requirement to utilize ever-safe plant to prevent accidental criticality, it is considered that fast reactor reprocessing could cost about twice as much per kg of heavy atoms (that is, core and all breeder fuel mixed) as for thermal reactor fuel.

Alternatives to using a process based on current thermal reactor reprocessing technology have been suggested. One is to use a process in which some of the fission products are allowed to remain with the separated plutonium in order to afford some measure of radiation protection against misappropriation. Such a process – known as CIVEX – was discussed during the INFCE programme. Another suggestion is not to separate the uranium and plutonium completely. Known as coprocessing, the intention is to reduce the risk of weapon proliferation by always allowing some uranium to be mixed with plutonium. However, neither CIVEX nor coprocessing are likely to be developed to the commercial stage before the turn of the century and are not discussed further.

Associated with both thermal and fast reactor fuel reprocessing is the management and eventual disposal of the highly active wastes including the liquors containing the fission products.

To a close approximation the fission products produced from the fission of plutonium are the same as those from the fission of uranium. However, because of their high thermal efficiency, the quantity produced in fast reactors per unit of electricity generated should be smaller than in most thermal reactors. On the other hand, because of the higher burnup and short irradiation time of fast reactor fuel, together with the desirability that fast reactor fuel should be reprocessed after a shorter cooling time than thermal oxide fuel, the quantity of high-level radioactive liquor containing these fission products will be up to an order of magnitude higher for fast reactors per unit mass of fuel reprocessed. But, in the final analysis, the quantity of vitrified waste produced from these liquors depends mainly upon the quantity of fission products and so in general terms (i.e. ignoring variations due to other added chemical reagents) the volume of vitrified waste per unit of electricity will be the same for both types of reactors and hence it can be stated that the cost of vitrification and disposal of the fission products from fast reactors will be about the same as for thermal reactors.

8.4.5 Other operating costs

The personnel and materials required to operate and maintain a fast reactor power station are assessed to be about the same as required for a thermal reactor power station. In terms of cost per unit of electricity generated, it is therefore assumed that fast reactor and thermal reactor operating costs will be identical.

8.4.6 Decommissioning

It is envisaged that the decommissioning of nuclear power stations will be carried out in three stages. Stage 1 will involve defuelling the reactor over 3 to 5 years after final shutdown and putting the station into a care and maintenance condition. During the next five-year period, Stage 2 will involve reducing the plant to the minimum size possible by removing all non-radioactive plant external to the reactor biological shield. Finally, Stage 3, which could be carried out at any time after defuelling but which may involve delays up to 50 years, will involve the complete dismantling and disposal of all active plant and components.

Preliminary studies have indicated that the total cost of decommissioning nuclear generating stations and eventually returning the site to a green field state will equate to about 10% of the original construction cost in real terms when assessed as a single payment at the time the plant is taken out of commission. If this amount is added as a levy to the original capital cost, it should be discounted according to the required rate of return. For example, if the RRR is 5% per annum the levy becomes 3% of the capital cost if it is assumed that nuclear plants have an operational lifetime of 25 years. There is no evidence to suggest at the present time that the decommissioning levy will be different for thermal and fast reactors. On the other hand, it is normal to assume that the decommissioning cost of fossil fired power stations will be covered by the scrap value of the plant.

8.5 SINGLE STATION GENERATING COSTS

8.5.1 Components of electricity cost

The cost per unit electricity (kWh) is the value which allows equalization of the receipts from the sale of the electricity produced with the costs incurred during the whole life of a power station. It is usual to divide the costs incurred into three main categories: the first is the investment corresponding with the power station construction which is a so-called fixed cost, i.e. it does not depend upon the operation and the future electricity generation of the station. The second is the fuel used to produce the electricity. In general, this is a variable cost which is proportional to the production of electricity. However, in the case of a nuclear power station which must be provided with an initial charge in order to commence operation, the initial charge can be considered to be part of the investment charge, as it is in fact necessary to incur its cost before the start of electricity production. At the end of the generating station operating life, the reactor will contain a final fuel charge which may provide a credit to offset partly the cost of the initial fuel charge. The third main category includes the costs corresponding to the operation of the station, such as personnel and maintenance materials. These are partially fixed charges (personnel) and partially variable charges which are mainly proportional to production (e.g. maintenance materials).

Thus there is a whole series of costs and receipts extending over the life of a power station. If it is required to obtain a single indicator like a levelized cost per kWh to compare one station type with another, it is necessary to pass from this series of costs to a unique value. The discounting technique can be used for this.

8.5.2 Discounted generating cost

For a power station producing electricity, the cost per kWh, C, is defined as that which equalizes the discounted costs (or expenditures) with receipts. If E_n is the annual cost and R_n the receipt in year n, we will have, taking for example the reference as the year of reactor start-up:

$$\sum_{n=-t}^{N} \frac{E_n}{(1+r)^n} = \sum_{n=1}^{N} \frac{R_n}{(1+r)^n} = \sum_{n=1}^{N} \frac{CWL_n}{(1+r)^n} = CW \sum_{n=1}^{N} \frac{L_n}{(1+r)^n}$$

(8.1)

where W is the electrical power of the station, N is the life of the station, L_n is the number of equivalent full power hours of operation during year n ($8760 \times$ load factor in year n), t is the construction period before the reference date and r is the annual discount rate or required rate of return when all costs are expressed in real terms.

The quantity:

$$\sum_{n=1}^{N} \frac{L_n}{(1+r)^n} \quad (8.2)$$

is the number of operating hours discounted over the life of the power station. Rearranging we obtain:

$$C = \frac{\sum_{n=-t}^{N} \frac{E_n}{(1+r)^n}}{W \sum_{n=1}^{N} \frac{L_n}{(1+r)^n}} \quad (8.3)$$

as the discounted mean generating cost over the lifetime of the station.

An alternative representation of the same quantity can be obtained by multiplying both sides of Equation (8.3) by the number of operating hours discounted over the life of the power station. We then get the present worthed capitalized generating cost per unit of installed net capacity (C_{cap}) given by:

$$C_{cap} = \frac{\sum_{n=-t}^{N} \frac{E_n}{(1+r)^n}}{W} \quad (8.4)$$

8.5.3 Generating costs

To determine whether fast reactor generating costs will, overall, be of the same order as thermal reactor generating costs, it is necessary to take a view on all the components of cost enumerated in the previous section. Table 8.4 contains typical values for the components of capital and fuel cycle costs for a thermal reactor in 1980 money terms. The values quoted are plausible for either AGR or PWR ordered in the UK during the 1980s and 1990s. Against these, a range of factors for the corresponding fast reactor costs can be examined in the context of varying uranium prices, to establish relative generating costs recognizing that uncertainties in thermal reactor costs as well as fast reactor costs will influence the firmness of results obtained. A comparison of generating costs which emerges is given in Table 8.5.

Table 8.5 shows the sensitivity of thermal and fast reactor generating costs to changes in each major component. The figures illustrate the importance of capital costs. Because thermal reactors are expected to have to bear rising prices for uranium and enrichment as prices of competing fossil fuels go up and which are also adequate to encourage expansion of supply, fast reactors could have higher capital costs than thermal reactors and remain competitive. However, the allowable margin in excess capital cost is limited

Table 8.4 Thermal reactor costs (commissioning date 1990–2000, 1 January 1980 money)

Uranium	$30–100/lb U_3O_8	
Enrichment	$100–150/kg SW	
Capital cost including interest during construction	£1100/kW	(2–1.4*)
Levy for decommissioning	£33/kW	(2–1.4)
Fuel fabrication	£100/kg ha	(10–8 core; 1.5 radial breeder)
Fuel reprocessing	£450/kg ha	(2)
Waste treatment	£50/kg ha reprocessed	(1)
Operations and maintenance costs at 70% load factor	£12/kWyr	(1)
Credit for plutonium	Zero	
$ exchange rate	$2 = £1	

Figures in parentheses are multiplication factors for fast reactor.
*Because of inherent differences in the characteristics of thermal and fast reactors, it is likely that capital cost ratios will vary as absolute safety standards are evolved during the development of a nuclear programme.

Table 8.5 Thermal and fast reactor generating costs (commissioning date 1990–2000, 25 year life, 70% load factor, 5% per annum discount rate)

	Thermal (%)		Fast (%)	
	Low U&SW	High U&SW	Low factors	High factors
Construction including IDC and decommissioning	66	58	72 (65)	78
Initial core				
uranium	1	3	—	—
enrichment	1	1	—	—
fabrication	1	1	1 (1)	1
Replacement fuel				
uranium	3	10	—	—
enrichment	3	4	—	—
fabrication	3	2	6 (8)	5
Reprocessing and waste management	12	12	13 (16)	10
Last charge (net)	—	—	1 (1)	—
Other operating costs	10	9	7 (9)	6
Total	100	100	100 (100)	100
Total £/kW pw	1740	1950	2190 (1760)	2900
Total p/kWh sent out	2.02	2.26	2.54 (2.04)	3.37

Figures in parenthesis based on unit factor for FR capital – see *Table 8.4.

The liquid metal fast breeder reactor

and the fourth column in Table 8.4 indicates that early fast reactors are likely to exceed it. But, as indicated in Section 8.4, with further development based on manufacturing and operating experience, fast reactor construction costs should be brought within the required range.

Nevertheless, apart from the obvious aim to reduce fast reactor capital costs relative to those of thermal reactors and the compensating effect of rising fuel prices for thermal reactors, other actions are possible that will help to bring fast reactor generating costs more in line with those of thermal reactors when assessed on a single-station basis. These possibilities are all concerned with the fuel cycle.

As seen from the figures given in Table 8.5, the principal components of the fuel cycle cost in fast reactors are the cost of reprocessing discharged irradiated fuel and providing replacement new fuel. The annual requirement for both these quantities is inversely proportional to the burnup (i.e. irradiation or fraction of heavy atoms fissioned) achieved in the fuel. Hence, it follows that if the mean burnup of fuel is doubled, the requirement for reprocessing and replacement fuel is halved. In a large programme of fast reactors, when all advantages of scale have been realized, this would lead to a halving of replacement fuel and reprocessing costs. (In the early stages of a fast reactor this saving would be partly offset by a need to provide smaller fuel processing plants which would lead to higher unit costs.) Other savings might be possible by reducing fabrication and reprocessing costs and increasing fuel rating.

8.5.4 Variations in fast reactor generating cost

As stated in Section 8.4, early fast reactors are likely to use mixed uranium/plutonium oxide fuel having fuel pellet diameters of around 5 mm in subassemblies comprising 300 or so pins. It should be possible to operate such reactors with fuel pellet diameters of 7 mm or so and thereby reduce the number of pins in each sub-assembly to about 170, resulting in a reduction of fabrication costs by about 10%.

The figures given in Tables 8.3 and 8.4 are based on the premise that current research and development programmes on fast reactor fuel reprocessing have a target to reduce the cooling time between discharging fast reactor fuel from a reactor and reprocessing to 6 months. If the cooling time were to be increased to one year or more the residual radioactivity of the irradiated fuel would be reduced and result in the ability to transport more sub-assemblies together in irradiated fuel flasks and allow some decrease in reprocessing plant capital cost by reducing shielding requirements, off gas and heat removal systems. Overall, such procedures may, although the figure is most uncertain, reduce unit reprocessing costs by about 10%.

The present intention is that mixed uranium/plutonium oxide fuel will have a linear rating of around 40 to 50 W per mm of fuel pin length. It is conceivable that it may prove possible to increase the linear rating by as much as a factor of two without altering other dimensions within the sub-assembly and thereby reduce the size of the initial fuel inventory by one-half.

Applying all these reductions to the figures given in Table 8.5 results in a decrease in generating cost of 9 and 12% for an early and later fast reactor respectively. The fast reactor generating costs given at the foot of Table 8.5 then become 2.2 and 3.1 pence per kWh respectively which, if there is pressure on supplies of fresh uranium resulting in higher enriched uranium prices, offer the promise of being competitive with thermal reactors. Also, they compare favourably with the alternative of constructing coal fired generating stations with capital costs in the range £500 to £600 per kWe which are expected to have discounted mean generating costs in excess of 3 pence per kWh assuming the real cost of coal rises at an average rate of 3% per annum from now to the end of the century [4]. Finally, as mentioned in Section 8.4, some safety standards might slowly change as the nuclear power programme evolves and it is possible that fast reactor generating costs might be fully competitive with thermal reactors even at the low uranium and separative work costs quoted in Table 8.4.

8.5.5 Disadvantages of single station generating cost comparisons

Although consideration of the physics of fast reactors may allow an increase in fuel rating and burnup as postulated above and future research and development may prove its practical feasibility, such a procedure changes the amount of plutonium required in new fast reactor fuel to enable it to operate for the required time. Also, altering fuel pin diameters can affect the amount of plutonium in the core whilst increasing the cooling time before reprocessing increases the amount of plutonium held up in irradiated fuel. Many of these 'improvements' in single station generating costs tend to increase the amount of plutonium required for each fast reactor. In Table 8.5 no value was ascribed to the plutonium produced in thermal reactors and used in fast reactors. For a closed generating system comprising both thermal and fast reactors, this is a reasonable thing to do because crediting thermal reactors with the value of the plutonium produced and debiting fast reactors with the plutonium they require initially would be internal transfer payments which would cancel when considering the system as a whole.

Another result of some of the possibilities given above is to reduce the amount of excess plutonium produced (or bred) in a fast reactor over the amount it consumes in producing power. For example, increasing the burnup achieved in new fuel requires an increase in the amount of plutonium in that fuel to sustain that burnup and counteract the increase in neutron absorption in the additional build-up of fission products. This is brought about by increasing the enrichment of plutonium contained in the depleted uranium which acts as a fertile material. The net result is that not only is the inventory of plutonium required in the reactor increased, but the net amount of plutonium produced is decreased because fertile atoms have been replaced by fissile atoms and their fission products. Thus, whilst the actions described may well reduce the generating cost of a single fast reactor, the number of fast reactors that can be installed from a given stockpile of plutonium and the number of fast reactors that can be installed later from excess plutonium bred could both be significantly reduced. The net result is that a whole balanced system of thermal and fast reactors may require more uranium and incur a higher overall expenditure for the same electricity produced than might otherwise have been the case.

Comparison of alternative generating stations using single-station costs can therefore be very misleading, as the effects that certain variable parameters may have on the entire system are not evaluated. The effect of changes in fast reactor fuel parameters on total uranium requirements has just been mentioned. Allied to this is the likelihood that, as fast reactors are exploited on a world-wide basis, the pressure of demand on available supplies of uranium for thermal reactors will be reduced and so thermal reactors will benefit by a moderation in the rate of rise in uranium prices. This itself introduces a paradox into single station generating cost comparisons. Thermal reactor generating costs could well be less than fast reactor generating costs with only moderate uranium price increase assumptions. However, moderate uranium price increases may only be applicable if fast reactors are assumed to be extensively installed on a world-wide basis.

Another assumption inherent in the figures given in Table 8.5 is that thermal reactor spent fuel will be reprocessed and the cost will be set against the electricity generating cost of thermal reactors. However, in the absence of fast reactors it may be economically preferable to delay reprocessing and merely store thermal spent fuel during the interim. As mentioned before, the economic case for reprocessing depends on the relative costs of storage, reprocessing, waste disposal and uranium/enrichment price; the latter could well be influenced by policy decisions on the fast reactor.

Finally, single station generating costs generally make some fixed assumption on lifetime load factors. In a real generating system stations will operate the varying load factors depending upon the load duration demanded and the ranking of the station within the merit order. The position of a station in the merit order will be influenced by the types of station already within and to be added to the system, together with their associated fuel costs. As already seen, the latter can themselves be influenced by the choice of station type.

It is concluded that single station generating cost comparisons are not wholly helpful in assessing the true consequences of following alternative fast reactor strategies; this is better done by considering a whole generating system and this aspect will be discussed next.

214 The economics of advanced converters and breeders

8.6 SYSTEM COSTS

8.6.1 Plutonium production, inventory and doubling time

Of the fresh fissile fuel produced in a fast reactor, an equal amount to the primary fuel destroyed is required for recycle into the reactor to maintain operation, leaving an excess for use in new plants. This excess over unity before deducting reprocessing losses is known as the 'breeding gain' and, as hinted at in Section 8.5, is dependent upon factors of fuel design such as rating, geometry and burnup, as well as the amount of neutron-absorbing materials in the reactor core. Following reprocessing of both irradiated core and breeder fuel, the important quantity is the excess plutonium production over that required to maintain operation.

The other important consideration of fast reactor logistics, and so eventually the economics of a complete nuclear generating system, is the amount of fissile material needed for a reactor to produce a given power output at a given load factor, i.e. its inventory. This inventory comprises two parts – fissile material present within the reactor itself and fissile material outside the reactor in the fuel processing parts of the cycle, which comprise irradiated fuel cooling, transport, reprocessing, refabrication into fresh fuel and transport back to the reactor. The former depends particularly upon the fuel rating and burnup achieved. The latter amount is dependent upon the rate fuel passes through the reactor (i.e. upon burnup and load factor) and also in turn upon the time taken to pass through the various stages of the fuel processing parts of the cycle.

The time taken for a fast reactor to produce enough plutonium to provide the total (in-pile and out-of-pile) inventory required by a new reactor is known as the linear doubling time. This is proportional to total plutonium inventory and inversely proportional to its net plutonium production, the latter as indicated above depending upon breeding gain and the amount of plutonium not recovered from process residues together with the fast reactor load factor.

Both a decrease in specific inventory and an increase in net plutonium production lead to a reduction in doubling time and a small specific inventory and a low doubling time are desirable attributes. A small initial plutonium inventory is important at the beginning of a fast reactor programme since, at a time when plutonium stocks are being used, the design with the smallest inventory will allow more fast reactors to be fuelled from the stock. This is of immediate importance to uranium requirements because, over the lifetime of each GWe of fast reactor capacity, there will be a reduction in the need for some 3000 to 4000 tonnes of uranium compared with using a similar thermal reactor capacity on a once-through fuel cycle. Subsequently, shorter doubling times will become increasingly important if thermal reactors are to be phased out of an expanding nuclear programme.

But a short doubling time depends critically upon the performance of the fast reactor fuel processing plant. In Fig. 8.3 the bottom left-hand corner represents the doubling time that a typical early mixed oxide fuelled commercial fast reactor (Table 8.2) would have if the plutonium held in process plant waste residues is 0.5% of total throughput and the time taken to return plutonium in new fuel to the reactor after its discharge in irradiated fuel is 9 months. Under these circumstances the linear doubling time would be about 30 years and would support a growth rate of fast reactor capacity of around 3% a year. The longer the time taken to return plutonium back to the reactor after discharge, the larger the out-of-pile inventory of plutonium in cooling ponds, reprocessing plants and fuel fabrication plants, and so the longer the doubling time. In addition, as all the plutonium in irradiated fuel must be reprocessed before recycle, the excess plutonium production decreases significantly as the percentage of total throughput held in process plant residues increases, thereby further lengthening doubling time.

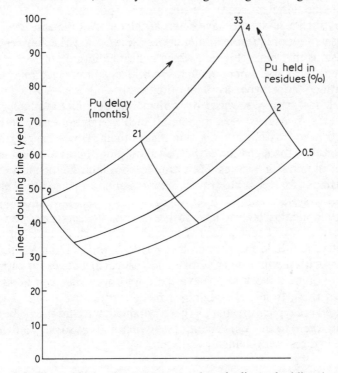

Fig. 8.3 Effect of fuel cycle parameters on breeder linear doubling time.

In addition, as already mentioned, fast reactor inventory and excess plutonium production and hence doubling time, is affected by such matters as fuel geometry, fuel rating, fuel burnup, etc. and it is clear that the determination of a fast reactor design to achieve the cheapest nuclear electricity generating system is a complex matter into which many factors enter and that the outcome must be a compromise to secure an optimum, as features which lead to increases in breeding gain (and so decrease doubling time) sometimes tend to increase total inventory (and so increase doubling time).

These many issues which are all relevant to fast reactor economics can only be adequately judged in the context of the complete electrical generation system of which the reactor forms part. A procedure is used in which the performance of a system for one nation or group of nations, or the whole world, is assessed over many decades. The basis of the procedure used, which was outlined in Section 8.4 and described fully in reference [5], is to assume a particular progression of future electricity demands and to test the effect of total system expenditure and uranium consumption (or any other relevant criterion of choice) as different power programmes, fast reactor designs or aspects of their performance including the performance of the associated fuel processing plants, are incorporated.

Many such assessments have been carried out over the past 20 years or so and it is neither practicable to summarize these nor to write down categorical conclusions that have emerged which indicate a future policy that must be followed. Each assessment made assumptions of future electricity demand, generating station types available (i.e. fossil – oil, coal, gas; or nuclear – thermal, fast with gas, water or sodium cooling), fuel availability, costs, plant performance, etc. and in general, such assumptions vary both in time and with who or what organization makes them. However, certain trends have emerged through the years which allow judgements to be made about a number of important issues. For example, they indicate the most economic proportions for possible future components of the total electricity generating system and the operating regime, i.e. high, mid or low merit, each component is likely to have to meet. Also for each type of power plant, preferred areas of technological improvement such as fuel burnup, fuel geometry, fuel held in process plant residues, process hold-up times, etc. together with the financial and/or fuel logistical incentives for carrying out the development work to achieve these improvements, can be identified. Some of these trends are examined further in relative terms. The results presented are typical of what has been obtained, but care should be taken in applying them to the assessment of new situations as some relativities may change in detail with changing assumptions.

8.6.2 Possible nuclear shares in electricity programmes

Each assessment of alternative nuclear strategies must make assumptions of future electricity need. Allowing for variation in system load factor and plant margin, these electricity needs can be translated into required capacity. The load factor of each station may then be calculated throughout

The liquid metal fast breeder reactor

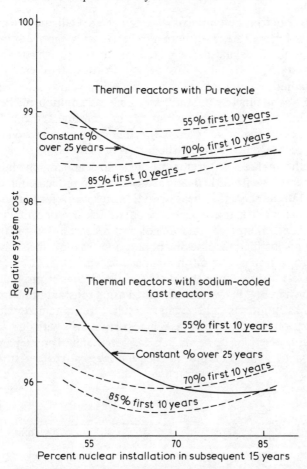

Fig. 8.4 Variation of system cost with proportion of nuclear capacity.

its life with reference to other stations in the system by stacking them in merit order (which depends upon operating costs rather than capital costs) under a load duration curve which may vary from year to year.

Alternative nuclear strategies have often been assessed on the basis that from a certain year, a proportion of all new plant is assumed to be nuclear. To justify the use of a constant nuclear proportion in any programme assessment, trial calculations are normally carried out using variant nuclear percentage assumptions. The results of such a test are indicated in Fig. 8.4 where relative total present worth system costs over a 25-year period are shown for two strategies as a function of percentage of nuclear plant. During a first 10-year period, the nuclear component was assumed to be 55, 70 or 85% of the total new installations. Each alternative was followed by a further 15-year period of either 55, 70 or 85%.

The curves of total system costs shown in Fig. 8.4 fall into two groups. The upper set is for a strategy of thermal reactors with plutonium recycle; the lower set is for a balanced programme of thermal reactors and sodium-cooled fast reactors. Total relative present worth system cost for the whole period studied is plotted for each initial 10-year nuclear percentage assumption as a function of the percentage assumption in the following 15-year period. These are shown as the broken lines on the figure. The continuous lines give the variation of system cost as a function of constant nuclear percentage through the whole 25-year period.

For the thermal reactor strategies, a high nuclear component is indicated in the early years with little financial advantage in reducing the component below 70% in the second 15-year period. In any event, for both thermal and fast reactor strategies, the dependence upon nuclear proportion is fairly flat. It is concluded, therefore, that a choice of a constant 70 to 80% nuclear component is justified for assessment purposes, as the effects of varying the nuclear proportion are small in comparison with other uncertainties and quite unlikely to influence the choice of development programmes. The curves shown in Fig. 8.4 were determined using constant uranium and fossil fuel cost assumptions and similar results are obtained with varying assumptions on fuel costs. Increases in fossil fuel costs with time tend to bias the nuclear percentage upwards. Increases in uranium and separative work costs will act in the opposite manner for thermal reactor strategies but

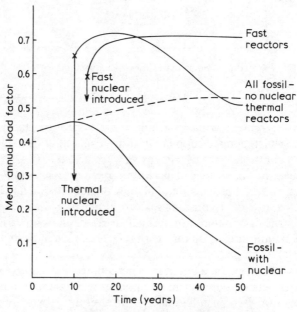

Fig. 8.5 Variation of mean annual load factors.

providing fast reactors are introduced early, fast reactor strategies are little affected by increasing uranium costs.

Using such a constant split in nuclear and fossil components in an installation programme, Fig. 8.5 shows how the mean annual load factor of each type of station varies with the strategy adopted. The solid lines show the variation in mean annual load factor for each type of station in a fossil/thermal/fast reactor strategy. The dotted line shows the mean annual load factor of all plant in a wholly fossil-fuelled programme. Because of the consequences of merit order operation, thermal reactor stations, which have lower operating costs than fossil-fuelled stations, will operate on base-load and have high mean annual load factors when first introduced. Once the base-load is fully supplied by thermal reactors, their mean annual load factor will begin to drop. Similarly, on introduction, fast reactors will occupy base-load, thereby depressing the thermal reactor load factor further. A consequence of operating at low load factors is to reduce uranium requirements for thermal reactors. At the same time these same reactors must be designed to have a capability for load following. The mean load factor of the fossil-fuelled plant is seen to fall rapidly as it is pushed further down the merit order by the nuclear plant. In the non-nuclear strategy, the

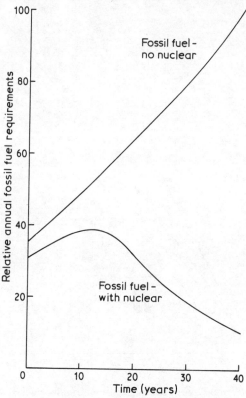

Fig. 8.6 Alternative US fuel requirements.

220 *The economics of advanced converters and breeders*

mean annual load factor of all the fossil plant remains more or less constant, varying only slowly with changes in system load factor and/or plant margin.

The effects on fossil fuel requirements for the strategies assumed in Fig. 8.5 are indicated in Fig. 8.6. On the assumptions of increasing electricity demand and no nuclear component, fossil fuel requirements rise steadily each year. With the introduction of a nuclear component within the same overall electricity programme, fossil fuel requirements rise slowly to a maximum and then drop as the percentage of the nuclear share of electricity production rises.

8.6.3 Uranium requirements

Figure 8.7 shows the relative annual quantity of uranium required for all thermal and mixed thermal and fast reactor strategies. The uppermost curve

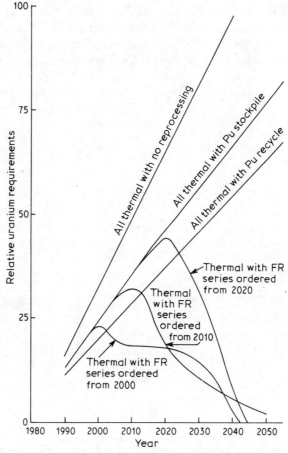

Fig. 8.7 Annual uranium requirements for alternative nuclear strategies.

indicates the quantity of uranium for an all-thermal reactor programme operating on the once-through fuel cycle, i.e. with the storage of spent irradiated fuel. The next two curves indicate the quantity of uranium for all thermal reactor programmes when it is assumed that spent fuel is reprocessed to recover both the uranium and plutonium. The upper of these two curves shows the uranium requirement when the uranium extracted from spent fuel is recycled but plutonium is stock-piled; the lower curve shows the further reduction in uranium requirement with plutonium recycle. The lowest curve shows the requirements for uranium if a series ordering of fast reactors is commenced at the turn of the century. Other curves show the effect of delays in the fast reactor programme. With the early fast reactor programme, annual uranium requirements would rise to a maximum of 23 relative units around the year 2000 and then fall steadily to zero before 2050. With 10 and 20 year delays in the series ordering of fast reactors, maximum annual demands rise to around 30 and 45 units respectively. Without fast reactors the annual requirement for uranium will rise continually and will be within the range 45 to 80 units by 2030, depending upon the thermal reactor

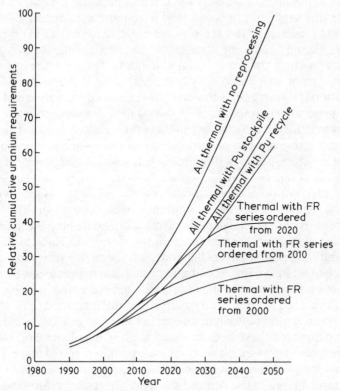

Fig. 8.8 Cumulative uranium requirements for alternative nuclear strategies.

fuel cycle adopted. Cumulative uranium requirements are shown in relative terms in Fig. 8.8. The important point to note here is that with fast reactors the need to mine additional uranium for an on-going nuclear programme can be avoided for several centuries, whereas with an all-thermal reactor strategy commitments to obtain fresh uranium continue so long as the reactors are installed and operated. A further point to be noted is that whereas Fig. 8.4 shows only a 2 to 3% saving in total present worth system expenditure due to the introduction of fast reactors into a nuclear strategy, Fig. 8.8 indicates that the saving in total uranium requirements can be a factor of 2 to 4, 30 to 50 years after their fullscale introduction. Obviously, the magnitude of economic saving due to the introduction of fast reactors will depend upon many factors, including the relative costs of thermal and fast reactors, together with the associated fuel cycle costs, but it seems fairly certain that in a world with finite uranium resources the demand on these resources will be greater in a world without fast reactors than in a world with fast reactors and it is to be expected that uranium prices will rise more quickly in the former case. It follows that the true economic benefit arising out of the introduction of fast reactors can never be determined. Some workers in the nuclear industry see this as a paradox; they argue that it is possible that uranium prices will tend to stabilize at relatively low levels if significant numbers of fast reactors are installed soon after the turn of the century, thereby reducing demand on uranium and hence avoiding the necessity to mine low-grade uranium deposits. On the other hand, if fast reactors are not introduced early in the next century, demands on uranium for all-thermal strategies of the same total installed capacity could be high and its price will rise rapidly, and it would then already be too late for fast reactors to influence the uranium market to any marked degree. This implies that fast reactors need to be introduced and established before the turn of the century when, perhaps, they are not demonstrably economic. In the final analysis, it could be that the choice between one reactor strategy and another may not be entirely a matter of immediate economic saving that might be achieved, but be an insurance policy against the risk of high uranium costs or the strategic desire to minimize future uranium imports and hence dependence on foreign uranium suppliers by utilizing to the fullest extent the 'waste' uranium arising from early thermal reactor programmes.

The changes in system expenditure or uranium requirements considered so far have all assumed that the fast reactor fuel-cycle processing operations will be developed to achieve a 9 month out-of-pile hold-up time and 0.5% or so diversion of total plutonium throughput to waste residues. Also the fast reactor parameters have been assumed as given in column 2 of Table 8.2. The effect of changes in fast reactor core parameters and of fuel cycle plant performance on system requirements are investigated in the Appendix.

The results given in the Appendix indicate that poor performance of fast reactor fuel process plants can increase uranium requirements significantly by reducing the fast reactor component within a given nuclear programme. Alternatively, if uranium supplies are severely restricted, poor fast reactor process plant performance could lead to a small nuclear programme because the number of thermal and fast reactors would both be limited. It is possible to counteract to some extent such adverse effects if large nuclear programmes are required whilst, at the same time, uranium requirements are to be kept to a minimum. One possibility is to introduce more advanced fuels such as carbide which would provide a higher breeding gain, as indicated in column 3 of Table 8.2. With good fuel-plant performance the doubling time for the carbide fuelled reactor is shown as around 14 years; this would rise to around 30 years if the plutonium out-of-pile time and quantity held in waste residues was to rise to 33 months and 4% respectively Such a doubling time is comparable with the best achievable with oxide fuel and hence, in the longer term, it might be expected that the total uranium requirements would be similar. But relatively little work has been done on proving carbide fuels in fast reactors and even less on the difficult problems associated with fabrication and reprocessing. During fabrication these problems arise on account of the pyrophoric nature of uranium/plutonium carbide powders, whilst during reprocessing this material has been found extremely difficult to dissolve. Thus their costs are very uncertain.

A second alternative is to use ^{235}U enrichment in fast reactors at times of plutonium shortage instead of continuing or reintroducing thermal reactors. Parameters for such a reactor are shown in column 4 of Table 8.2 from which it is seen that the plutonium production after reprocessing is almost a factor 4 higher than the net production from the corresponding oxide fuelled reactor enriched with plutonium. The production of ^{235}U required to fuel these reactors results in increased uranium consumption but an advantage is gained later as a result of the more rapid introduction of fast reactors. The overall result on annual uranium requirements is shown in Fig. 8.9. The lowest curve in Fig. 8.9 shows the relative annual uranium requirements for a strategy of thermal and fast reactors in which plutonium-enriched oxide fuelled fast reactors are introduced from around 1990, at first slowly, but after 2005 to the extent of plutonium availability. With a good fuel-plant performance (i.e. plutonium turn-round time of 9 months and plutonium held in residues 0.5%) uranium requirements rise to a peak around the year 2010 and then fall to zero by 2050. The next curve shows the uranium requirements for the same strategy in which the fast reactor fuel plant performance is such that the turn-round time has increased from 9 to 33 months and the plutonium held in residues from 0.5 to 4%. The peak annual uranium requirement has now doubled and there would still be a large commitment to further uranium supplies after the middle of the next century. The lower broken curve illustrates the effect of introducing ^{235}U enrichment in fast reactors at times of plutonium shortage. This shows quite clearly that although the maximum annual uranium requirement is about the

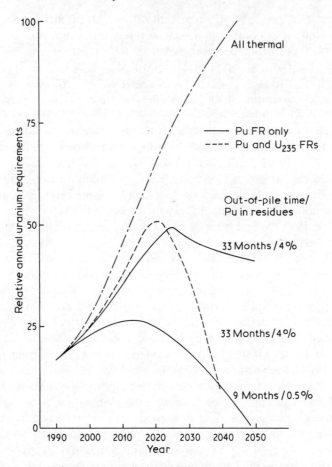

Fig. 8.9 Annual uranium requirements: Pu and ^{235}U FRs.

same, it quickly decreases to zero by around the middle of the next century in spite of the poor performance of the fast reactor process plants. The uppermost curve on the figure indicates the annual uranium requirements of an all-thermal reactor programme having the same installed capacity. Figure 8.10 shows the same information plotted on a cumulative basis and clearly indicates the advantage of using ^{235}U enrichment in fast reactors to restrict total uranium requirements. The use of ^{235}U enrichment under these circumstances also yields a saving in the total separative work required for enriching fuel (for both thermal and fast reactors) and as it can be anticipated that the capital, fuel fabrication and reprocessing costs for fast reactors using ^{235}U enrichment will be very similar to those using plutonium enrichment, an economic benefit should result.

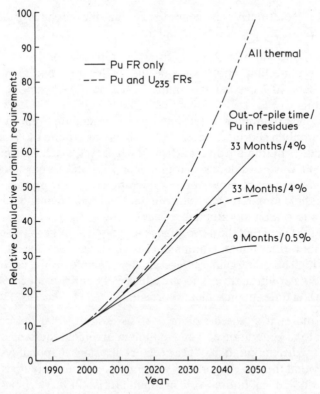

Fig. 8.10 Cumulative uranium requirements: Pu and ^{235}U FRs.

8.6.4 System analyses – conclusions

The main conclusions that have emerged from economic and logistic system analyses, as outlined in the preceding paragraphs and in the Appendix, are:

1. Minimum system expenditure will be achieved if around 80% of all new generating capacity is nuclear rather than fossil fuelled.
2. Total uranium requirements will be kept to the minimum if plutonium is retained for use in fast reactors, rather than recycling it in thermal reactors. Such a strategy will have a lower total system generating cost than an all-thermal reactor strategy. This is because the lower uranium demand will reduce uranium price rises below what they would otherwise have been and so help to stabilize the costs of thermal reactor generation in addition to the direct savings of uranium provided by the fast reactor.
3. Given the strategy outlined in (2), the achievement of a minimum system expenditure is not affected strongly by fast reactor core parameters except perhaps through the mechanism of uranium price. To minimize total

uranium requirements and hence to minimize the pressures of demand on the available supply, it is beneficial to:

(i) Introduce fast reactors as early and as rapidly as possible.
(ii) Use plutonium enriched oxide fuel with 5 to 7 mm diameter fuel pellets, 10% maximum burnup and around 50 W/mm peak pin linear rating.
(iii) Reduce plutonium out-of-pile time to around one year by the time plutonium shortages occur, i.e. probably by around the year 2010.
(iv) Reduce plutonium diverted from the fuel cycle and held in processing plant waste residues to around 1 or 2% total throughput for fuel logistic reasons. However, considerations of waste management and possible long-term environmental impact could demand this to be as low as reasonably attainable – say 0.5%.
(v) In the event that the performance suggested in (iii) and (iv) cannot be achieved, uranium requirements can be reduced by the introduction of high breeding gain fast reactors, for example, by the use of carbide fuel. An alternative is to introduce ^{235}U enrichment in oxide fuel during times of plutonium shortage.

Furthermore, if a closed system analysis indicates that the benefits of minimum total generation cost and minimum uranium requirements can be achieved by the exploitation of fast reactors in a single national system, it is very likely that the same arguments will apply to a group of nations or to the whole world. Hence, future world uranium prices will vary depending upon the choice of reactor strategy in the world as a whole. It follows that some nations or regions could deploy fast reactors to the full by the establishment of a trade in irradiated thermal reactor fuel containing plutonium with those nations or regions wishing to deploy only thermal reactors for as long as possible.

8.7 CONCLUDING REMARKS

The general conclusion is drawn that to utilize fully uranium resources available from low cost deposits, and to provide a source of power which will be economic compared with other alternatives, the option to use stocks of plutonium and depleted uranium arising from thermal reactor programmes in fast reactors should be developed. The full deployment of the fast reactor option will take some considerable time because of the practicable penetration rate that might be industrially achievable, together with the large investment involved in both reactor and fuel cycle plants. At the time such full deployment is implemented, the pressure on other fuel resources may well be great, and if the fast reactor is to be most effective in meeting energy demands, it is necessary to complete the demonstration and development phases in good time. It is therefore suggested that the construction of a commercial demonstration fast reactor, together with its associated fuel servicing plants, should proceed as early as practicable even though, standing alone, it may not be immediately economic in its own right. In this way the option of recycling uranium and plutonium in fast reactors will be available as an established technology for the generation of economic electricity when the effects of depletion of the gaseous and liquid carbon fuels begin to bite.

REFERENCES

[1] *Assessment studies on plutonium recycle in CANDU reactors* (29 November 1978) INFCE/DEP/WG4/69. Canadian Paper to INFCE.
[2] INFCE (1980) Vol. 8, *Advanced Fuel Cycle and Reactor Concepts*.
[3] INFCE (1980) Vol. 9, *Summary Volume*.
[4] CEGB Annual Report (1979/80) *Appendix 3*.
[5] Iliffe, C.E. (June 1973) *Discount G: a digital computer code for assessing the economics of nuclear power*, TRG Report 2285(R).

APPENDIX

Variation in system expenditure and uranium requirement with fast reactor core parameters

Changes in fast reactor core parameters that might reduce the generating cost of a single reactor will not necessarily lead to an overall reduction in total system expenditure on a whole nuclear generating system because such changes might adversely affect reactor doubling time and hence the ability of a fast reactor to penetrate a nuclear sector. In Fig. 8A.1, relative total system present worth expenditure is shown as a function of oxide fuelled fast reactor fuel pellet diameter for various values of fuel density, fuel burnup and fuel pin linear power rating, with a core height of 1 m and axial blanket thickness of 0.4 m.

Most of the curves pass through a minimum system cost at a fuel diameter of just over 6 mm. Increasing fuel density whilst keeping the fuel linear rating and burnup constant reduces system expenditure slightly. The principal effect here is that for a given linear pin rating and burnup a certain quantity of fissile plutonium is required at the start of fuel irradiation life. As the density of the fuel is increased, so the amount of fertile material rises and so the breeding gain of the reactor improves, which results in a lower doubling time and, in turn, a greater penetrating power of the fast reactor into the system, and a lower total uranium requirement. Increasing the irradiation of fast reactor fuel appears from Fig. 8A.1 to be generally effective in reducing system costs. Not only is the amount of fuel to be processed reduced by increasing burnup, but also the out-of-pile plutonium inventory is significantly reduced and hence, within the range considered,

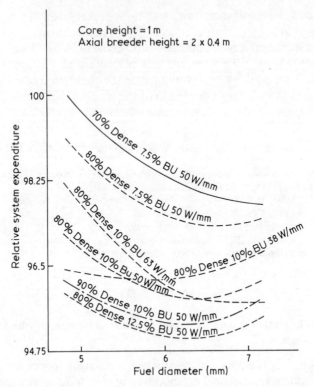

Fig. 8.A.1 Change in relative system expenditure with fast reactor fuel diameter–oxide.

the penetrating power of the fast reactor is improved because of the reduction in doubling time. However, as burnup is increased, the required new fuel plutonium enrichment must be increased and so the in-pile inventory also is increased. This, in turn, affects the breeding gain in the reactor and if irradiation is increased too much, the doubling time of the fast reactor will begin to increase rapidly and result in rising system costs due to additional uranium for a larger thermal reactor section. Figures 8A.2 and 8A.3 indicate the effect of increasing core burnup on breeding gain, inventory and doubling time. From Fig. 8A.2 it is apparent that an increase in burnup leads (1) to a decrease in the out-of-pile inventory due to a longer residence time in the core, and (2) to an increase in in-pile inventory, mainly due to the increase required in core enrichment. These opposite effects lead to a minimum in the total inventory required at about 15 to 20% maximum core burnup. The breeding gain decreases with increase in core burnup, going to zero around 40% burnup, which has a very marked effect on the reactor

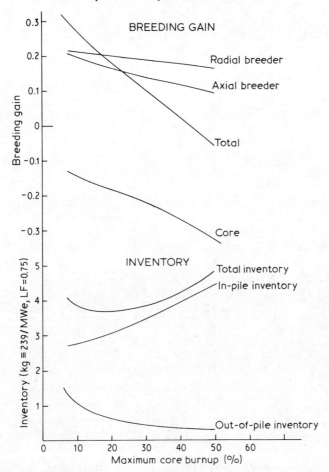

Fig. 8.A.2 Breeding gain and inventory versus maximum core burn-up.

doubling time as seen in Fig. 8A.3. Finally, Fig. 8A.1 indicates the effect of changing linear pin rating on system expenditure. At any given pin diameter, increasing linear rating decreases the size of the core required and the initial inventory of plutonium, whilst at the same time the plutonium enrichment required in the fuel is increased. The latter has an effect on the breeding gain which is decreased, and affects the doubling time in an opposite sense to that due to the change in inventory. With small values of fuel pellet diameter, the mass rating will be higher at any given linear pin rating than for higher values of fuel pellet diameter, and it is seen that for a pellet diameter of 5 mm, system expenditure is decreased if the linear rating is decreased. At a value of 7 mm for fuel pellet diameter, the opposite is seen, i.e. system expenditure decreases as the linear pin rating is increased.

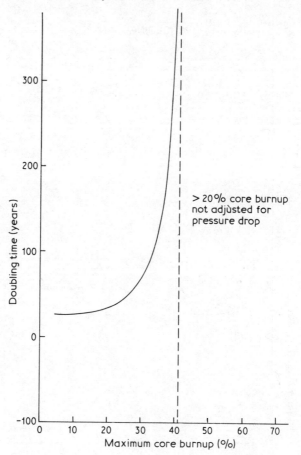

Fig. 8.A.3 Doubling time versus maximum core burn-up.

As already indicated, varying fast reactor core parameters such as pin diameter and rating varies plutonium inventory and production and hence, doubling time. These in turn affect the rate at which the number of fast reactors can grow and thus for a given nuclear power programme, the number of thermal reactors required in support and hence the total need for natural uranium. Figure 8A.4 shows the relative cumulative use of natural uranium for oxide fuelled reactors as a function of fuel pellet diameter and fuel pin linear rating for a given maximum fuel burnup and core height and fuel density. As with total system expenditure, the curves pass through somewhat flat minima, but it is seen that a 'best' fuel is one having a fuel pellet diameter of around 6 mm and a linear rating in the range 50 to 60 W/mm.

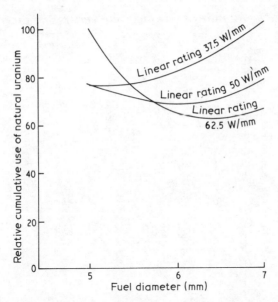

Fig. 8.A.4 Change in uranium requirements with fuel diameter and linear rating–oxide.

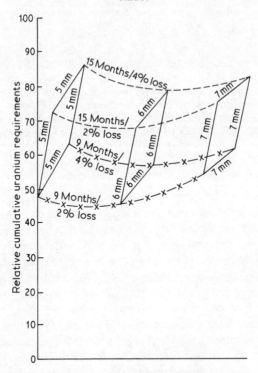

Fig. 8.A.5 Variation in relative uranium requirements with fuel pin diameter, plutonium out-of-pile hold-up time and plutonium held in process plant waste residues – oxide fuel; where ———— is 15 months delay, —x— is 9 months delay, and ———— is fuel diameter; for 80% theoretical density, linear rating 50 W/mm, core height 1 m, axial breeder height 2 × 10.4 m and core burn-up 10.0%.

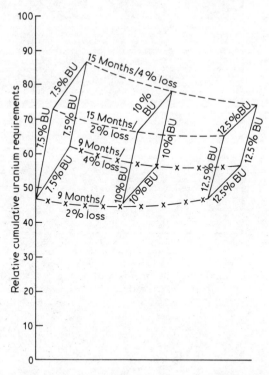

Fig. 8.A.6 Variation in relative uranium requirements with fuel burn-up, plutonium out-of-pile hold-up time and plutonium held in process plant waste residues – oxide fuel; where — — — is 15 months delay, —x— is 9 months delay, and ——— is burn-up; for 6 mm fuel diameter, 80% density, linear rating 50 W/mm, core height 1.0 m and axial BR 2×0.4 m.

The changes in system expenditure or uranium requirements with variations of specific inventory and breeding gain due to changes in the various fast reactor parameters considered so far have all assumed that the fast reactor fuel cycle processing operations will be developed to achieve a 9 month out-of-pile hold-up time and 0.5% or so diversion of total plutonium throughput to waste residues. Since out-of-pile inventory varies with process hold-up time, and net plutonium production varies with process losses to waste residues, the importance of varying these two characteristics of fuel cycle performance must also be assessed. Figure 8A.5 indicates the variation in uranium requirements with fuel pin diameter for oxide fuelled cores for a given fuel burnup, rating, fuel density and core geometry. Whilst these results do not indicate a strong preference for a change in the selection of the 'best' fuel pellet diameter (i.e. around 6 mm) they do demonstrate that the effects of a poor performance in fuel cycle processing can lead to large increases in uranium requirements compared to small changes in fuel pellet diameter. The interaction of fuel burnup with process plant performance is indicated in Fig. 8A.6 for oxide fuel of a given fuel pellet diameter, rating, fuel density and core geometry. As might be expected, the worse the performance of the reprocessing plant, the greater the incentive to achieve a higher burnup to minimize the adverse effect on total uranium requirement.

9

J. Baumier and L. Duchatelle
THE CASE FOR DEVELOPING THE LIQUID METAL FAST BREEDER REACTOR IN FRANCE

9.1 INTRODUCTION

In the past, oil was abundant and cheaper than almost all other forms of energy for all purposes, including for the production of electricity. However, as we are all aware, its price has since rocketed, and will certainly continue to rise at a rate faster than inflation. The underlying cause of this is that discoveries of new oil are not keeping pace with demand, resulting in a continuous depletion of reserves expressed as years of current production. To reduce their energy dependence and balance of payments deficits, while continuing to satisfy the growing demand for electricity, many countries have therefore decided to drastically limit use of oil in electricity production and switch to other sources of energy. Although the main replacement source will in some cases be coal, which is extremely abundant, the cheapest source of energy for generation of electricity is usually nuclear fission. For this reason, and despite the general slow-down in world-wide nuclear power plant construction in the last few years, the last quarter of the 20th century will almost certainly see increasing use of nuclear energy to produce the electricity needed to sustain economic growth, maintain standards of living in the industrialized nations, and create the conditions for improving the quality of life in the developing countries.

Most of the world's nuclear electricity is presently produced by light water reactor (LWR) power plants fuelled with slightly enriched uranium, and using pressurized water reactors (PWRs) or boiling water reactors (BWRs). Between now and the end of the century, existing and planned LWRs will consume large quantities of uranium, but the world's uranium reserves will remain limited, especially as LWRs burn only enriched product, having no use for the large amounts of depleted uranium produced at enrichment plants. In consequence, the selling price of uranium will almost certainly start rising above its present abnormally low level*. Indeed, the rate of price

---

*The 1980 uranium price, of around $30 per lb, is at best only just above the ex-mine cost, and many uranium mines are currently unprofitable.

increase may well be comparable to that of oil over the last ten years. If this occurs, the total cost of electricity generated by LWRs (and other types of thermal reactor) will also rise steeply, and it will become advantageous to deploy fast breeder reactors (FBRs) on a large scale. In effect, the cost of electricity generated with FBRs will then be equivalent to, or less than, that produced by LWRs, because:

1. FBR electricity costs are far less dependent on uranium prices, as they use only depleted uranium, the otherwise largely valueless byproduct of uranium enrichment for LWRs[†].
2. The fissile fuel for FBRs is plutonium. Initially, this plutonium can be recovered when reprocessing spent LWR fuel, and fuelling FBRs is in fact by far the most satisfactory way of using the fissile plutonium produced by LWRs. From then on, the FBRs produce more plutonium than they burn, converting low-value uranium-238 to high-value plutonium.

Of course, the cost of electricity from FBR plants is influenced by the selling price of plutonium. However, unlike for uranium, the price of plutonium should not rise too steeply, because it may be predicted reasonably that plutonium from LWRs will be abundant, whereas the demand for plutonium for FBRs will be quite low, being basically limited to first core loads, after which the FBRs produce their own plutonium.

To substantiate the above statements, we analyse herein the economic aspects of electricity generation by FBR plants, with particular reference to FBRs planned in France during the 1980s, and in comparison with French PWR units.

Section 9.2 outlines the bases of comparison between FBRs and PWRs, and describes methods used to calculate costs.

Section 9.3 describes the FBRs and PWRs concerned, defines and explains the concepts of breeding gain and doubling time for FBRs, and calculates the energy extracted from uranium by the two types of reactor.

Section 9.4 contains the results of analysis and FBR/PWR comparison for capital, operating, fuel cycle and total generating costs, and draws conclusions for the competitivity of FBRs.

Finally, Section 9.5 gives general conclusions on the advisability of large-scale FBR deployment.

9.2 BASES OF COMPARISON

The cost of electricity generated with thermal reactors is a function, essentially, of capital (first) costs and operating costs for the power plants themselves and for associated fuel cycle plants. Comparing the total costs of

[†] World reserves of depleted uranium will reach 500 000 tonnes sometime between 1985 and 1990. Ignoring the question of plutonium availability, this is sufficient for production of some 600 billion tonnes of oil equivalent (toe) of energy using FBRs, a figure comparable to the energy value of the world's total oil, gas and coal reserves at present extraction costs.

electricity from various types of thermal reactor plant (PWR, BWR, HWR, etc.) is not overly difficult, provided the fuel cycle is open.

However, for FBRs the fuel cycle is closed, and cost comparisons between FBRs and LWRs (or other types of thermal reactor) are more complicated. Use of plutonium FBRs increases the value of plutonium produced in PWRs, and hence reduces LWR fuel-cycle costs. Inversely, of course, FBR fuel-cycle costs are increased.

9.2.1 General

After the first units of each type, for which costs are necessarily atypical, the construction costs of the FBR and PWR units considered herein will remain roughly constant in real (present value) terms. Although basically determined by the rate of growth of electricity consumption, the rate of deployment of FBRs will then also be accelerated by the economic advantages to be gained from replacing obsolete PWRs by FBRs, and the only true limitation on the number of FBR plants will be plutonium availability. However, until this stage is reached, when the main advantage of FBRs over PWRs will be their lower fuel cycle costs, investment decisions must be at least partly based on the total costs of electricity generated by the first units of each type.

Herein, these costs are evaluated for the new French 1300 MWe PWR design, based on the four units under construction at the Paluel site, and for a first series of two, four or six French 1500 MWe class FBR units, also supposed all installed at a single site. All costs are indicated in French francs, based on current economic conditions, using a present-worth technique to levelize costs over the entire operating lives of the power plants and associated fuel cycle plants.

It should be noted that any economic comparison of this sort between FBRs and PWRs is inevitably weighted in favour of PWRs, because they are already in series production (about 5 units per year in France) and benefit from a much longer period of industrial experience, all of which reduces costs. This built-in advantage is even greater for fuel-cycle costs than for power plant construction, because French uranium enrichment facilities are sized to supply the equivalent of about eighty 1000 MWe LWR units, and French PWR fuel fabrication and LWR fuel reprocessing plants are sized to serve 35 units of the same size. In other words, PWR fuel cycle costs benefit from considerable economies of scale, not applicable to the first FBRs.

9.2.2 Levelizing and cost calculation methods

Several scenarios can be envisaged for the future of nuclear power in a country like France, ranging from exclusive use of PWRs to construction of FBRs only beyond a certain date. Three such scenarios are examined below,

The case for the liquid metal fast breeder reactor in France

in Section 9.2.2 (a–c). In each case, cost levelizing formulae are given to estimate the present value of future expenditures and then determine the mean levelized cost of each kWh of electricity generated with each type of reactor.

The mean levelized cost of electricity ($\overline{\text{cost}}_{EL}$) is the sum of the mean levelized power plant cost ($\overline{\text{cost}}_{PP}$) and the mean levelized fuel cycle cost ($\overline{\text{cost}}_{FC}$):

$$\overline{\text{cost}}_{EL} = \overline{\text{cost}}_{PP} + \overline{\text{cost}}_{FC} \tag{9.1}$$

In greater detail, the $\overline{\text{cost}}_{EL}$ is a function of capital (first) costs, operating costs and fuel-cycle costs, levelized over the entire operating life of the plant considered. Herein, the base date for levelizing is the start of commercial operation of the plant. As stated, all costs are in French francs, estimated on the basis of current economic conditions. The discount rate employed in determining the present values of future expenditure is 9%.

Expressing this as a formula:

$$\sum_{j=0}^{j=n-o} \frac{\Omega_j \times \overline{\text{cost}}_{EL_j}}{(1+i)^j} = \sum_{j=m-o}^{j=0} \frac{(\text{direct capital costs})_j}{(1+i)^j}$$

$$+ \sum_{j=o}^{j=n-o} \frac{(\text{operating costs})_j}{(1+i)^j}$$

$$+ \sum_{j=m-o}^{j=n-o} \frac{(\text{fuel-cycle costs})_j}{(1+i)^j} \tag{9.2}$$

where: i = discount rate (9%), m = date of first capital outlay or date of first fuel cost, 0 = start of commercial operation, n = economic lifetime of plant or fuel cycle, and Ω_j = electricity produced in period j.

It follows that:

$$\overline{\text{cost}}_{EL} \times \sum_{j=o}^{j=n-o} \frac{\Omega_j}{1.09^j} = \sum_{j=m-o}^{j=o} \frac{CA_j}{1.09^j} + \sum_{j=o}^{j=n-o} \frac{OP_j}{1.09^j}$$

$$+ \sum_{j=m-o}^{j=n-o} \frac{FC_j}{1.09^j} \tag{9.3}$$

where: CA_j = capital costs in period j, OP_j = operating costs in period j, and FC_j = fuel-cycle costs in period j.

Hence, the mean levelized cost of electricity is:

$$\overline{\text{cost}_{EL}} = \frac{\sum_{j=-m-o}^{j=o} \frac{CA_j}{1.09^j} + \sum_{j=o}^{j=n-o} \frac{OP_j}{1.09^j} + \sum_{j=m-o}^{j=n-o} \frac{FC_j}{1.09^j}}{\sum_{j=o}^{j=n-o} \frac{\Omega_j}{1.09^j}} \quad (9.4)$$

The above definition of $\overline{\text{cost}_{EL}}$ is intended essentially for comparison purposes and is far from the total cost of producing and distributing electricity. In particular, it does not include the costs of transmission and distribution infrastructures, nor the utility's tax payments, social insurance contributions, R&D outlays, etc., and is in no way equivalent to an electricity selling price capable of generating profits and sufficient cash-flow for investment.

The assumed discount rate of 9% is quite high. In this respect, it should be pointed out that setting high rates of return on investment is more inhibitive to the development and deployment of FBR plants than PWR plants, because they are more capital-intensive.

(a) *Scenario one – PWRs only*

Producing electricity with a PWR plant involves fuel-cycle costs throughout the life of the plant, related to natural uranium procurement (U), uranium conversion (C), enrichment (E), fuel fabrication (FF), reprocessing and radwaste storage (R), and plutonium storage (PuS). On the other side of the balance sheet, there is a credit for enriched uranium recovered by reprocessing (U').

Levelizing then gives the mean levelized cost for each of these operations, calculated over the lifetime of the corresponding fuel cycle plants, assumed herein to be 15 years. Each such levelized cost is then multiplied by the quantities involved and these products are summed to give the mean levelized fuel-cycle cost per kWh of electricity.

The formula used is detailed in the Appendix. It includes the specific costs of plutonium storage, assuming that there is no plutonium recycle in the PWRs.

(b) *Scenario two – PWRs plus FBRs*

This case differs from the preceding one, in that it is only the plutonium which is not used in the FBRs which is stored. Also, when calculating PWR-related costs, an additional credit must be allotted for the plutonium

The case for the liquid metal fast breeder reactor in France

used in FBRs (Pu′), and this credit increases as the number of FBRs grows. In other words:

PWR $\overline{\text{cost}}_{FC}$ with FBRs = PWR $\overline{\text{cost}}_{FC}$ without FBRs − difference in plutonium storage costs − credit for the plutonium used in the FBRs

The formula used to calculate $\overline{\text{cost}}_{FC}$ for the PWRs is given in the Appendix.

As far as the FBRs are concerned, $\overline{\text{cost}}_{FC}$ is a function of the costs of plutonium (Pu′), depleted uranium (U′$_0$), fabricating the fissile fuel assemblies (F_{F+AB}), which include the axial fertile blanket, and the radial fertile fuel assemblies (F_{RB}), reprocessing the fissile fuel assemblies (R_{F+AB}), and storing the radial fertile fuel assemblies (S_{RB}).

The formula for $\overline{\text{cost}}_{FC}$ for the FBRs is shown in the Appendix: it includes credits for recovered uranium (U″) and plutonium (Pu″).

(c) *Scenario three – mainly FBRs*

In this case, the total plutonium breeding gain is sufficient to satisfy the growth in electricity demand using just FBRs, and the PWRs are progressively phased out. The FBR radial fertile fuel assemblies are reprocessed (in Scenario one they were stored). The $\overline{\text{cost}}_{FC}$ for the FBRs, calculated as in the Appendix, tends to decrease, because the price of plutonium drops due to reduced demand for new plutonium, and because of the increased size of FBR fuel cycle plants and consequent economies of scale.

9.3 TECHNICAL ASPECTS

9.3.1 **French fast breeders**

The fast breeder reactors developed in France are of the pool type: i.e. all primary components (including the core) are housed inside a single, very large, sodium-filled vessel. This primary sodium is circulated through the core, to remove heat from the latter, then this heat is transferred to secondary sodium coolant loops. These loops include steam generators, at which the secondary sodium in turn gives up its heat to produce steam. Finally, the steam drives a turbogenerator to produce electricity.

The reactor core can be of the homogeneous or the heterogeneous type:

1. In a homogenous design, the centre of the core consists exclusively of fissile fuel assemblies, each consisting of a number of fuel pins containing a mixture of oxides of depleted uranium and fissile plutonium (14%) in the centre, and shorter lengths of depleted uranium only at the top and bottom (axial fertile blanket). This central section is surrounded by assemblies containing only depleted uranium (radial fertile blanket),

240   The economics of advanced converters and breeders

2. In a heterogeneous design, the centre of the core is larger, containing a mixture of fertile fuel and fissile fuel assemblies, the latter's mixed oxide fissile section including roughly 20% plutonium.

In all cases, the power density of an FBR core is very high (typically 500 kW per litre), and its volume is therefore about a fifth that of a PWR for a reactor of the same capacity.

The following sections indicate the main characteristics of French FBRs that are relevant to the present analysis.

(a) *The Superphénix (SPX 1) power plant at Creys-Malville*

- Net electrical output = 1200 MWe
- Thermal efficiency = 41%
- Mass of U and Pu in mixed oxide fissile fuel = 34.4 tonnes
  Total Pu mass = 5.9 tonnes
  ^{239}Pu equivalent mass of Pu = 4.7 tonnes
  Ratio of ^{239}Pu equivalent mass to total Pu mass = 0.81
- Fissile fuel assembly in-core residence time = 640 EFPD
- Discharge burnup = 70 000 MWd/T
- Mass of axial fertile material = 21.4 tonnes
- Mass of radial fertile material = 51.5 tonnes
- Proportion of core assemblies replaced at each refuelling shutdown:
  Fissile fuel assemblies = 1/2
  Radial fertile fuel assemblies = 1/4
- Breeding gains:
  Fissile (mixed oxide) fuel = −0.122
  Axial fertile blanket = +0.196
  Radial fertile blanket = +0.167
- Total breeding gain = +0.241

Breeding gain is defined and explained in Section 9.3.3 below.
EFPD: Effective full power days.

(b) *1500 MWe class post-Superphénix (SPX 2) power plants*

- Net electrical output = 1450 MWe
- Thermal efficiency = 41%
- Mass of U and Pu in mixed oxide fissile fuel = 35.2 tonnes
  Total Pu mass = 6.4 tonnes
  ^{239}Pu equivalent mass of fuel = 5.1 tonnes
  Ratio of ^{239}Pu equivalent mass to total Pu mass = 0.81
- Fissile fuel assembly in-core residence time = 960 EFPD*
- Discharge burnup = 120 000 MWd/T*
- Mass of axial fertile material = 18.3 tonnes

The case for the liquid metal fast breeder reactor in France

- Mass of radial fertile material = 62.4 tonnes
- Proportion of core assemblies replaced at each refuelling shutdown:
  Fissile fuel assemblies = 1/3
  Fertile fuel assemblies = 1/6
- Breeding gains:
  Fissile (mixed oxide) fuel = −0.148
  Axial fertile blanket = +0.191
  Radial fertile blanket = +0.203
- Total breeding gain = +0.246

*From 1992.

Note: The fissile fuel assemblies contain the axial fertile fuel (depleted uranium) as well as the fissile mixed oxide fuel (depleted uranium plus plutonium).

9.3.2 French PWRs

The French 1300 MWe class PWR power plants considered herein have the following characteristics:

- Reactor type = 4-loop PWR – modified Westinghouse design
- Net electrical output = 1275 MWe
- Thermal efficiency = 34%
- Mass of enriched uranium in core = 104 tonnes
- Uranium enrichment, average = 3.2%
- Fuel in-core residence time = 900 EFPD
- Discharge burnup, average = 33 000 MWd/T
- Proportion of fuel assemblies replaced at each refuelling shutdown = 1/3

9.3.3 Breeding

In an FBR core, each fission releases about three neutrons. On average, one of these neutrons is captured by a fissile isotope of plutonium, splitting the latter and sustaining the chain reaction. The other two neutrons are captured by the fertile material (depleted uranium) in and around the fissile core, or by structural and shielding elements, etc. Neutrons captured by structural/shielding materials are wasted, and the reactor is therefore designed to minimize such captures and hence maximize the number of neutrons captured by ^{238}U, converting the latter to fissile plutonium. Thus, each time the core's fissile plutonium atom inventory is reduced by one (due to fission), more than one fertile atom of ^{238}U is converted to a fissile plutonium atom. In other words, the reactor *breeds* more plutonium than it consumes.

The newly created fissile atoms consist in fact of several isotopes of plutonium, with differing nuclear properties, including the mean number of neutrons produced per fission. For calculational purposes, the produced mass of each plutonium isotope is therefore multiplied by a suitable weighting coefficient to convert to an equivalent mass of ^{239}Pu. These individual equivalent masses are then summed to give the total produced ^{239}Pu *equivalent mass* of plutonium, which is used to calculate the *total breeding gain (TBG)*, defined by the following formula:

$$TBG = \frac{\text{produced }^{239}\text{Pu equivalent mass} - \text{consumed }^{239}\text{Pu equivalent mass}}{\text{total fissioned mass (Pu plus U)}} \quad (9.5)$$

The denominator can be converted to an energy equivalent form, using the approximate formula (see Section 9.3.4(a)):

Fissioning 1 g of heavy atoms releases about 1 MWd of thermal energy

The numerator represents the ^{239}Pu equivalent mass of 'bred' plutonium. Hence:

$$TBG = \frac{^{239}\text{Pu equivalent mass of Pu bred during time } t}{\text{thermal energy produced during time } t \text{ (in MWd)}} \quad (9.6)$$

The ^{239}Pu equivalent mass of bred plutonium (m) is then:

$$m = TBG \times \frac{P}{\eta} \times EFPD \text{ (g)} \quad (9.7)$$

where: P = electrical rating of power plant in MWe, η = thermal efficiency of power plant, and $EFPD$ = effective full power days of operation. Alternatively:

$$m = TBG \times \frac{P}{\eta} \times 365 \times f \text{ (g/yr)} \quad (9.8)$$

where f is the plant load factor.

The concept of *plutonium mass in the fuel cycle* is then used to relate the characteristics of the particular power plant to those of the fuel cycle, which is necessarily closed for an FBR:

1. Plutonium produced in the reactor is recycled after reprocessing and new fuel fabrication,
2. The only material that exits the cycle (apart from wastes) is the bred excess plutonium, and the only material that enters the cycle is depleted uranium (1 tonne per GW per annum), converted in the reactor to replace the burnt plutonium.*

---

*Depleted uranium is far from rare, and its use in FBRs is its only foreseeable large-scale use. It should be noted that no enriched uranium is employed in FBRs.

The flux of fissile material (mass divided by residence time) is constant for all stages of the fuel cycle. If the cycle is divided into just an 'ex-reactor' and an 'in-reactor' part, then the plutonium in-reactor mass and residence time (M_{IR} and T_{IR}) and ex-reactor mass and residence time (M_{ER} and T_{ER}) are related by the following formula:

$$\frac{M_{IR}}{T_{IR}} = \frac{M_{ER}}{T_{ER}} \qquad (9.9)$$

The total mass of plutonium in the cycle (M_C) is by definition the mass necessary for operation of a reactor, and:

$$M_C = M_{IR} + M_{ER} = M_{IR}\left(1 + \frac{T_{ER}}{T_{IR}}\right) \qquad (9.10)$$

For a reactor, the *linear doubling time* (T_{LD}) in years is the time necessary to obtain an additional mass of plutonium in the fuel cycle sufficient for the operation of another identical reactor. In other words:

$$T_{LD} = \frac{M_C}{m'} \qquad (9.11)$$

where m' is the net mass of excess plutonium bred annually:

$$m' = m - \text{losses in the cycle} = m - \delta m$$

If, for example, $\delta m = 2\%$ (a maximum value, based on French experience), then $m' = 0.98\,m$, and we have:

$$T_{LD} = \frac{M_{IR}\left(1 + \dfrac{T_{ER}}{T_{IR}}\right)}{(TBG \times P/\eta \times 365 \times f) - \delta m} = \frac{M_{IR}\left(1 + \dfrac{T_{ER}}{EPFD/365f}\right)}{TBG \times P/\eta \times 365f \times 0.98}$$

$$= \frac{M_{IR}\left(\dfrac{1}{365f} + \dfrac{T_{ER}}{EPFD}\right)}{TBG \times P/\eta \times 0.98} \qquad (9.12)$$

It should be noted that, whereas TBG is a function only of the particular characteristics of each reactor, T_{LD} also depends on fuel-cycle characteristics (ex-reactor residence time and in-cycle losses).

T_{LD} is the linear doubling time calculated for one reactor. We can also define a *compound doubling time* T_{CD} for a group of FBRs, calculated from the formula:

$$T_{CD} = 0.693\,T_{LD} \qquad (9.13)$$

Finally, the total breeding gain (TBG) for a reactor can be adjusted by varying the characteristics of the radial and axial fertile blankets. For

244   The economics of advanced converters and breeders

example, its maximum value can be adjusted so that, for an increasing number of FBRs, the plutonium stock remains close to zero.

9.3.4 Energy extracted from uranium in PWRs and FBRs*

FBRs extract more energy from uranium than PWRs. This is demonstrated below by calculating the quantities of energy extracted in PWRs (with and without plutonium recycle) and FBRs.

(a) *Energy released by fission*

Fissioning the nucleus of a heavy atom releases about 200 MeV of energy, most of which is the kinetic energy of the fission fragments, the remainder consisting of neutrons, beta radiation and gamma radiation. Roughly 90% of this energy is released in the fuel, and 10% in the moderator, reflector and biological shielding. The end result is that approximately 200 MeV of energy appears as heat.

As $1 \text{ eV} = 1.6 \times 10^{-12}$ ergs $= 1.6 \times 10^{-19}$ joules, we have:

$$200 \text{ MeV} = 3.2 \times 10^{-11} \text{ J} = 3.2 \times 10^{-11} \text{ W/s}$$

$$= \frac{3.2 \times 10^{-14}}{3600} \text{ kW/h} = 0.89 \times 10^{-17} \text{ kWh (thermal)}$$

Also, 235 g of ^{235}U contain 6.023×10^{23} atoms, and the number of fissile nuclei per gram of ^{235}U is thus:

$$\frac{6.023 \times 10^{23}}{235} = 2.56 \times 10^{21}$$

It follows that fissioning 1 g of ^{235}U releases:

$$2.56 \times 10^{21} \times 0.89 \times 10^{-17} = 22\,800 \text{ kWh (thermal)} = 0.95 \text{ MWd}$$
$$= \text{approximately 1 MWd} \qquad (9.14)$$

(b) *Energy extracted from uranium by PWRs without plutonium recycle*

When enriched uranium is consumed in a PWR, some of the non-fissile uranium-238† in the fuel is converted to plutonium, and part of this plutonium is then burned in the reactor. For a PWR of the type considered:

1. Average initial fuel enrichment = 3.2% (i.e. 32 g of ^{235}U per kg of U).
2. Average discharge fuel enrichment = 0.95% (i.e., 9.5 g of ^{235}U per kg of U).

Economie de l'Energie nucléaire by J. Baumier (Instn) Saclay, cours de Genie Atomique.
†In fact, in a PWR, about 9% of all fissions are those of ^{238}U atoms (provoked by fast neutrons). This percentage is virtually constant, hardly varying with burnup.

The case for the liquid metal fast breeder reactor in France

3. Discharge burnup = 33 000 MWd/T.

As a first approximation we have:

$$(32 - 9.5) \text{ g/kg} \times K = \text{MWd/kg}$$

Hence, $K = 1.46$ MWd/g. In other words, *on average*, and with a discharge burnup of 33 000 MWd/T, a total of 1.46 MWd of energy is produced per gram of ^{235}U consumed in the reactor, a value much higher than the 0.95 MWd/g obtained from total fissioning of 1 g of ^{235}U. It follows that a considerable part of the energy produced in a PWR is due to plutonium fission.

PWR fuel is enriched in plants whose feed is natural uranium containing 0.7% of ^{235}U (7 g per kg of U), and which operate with a tails assay of around 0.2% (i.e. the depleted uranium byproduct of enrichment contains about 0.2% of ^{235}U).

From the law of conservation of mass, we can thus write:

$$FN_o = PN_P + WN_W \qquad (9.15)$$
$$F = P + W \qquad (9.16)$$

where: F = mass of natural U input to the enrichment plant, P = mass of enriched U produced, W = mass of depleted U (tails), N_P = degree of enrichment of the enriched U, N_o = degree of enrichment (% ^{235}U) of the natural U feed, and N_W = tails assay.

Eliminating W, we have:

$$\frac{P}{F} = \frac{N_o - N_W}{N_P - N_W} = \frac{0.71 - 0.2}{3.2 - 0.2} = 0.17$$

One kg of natural U thus gives 170 g of 3.2% enriched U at a tails assay of 0.2%. With a discharge burnup of 33 MWd per kg, the total energy produced in a PWR per kg of natural U (i.e. per 0.17 kg of 3.2% enriched U) is thus:

0.17 (kg enriched U per kg natural U)×33 (burnup−MWd per kg enriched U)×0.335 (thermal efficiency of the power plant−MWe per MWth)×1000 (kWe per MWe)×24 (hours per day)

$$= 45\,100 \text{ kWeh per kg of natural U.}^*$$

(c) *Energy extracted from uranium by PWRs with plutonium recycle*

As indicated above, with a discharge burnup of 33 000 MWd/T, a PWR consumes 22.5 g of ^{235}U per kg of U. However, not all these ^{235}U nuclei are

---

*When considering total energy balances (for comparison with other types of reactor, such as those using natural rather than enriched uranium as fuel), this value must be decreased by the amount of electrical energy needed to enrich the uranium, namely about 1600 kWh.

fissioned, some being converted by neutron capture. The number of fissioned nuclei is equal to the total number of nuclei consumed multiplied by the ratio of the fission and absorption cross-sections, which has the average value of 0.85 for ^{235}U. Thus, out of the 22.5 g of ^{235}U consumed, 19.1 g (22.5 × 0.85) are fissioned, producing 19.1 × 0.95 = 18.14 MWd of energy.

As already noted, an almost constant percentage of fissions (and hence energy produced) corresponds to ^{238}U fissioning by fast neutrons. This is 9%, and the corresponding amount of energy at a discharge burnup of 33 000 MWd/T is thus 0.09 × 33 = 3 MWd/kg. The additional 11.9 MWd of energy produced per kg of fuel (33 − 18.14 − 3) is due to fissioning of part of the plutonium produced in the reactor. Even so, not all the produced plutonium is consumed in the reactor, and the remainder can be recovered during reprocessing, then recycled in thermal reactors (e.g. PWRs) along with recovered uranium.

In this case, the total amount of energy that can be produced with PWRs is about 75 000 kWeh per kg of natural U.

(d) *Energy extracted from uranium by FBRs*

Sections 9.3.4(b) and (c) show that PWRs extract only between 0.6 and 0.7% of the energy in natural uranium without plutonium recycle, and about 1% with recycle.

If the plutonium is used instead to fuel FBRs, and making due allowance for losses during reprocessing, fuel fabrication, etc., then as much as 60% of the energy can be extracted from the uranium, namely:

600(g/kg)×0.95(MWd/g)×0.40(thermal efficiency of the FBR plants)× 24(hours per day)

= about 5 400 000 kWeh per kg of natural U.

In other words, using FBRs increases the amount of energy extracted from the natural uranium by a factor of 120 over PWRs without plutonium recycle (5 400 000/45 700), and by a factor of 72 over PWRs with plutonium recycle (5 400 000/75 000).

9.3.5 **Typical doubling times** (see Section 9.3.3)

(a) *Doubling time for a 1500 MWe class FBR with homogeneous core*

$$M_{IR} = 5.12 \text{ tonnes } (^{239}\text{Pu equivalent mass}).$$
$$T_{ER} = T_{CL} + T_{RP} + T_{FF} \qquad (9.17)$$

where: T_{CL} = cooldown time between discharge from the reactor and

The case for the liquid metal fast breeder reactor in France   247

reprocessing (1 year), T_{RP} = reprocessing time (0.35 years), and T_{FF} = new fuel fabrication time (0.35 years).

Thus, $T_{ER} = 1+0.35+0.35 = 1.7$ years.

$EFPD = 960$ for 120 000 MWd/T discharge burnup (about 220 dpa), from 1992.

Note: Target discharge burnup for this class of FBRs is 150 000 MWd/T (about 275 dpa).

Total breeding gain $(TBG) = 0.25$. Hence:

$$\text{Linear doubling time } (T_{LD}) = \frac{5.12 \times 10^6 \times \frac{1}{365 \times 0.7} \times \frac{1.7}{960}}{0.25 \times \frac{1450}{0.41} \times 0.98} = 34 \text{ years} \quad (9.18)$$

Compound doubling time $(T_{CD}) = 0.693 \times T_{LD} = 24$ years   (9.19)

Note: T_{LD} and T_{CD} are calculated above from Equations (9.12) and (9.13) and the reactor characteristics in Section 9.3.1(b)

(b) *Doubling time for a 1500 MWe class FBR with heterogeneous core*

In this case, the total breeding gain can be as high as 0.4, giving a compound doubling time of $24 \times 0.25/0.4 = 15$ years.

9.4 COMPARATIVE COSTS OF FBR AND PWR POWER PLANTS

This comparison is based on:

1. The four French 1300 MWe class PWR units under construction at the Paluel site. Planned dates of start of commercial operation are respectively September 1982, February 1983, March 1984 and June 1985.
2. One, two or three pairs of 1500 MWe class FBR units, all grouped at the same site, and starting commercial operations at 18-month intervals, beginning in June 1989. These plants are presently at the conceptual design stage, and detailed design has not yet commenced: all cost data herein are thus no more than good estimates, based on current forecasts (the current tendency is towards a reduction in FBR costs).

In general, all costs are based on 1979 economic conditions. However, FBR/PWR relative values are also valid for 1980.

9.4.1 Power plant capital costs

Table 9.1 shows the capital cost breakdown on a percentage basis for:

1. The 1200 MWe class Superphénix FBR plant at Creys-Malville (SPX 1).
2. A first series of two or three pairs of 1500 MWe class post-Superphénix FBR units (SPX 2).
3. The four 1300 MWe class PWR units at Paluel.

Table 9.1 Percentage breakdown of PWR and FBR power plant capital costs

	SPX 1 (1200 MWe) %	SPX 2 (1500 MWe) first series of 2 or 3 pairs %	Four Paluel 1300 MWe PWR units %
Direct capital costs – design, equipment site erection, etc.			
NSSS – % of total direct capital costs	68	62	40
Balance of plant – % of total direct capital costs	32	38	60
Total direct capital costs – % of total capital investment	69.5	73	75
Interest payments – % of total capital investment	24	21.5	20
Architect–engineer and pre-operating costs – % of total capital investment	6.5	5.5	5
Total capital investment	100	100	100

The ratio of the average levelized capital cost per KWh for FBRs and PWRs is given in Table 9.4, along with the corresponding ratios for operating and fuel-cycle costs.

9.4.2 Fuel-cycle costs

As explained in Section 9.2.1, the comparison between FBR and PWR fuel cycle costs is quite heavily weighted in favour of PWRs. In addition, we have calculated the average levelized fuel-cycle cost for PWRs as per the Appendix, supposing that the plutonium produced in PWRs is used in FBRs, thus eliminating Pu storage costs from PWR fuel-cycle costs, and also maximizing the Pu credit in the latter.

The first stage in calculation of fuel cycle costs is to define the input cost data for nuclear materials and fuel-cycle operations, for each type of reactor. When those are common to both reactor types (FBR and PWR), then the ratio of the costs is indicated, rather than absolute values.

The case for the liquid metal fast breeder reactor in France

(a) *Input cost data*

1. Price of yellowcake (uranium concentrate): 520 F/kg or 1040 F/kg, the latter figure being chosen to bring out the effects of a doubling in U prices by the 1990s, which is by no means unlikely, the current price only just assuring profitability for the operators of uranium mines and mills.
2. Conversion to UF_6: 23 FF per kg of U.
3. Enrichment: 605 FF per SWU (Separative Work Unit).
4. Plutonium: 60 FF per gram in 1980 – increasing linearly to 100 FF/g by the year 2000.
5. Depleted uranium: 70 FF per kg.
6. Fuel fabrication and reprocessing: costs are indicated as ratios of FBR costs to corresponding PWR costs, based for PWRs on a large number of reactors, and for FBRs on two or six SPX 2 (1500 MWE class) units:
   (i) PWR fuel fabrication: 1.
   (ii) Fabrication of fissile fuel assemblies for FBRs (including axial fertile blankets): 5.4 or 4.4.
   (iii) Fabrication of radial fertile fuel assemblies for FBRs: 1.12.
   (iv) Reprocessing of PWR fuel: 1.
   (v) Reprocessing of FBR fissile fuel assemblies (including the axial fertile blanket): 7.6 or 4.7.
   (vi) Reprocessing of FBR radial fertile fuel assemblies: 1.6.

(b) *Other input data*

The characteristics of the FBR and PWR power plants are as defined in Sections 9.3.1 and 9.3.2.

The cool-down time between discharge from the reactor and reprocessing is assumed to be 12 months for FBR fuel and 36 months for PWR fuel.

Discharge burnups are taken as:

1. FBRs, until 1992: 105 000 MWd/T.
2. FBRs, from 1992: 120 000 MWd/T.
3. PWRs: 33 000 MWd/T.

The number of full power operating hours per reactor per annum is assumed to be:

1. First year of commercial operation: 4400 hours.
2. Second and third years: 5300 hours.
3. Fourth to twenty-first years: 6200 hours.

The assumed reactor operating lifetime is 21 years.

(c) *Results of calculation*

The data above and in Section 9.4.2(a) have been used to calculate PWR fuel cycle costs and FBR fuel cycle costs (two, four or six pairs of FBRs). The results of these calculations are given in Tables 9.2 and 9.3, as a breakdown of total fuel cycle costs on a percent basis.

Table 9.2 Breakdown of PWR fuel cycle costs

	%
Yellowcake	40
U conversion	2
Enrichment*	34
Fuel fabrication†	11
Reprocessing†	22
Plutonium credit	−4
Uranium credit + enrichment credit	−5

*Based on an enrichment plant sized to supply the equivalent of eighty 1000 MWe PWR units.
† Based on fuel fabrication and reprocessing plants sized to serve the equivalent of thirty-five 1000 MWe PWR units.

The breakdowns in Tables 9.2 and 9.3 have been used to calculate the average levelized fuel-cycle costs for FBRs (four or six SPX 2 type units) and PWRs. The corresponding FBR/PWR cost ratios are shown in Table 9.4.

Table 9.3 Breakdown of FBR fuel-cycle costs

		SPX 2		
	SPX 1* %	One pair %	Two pairs %	Three pairs %
Plutonium	30.5	38	46.5	53
Depleted uranium	0.5	0.4	0.5	0.6
Fabrication of fissile fuel assemblies	24.5	21	22	23
Fabrication of radial fertile fuel assemblies	2.5	3.5	4.2	4.7
Reprocessing of fissile fuel assemblies	62.1	61	55.5	51
Storage of radial fertile blanket	1	1	1	1
Plutonium credit	−21	−24.5	−30	−33
Uranium credit	−0.1	−0.2	−0.2	−0.3

* In association with two pairs of type SPX 2 units.

Table 9.4 Ratios of average levelized FBR costs to average levelized PWR costs

	SPX 1	SPX 2 (third pair of the first series)
Total capital costs	2.1	1.30
Power plant operating costs	1.69	1.10
Fuel cycle costs	1.79	0.95
Total electricity generating cost ($\overline{\text{cost}}_{EL}$)	2.02	1.15

9.4.3 Power plant operating costs

Preliminary comparisons based on FBR and PWR units of comparable size (Phénix 250 MWe FBR and Chooz 300 MWe PWR) suggest that their operating costs differ little in real terms. For this reason, we have assumed that the average levelized operating costs of FBRs will progressively decrease, and that the corresponding FBR/PWR cost ratios will progressively approach unity, as shown in Table 9.4.

9.4.4 Total electricity generating costs per kWh

The average levelized cost elements whose breakdowns are indicated in Sections 9.4.1 to 9.4.3 have been summed to give the total average levelized cost per kWh of generated electricity ($\overline{\text{cost}}_{EL}$). Table 9.4 shows the calculated ratios of the average levelized costs for FBRs and PWRs.

Thus, for example:

1. Fuel cycle $\overline{\text{costs}}$ for SPX 1 (1200 MWe Supérphénix unit at Creys-Malville) are 1.79 times those for the 1300 MWe class PWR units at Paluel, but when the second or third pair of 1500 MWe class FBR units (SPX 2) come into service, the FBR $\overline{\text{cost}}_{FC}$ becomes less than the PWR $\overline{\text{cost}}_{FC}$.
2. The total cost of electricity generated with the FBR plants progressively decreases in comparison with the corresponding PWR costs (from 2.02 times PWR $\overline{\text{cost}}_{EL}$ to just 1.15 times).

It may be noted first that increasing the number of FBRs reduces FBR capital costs, and hence the cost of electricity.

Secondly, increasing the number of FBRs also reduces FBR fuel cycle costs for all the FBRs in operation, again decreasing electricity costs, and the per-kWh electricity costs for a given number of FBRs always lie on the corresponding fuel cycle isocost value.

252 The economics of advanced converters and breeders

At 1979 uranium prices, FBRs become almost competitive with PWRs (to within 15%) as soon as the second or third pair of SPX 2 type FBRs come into service, and a doubling of the natural uranium price would assure complete competitivness of the FBRs.

9.5 CONCLUSIONS

This analysis has shown that fast breeder reactor power plants could quite rapidly produce electricity at costs comparable with PWRs. In effect, the difference between FBR and PWR electricity costs is just 10 to 15% at current uranium price levels, as soon as the third pair of 1500 MWe class FBR units comes on stream. This suggests that FBRs should be widely deployed, and that this decision should be taken soon, if only to help limit to reasonable values the rise in uranium prices that is almost certain to occur before the end of the century, and thus also help limit the total costs of generating the nation's electricity. If, however, for lack of a sufficiently long-term and even medium-term view, it is decided not to do this, then there will be no way of mastering uranium prices.

It is in the interest of the community as a whole that decisions be made on the basis of cost-benefit analyses, and it is evident that the relatively small additional costs of producing electricity during the first years of FBR construction and deployment will be more than amply compensated for by the savings made as soon as the first major uranium price increases occur.

Insofar as the economic position of FBRs with respect to PWRs is concerned, we are in many ways in the same situation as that of PWRs with respect to oil before the 1973 Arab–Israeli war.

These conclusions apply to all the world's major industrial nations, not only to France. It is thus essential to convince political leaders and the general public of the fundamental importance of fast breeder reactors for satisfying the ever-growing energy needs of mankind.

APPENDIX

Formula for calculating average levelized fuel-cycle costs for PWRS. Scenario one – PWRs only (no FBRs)

$$\mathrm{PWR}\,\overline{\mathrm{cost}}_{FC} = \sum_{j=m-o}^{j=n-o} \left[\frac{(U \times \mathrm{price}\,U)_j}{1.09^j} + \frac{(C \times \overline{\mathrm{cost}}\,C)_j}{1.09^j} + \frac{(E \times \overline{\mathrm{cost}}\,E)_j}{1.09^j} \right.$$

$$\left. + \frac{(FF \times \overline{\mathrm{cost}}\,FF)_j}{1.09^j} + \frac{(R \times \overline{\mathrm{cost}}\,R)_j}{1.09^j} + \frac{(SPu \times \overline{\mathrm{cost}}\,SPu)_j}{1.09^j} - \frac{(U' \times \mathrm{price}\,U')_j}{1.09^j} \right]$$

$$\times \frac{1}{\displaystyle\sum_{j=0}^{j=21} \frac{\mathrm{Electricity\,production}_j}{1.09^j}}$$

The case for the liquid metal fast breeder reactor in France

Note: No allowance has been made for the costs of americium decontamination before the plutonium is reused. The Am is formed from ^{241}Pu and its quantity increases at the rate of 6000 ppm per annum. In other words, the longer the Pu is stored before use in FBRs, the higher these costs.

Formula for calculating average levelized fuel-cycle costs for PWRs. Scenario two – mix of PWRs and FBRs

PWR $\overline{\text{cost}}_{FC}$ with FBRs = PWR $\overline{\text{cost}}_{FC}$ without FBRs

$$-\frac{1}{\sum_{j=0}^{j=21}\dfrac{\text{Electricity production}_j}{1.09^j}}\left[\sum_{j=o}^{j=n-o}\frac{(\text{SPu}'\times\overline{\text{cost}}\,\text{SPu})_j}{1.09^j}+\frac{(\text{Pu}'\times\overline{\text{cost}}\,\text{Pu})_j}{1.09^j}\right]$$

Formula for calculating average levelized fuel-cycle costs for FBRs. Scenario two – mix of PWRs and FBRs

$$\text{FBR } \overline{\text{cost}}_{FC} \text{ with PWRs} = \left[\sum_{j=m-o}^{j=n-o}\frac{(\text{Pu}'\times\text{price Pu})_j}{1.09^j}+\frac{(U'_o\times\text{price }U'_o)_j}{1.09^j}\right.$$

$$+\frac{(F_{F+AB}\times\overline{\text{cost}}\,F_{F+AB})_j}{1.09^j}+\frac{(F_{RB}\times\overline{\text{cost}}\,F_{RB})_j}{1.09^j}+\frac{(R_{F+AB}\times\overline{\text{cost}}\,R_{F+AB})_j}{1.09^j}$$

$$\left.+\frac{(S_{RB}\times\overline{\text{cost}}\,S_{RB})_j}{1.09^j}-\frac{(\text{Pu}''\times\text{price Pu})_j}{1.09^j}-\frac{(U''_o\times\text{price }U'_o)_j}{1.09^j}\right]$$

$$\times\frac{1}{\sum_{j=0}^{j=21}\dfrac{\text{Electricity production}_j}{1.09^j}}$$

Formula for calculating average levelized fuel-cycle costs for FBRs. Scenario three – mainly FBRs (progressive phasing-out of PWRs)

$$\text{FBR } \overline{\text{cost}}_{FC} = \left[\sum_{j=m-o}^{j=n-o} \frac{(\text{Pu}' \times \text{price Pu})_j}{1.09^j} + \frac{(\text{U}'_o \text{ price U}'_o)_j}{1.09^j} \right.$$

$$+ \frac{(F_{F+AB} + \overline{\text{cost}} \, F_{F+AB})_j}{1.09^j} + \frac{(F_{RB} \times \overline{\text{cost}} \, F_{RB})_j}{1.09^j} + \frac{(R_{F+AB} \times \overline{\text{cost}} \, R_{F+AB})_j}{1.09^j}$$

$$\left. + \frac{(R_{RB} \times \overline{\text{cost}} \, R_{RB})_j}{1.09^j} - \frac{(\text{Pu}'' \times \text{price Pu})_j}{1.09^j} - \frac{(\text{U}''_o \text{ price U}'_o)_j}{1.09^j} \right]$$

$$\times \frac{1}{\sum_{j=0}^{j=21} \dfrac{\text{Electricity production}_j}{1.09^j}}$$

10

L.G. McConnell and L.W. Woodhead
THE ECONOMICS OF THE CANDU REACTOR

10.1 INTRODUCTION

The purpose of this chapter is to:

1. Discuss the cost of producing electricity from CANDU-Pressurized Heavy Water (PHW) nuclear generating units.
2. Present actual cost experience of CANDU units in Ontario Hydro.
3. Compare CANDU cost experience with fossil (coal) experience in Ontario Hydro.
4. Present projected CANDU and fossil cost data in Ontario Hydro.
5. Present cost estimate comparisons of CANDU and Light Water Reactors (LWR) in Ontario Hydro.

10.2 COST CRITERIA

The cost objective of Ontario Hydro is to produce and deliver electricity at the lowest long-term cost to Ontario customers, while satisfying the other rudimentary objectives: worker safety, public safety, environmental protection and reliability. If a comparison is made between two alternative types of generation, the degree to which all of those objectives are satisfied should be considered.

A comprehensive discussion of the CANDU-PHW type nuclear unit, including worker safety experience, public safety experience, environmental protection experience, and reliability experience, has been reported by the authors in a companion paper [1] (not included in this book).

The load of Ontario Hydro (as with most electrical utilities) varies with time. Loads peak in the daytime Monday to Friday when factories are busy and society is active. In Ontario, the loads are higher during the winter when temperatures are low.

The most economical generating system for Ontario Hydro is a mix of hydraulic generation, fossil generation, and nuclear generation. The majority of available economic hydraulic resources in Ontario has been developed. New loads must be met by alternative resources of which nuclear and coal are the primary options for the remainder of this century.

The Ontario Hydro system load factor is typically 68% (the ratio of average annual power to peak annual power). Fossil-fired generation is most economical for peak load requirements because of its lower capital and operating, maintenance and administrative (OM&A) costs. Nuclear generation is most economical for base-load application because its higher capital and OM&A costs are more than offset by the very low fuelling costs. This chapter is limited to a cost discussion for base-load generation in Ontario Hydro.

Cost evaluations for generation commitment decisions of Ontario Hydro are very complex, utilizing present value techniques, uncertainty analyses, load forecasts, reliability assessments, environmental impacts, etc., that are beyond the scope of this paper.

The total unit energy cost (TUEC) method is a simple and accurate indicator of the relative economics for base-load application, and is used in this paper.

10.2.1 Total unit energy cost (TUEC)

The cost of producing electricity from generating stations involves the following cost classifications:

1. The research and development of generation concepts.
2. The cost of building the stations.
3. The cost of operating and maintaining the stations.
4. The cost of fuelling the stations.
5. The cost of in-service modifications.
6. The cost and benefits associated with disposal of the stations at the end of their useful life.
7. Overhead costs to support the above cost classifications.

In addition, the cost of producing electricity must also consider:

1. The method employed for financing and amortizing the investments.
2. The interest rates applicable to the above classifications.
3. The lifetime assumed for the facilities.
4. The reliability of the stations to produce electricity.
5. The policies that are adopted concerning source of supply, taxes, regulations, etc.

The total unit energy cost (TUEC) is defined as the total annual cost of producing electricity ($)* divided by the total annual electricity energy produced (kWh).

$$\text{TUEC} = \frac{\text{Total annual cost}}{\text{Total annual electricity produced}}$$

---

* All costs expressed in this chapter are in Canadian dollars unless otherwise stated.

The economics of the CANDU reactor

In this chapter the research required to develop the generation concepts, and the costs and benefits of station disposal have been excluded. In the opinion of the authors, these exclusions do not have a serious effect on the absolute costs and relative costs of the generation alternatives, in the long-term, for a major programme.

The four cost components for the CANDU-PHW concept are:

1. Annual interest and depreciation of the capital cost.
2. Annual operation, maintenance, and administration cost.
3. Annual fuelling cost.
4. Annual heavy water upkeep cost.

The three cost components for the Light Water Reactor (LWR) concept and coal-fired stations are:

1. Annual interest and depreciation on the capital cost.
2. Annual operation, maintenance, and administration cost.
3. Annual fuelling cost.

The computation of the annual interest and depreciation cost depends upon four factors:

1. The initial capital cost and the capital modifications cost.
2. The interest rate.
3. The lifetime of the station.
4. The method of amortization of the initial capital cost and the capital modifications cost.

The initial capital cost includes:

1. The design and engineering cost.
2. The construction cost.
3. The commissioning cost.
4. The permanent in-reactor fuel charge.
5. The heavy water inventory.
6. Overheads.
7. Accumulated compound interest during construction.

The initial capital cost includes the permanent in-reactor fuel charge (one-half of the initial fuel charge) and the heavy water inventory. The initial 'dry' capital cost is identical to the initial capital cost except that the permanent in-reactor fuel charge, the heavy water inventory and commissioning are excluded.

The annual operation, maintenance, and administration cost includes:

1. Labour.
2. Materials.
3. Purchased services.
4. Interest on operating and maintenance inventories.
5. Overheads (including taxes).

The annual fuelling cost includes:

1. Fuel (quantity and price).
2. Interest on inventory.
3. Transportation.
4. Overheads.

The annual heavy water upkeep cost is comprised of two basic factors:

1. The cost of replacing any heavy water lost during operation.
2. The cost of upgrading any heavy water which becomes downgraded during operation (diluted with ordinary water).

The total unit energy cost (TUEC) is the sum of the unit energy cost (UEC) for each of the cost components. As an example, the fuelling unit energy cost is as follows:

$$\text{Fuelling unit energy cost} = \frac{\text{Fuelling annual cost}}{\text{Total annual electricity energy produced}}$$

The total unit energy cost is very dependent on the capacity factor* achieved.

The total annual electricity energy used to determine TUEC may be either the gross or net electricity produced. Ontario Hydro prefers to use net energy – TUEC (net). However, for some utilities only the gross production is published and TUEC (gross) is determined.

The specific capital cost is the total initial capital cost ($) divided by the net capacity (kW) and is expressed in dollars per kilowatt.

10.3 ONTARIO HYDRO COST COMPARISON – CANDU-PHW VERSUS FOSSIL (COAL)

The cost comparison between CANDU-PHW units and alternative sources of generation will depend upon many factors which are particular to the electrical utility making the comparison.

Nuclear fuel cost tends to be independent of the distance between the uranium source and the generating station because transport cost of nuclear fuel is small. In the case of coal, the transport cost is low if the generating unit is near the coal mine, but can be very high if the coal has to be transported a great distance.

The following data illustrates that the CANDU-PHW is very competitive within Ontario Hydro where economic hydroelectric resources have been almost fully developed and where coal must be transported a minimum of 800 kilometres. There are other locations in Canada in which coal-fired

---

*Capacity factor = $\dfrac{\text{actual energy produced}}{\text{perfect production}}$ for any specified period.

Table 10.1 Pickering/Lambton cost comparison, 1980. Pickering and Lambton net capacity factor: 82.6%*

	UEC m$/kWh (net)	
	Pickering NGS-A	Lambton TGS
Interest and depreciation	6.06	1.94
Operation, maintenance and administration	3.95	1.78
Fuelling	2.33	17.46
Heavy water upkeep	0.43	—
Total unit energy cost (net)	12.77	21.18
Station data	Pickering	Lambton
Capacity (maximum continuous rating) (MWe net)	4×515	4×495
In service	1971–1973	1969–1970
Initial capital cost (M$ Canadian escalated)	746.5	257.0
Specific capital cost ($/kW)	362.4	129.8
Economic lifetime (years)	30	30
Depreciation method	Straight line	Straight line
Interest rate (%)	9.7	9.7

* Assumes Lambton also operated at base-load with net capacity factor of 82.6%. Lambton actual 1980 net capacity factor was 58.0%.

generation is cheaper than CANDU-PHW where the generating unit is near the mine. The high current and projected cost of oil and gas makes their use uneconomical for base-load generation in Ontario Hydro.

More specifically, the following presentation compares the Ontario Hydro Pickering Nuclear Generating Station-A (NGS-A) with the Ontario Hydro Lambton Thermal Generating Station (TGS). The Pickering NGS-A comprises four 515 MW (net) nuclear units of the CANDU-PHW type. The Lambton TGS comprises four 495 MW (net) units which burn coal. Both stations were built at the same time, both are of modern design, and both stations are fully operational with good performance records.

For the year 1980, Pickering NGS-A had a net capacity factor of 82.6%. Table 10.1 illustrates the unit energy costs (UEC) of these two stations, and the following points should be noted:

1. The coal-fired capital cost is much lower than the nuclear capital cost.
2. The coal-fired OM&A cost is lower than the nuclear OM&A cost.
3. The nuclear fuelling cost is very much lower than the coal-fired fuelling cost.
4. The heavy water upkeep cost, which applies only to the nuclear, is only a small percentage (about 4%) of the total unit energy cost.
5. For base-load application, Pickering NGS-A had approximately one-half the total unit energy cost of Lambton in 1980.

During the period up to 1980, in-service capital modifications have been made to both Pickering NGS-A and Lambton TGS. These modifications are amortized on a remaining lifetime basis and are included in the comparison. It is expected that further capital modifications will be required from time to time to replace major components and meet new requirements. For example, the pressure tubes at Pickering NGS-A may have to be replaced and Lambton may have to be retro-fitted with SO_2 scrubbers to meet acid rain requirements.

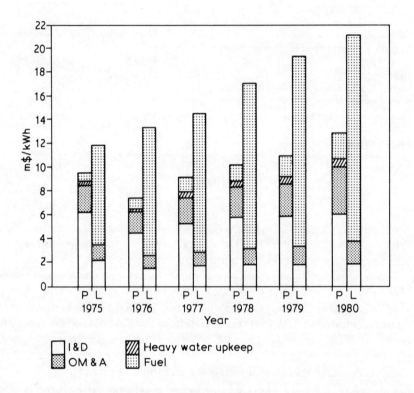

Fig. 10.1 Total unit energy cost components, thermal versus nuclear (1975–1980), where P is Pickering (nuclear) 4×515 MW(e) and L is Lambton (coal) 4×495 MW(e).

Figure 10.1 presents the Pickering versus Lambton unit energy costs (assuming Lambton operated at the same high capacity factors as Pickering) for each year 1975 to 1980 inclusive. This graph clearly shows the steadily increasing cost advantage of the nuclear plant due to the continuing inflation of coal costs. This is an example of the 'inflation-proof' characteristics of the CANDU-PHW. The graph also shows that the cost of heavy water upkeep is only a small component (about 4%) of the total unit energy cost.

10.3.1 Highlight

The base-load cost (TUEC) of the Pickering NGS-A has been consistently well below the cost of the Lambton TGS (coal-fired). This cost advantage is expected to increase as fossil fuels become more expensive.

10.4 CANDU COSTS VERSUS TIME

The actual or estimated TUEC for in-service stations and stations under construction will vary with time due to a variety of reasons including:

1. Escalation of labour and material costs.
2. Changes in interest rates.
3. Escalation of fuel costs.
4. Changes in design and operating requirements.
5. Changes in operating performance.
6. Competence and maturity of workforces (design, manufacturing, construction, operation).

The Pickering NGS-A and the Lambton TGS were built in the late 1960s and placed in service in the early 1970s. During the 1970s, high inflation caused capital, OM&A, and fuelling costs to be driven rapidly upwards. As a result, new coal-fired generating stations such as Nanticoke TGS (8×490 MW net) and new nuclear stations such as Bruce NGS-A (4×740 MW net) have higher capital costs.

In addition, the TUEC of the in-service coal-fired station, Lambton TGS (4×498 MW) and the in-service nuclear station, Pickering NGS-A (4×515 MW), are rising due to inflation in OM&A and fuelling costs.

Table 10.2 Bruce NGS-A 1980 costs. Net capacity factor = 86.7%

	UEC m$/kWh
Interest and depreciation	9.67
Operation, maintenance and administration	3.06
Fuelling	2.67
Heavy water upkeep	0.45
Total unit energy cost (net)	15.85
Station data	
Capacity (maximum continuous rating) (MWe net)	4×740
In service	1977–1979
Original capital cost (M$ Canadian escalated)	1961.1
Specific capital cost ($/kW)	662.5
Economic lifetime (years)	30
Depreciation method	Straight line
Interest rate (%)	9.7

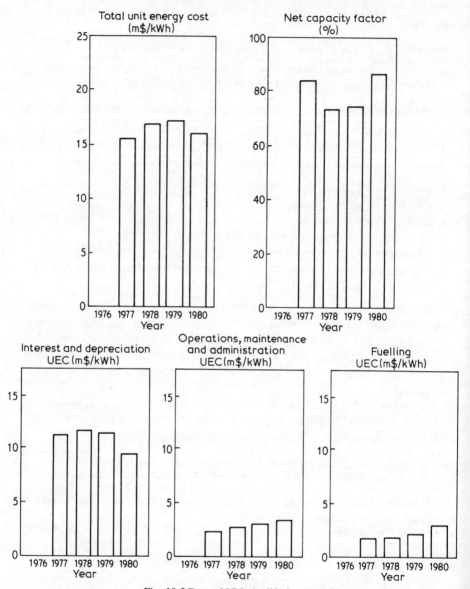

Fig. 10.2 Bruce NGS-A – lifetime trends.

The specific capital cost of Bruce NGS-A compared with that of Pickering NGS-A is affected by three major factors:

1. Bruce NGS-A has lower costs due to larger unit size.
2. Bruce NGS-A has higher costs due to new regulatory requirements.
3. Bruce NGS-A has much higher costs due to inflation of labour and materials.

The economics of the CANDU reactor

Table 10.3 Nuclear capital cost data (net)

Station	Net Capacity (MWe)	Initial capital cost (M$)	Specific cost ($/kW)	Dry* capital cost (M$)	Specific Dry* capital cost ($/kW)	Year in service
Actual						
Pickering NGS-A	2060	746.5	362.4	565.7	274.6	1971–1973
Bruce NGS-A	2960	1961.1	662.5	1498.9	506.4	1977–1979
Estimated						
Pickering NGS-B	2064	3097.3	1500.6	2325.6	1126.7	1983–1984
Bruce NGS-B	3024	4579.0	1514.2	3400.0	1124.3	1983–1987
Darlington NGS	3524	6639.5	1884.1	5490.5	1558.0	1988–1991

*Dry capital costs exclude heavy water, fuel, and commissioning.

The result is that the Pickering NGS-A specific capital cost was 362.4 $/kW (net) and Bruce NGS-A was 662.5 $/kW (net). Pickering NGS-A came into service between 1971 and 1973; Bruce NGS-A between 1977 and 1979.

Table 10.2 shows Bruce NGS-A Unit Energy Costs in 1980, while Fig. 10.2 shows the lifetime trends (1977 to 1980).

Table 10.3 presents the actual initial capital cost of Pickering NGS-A and Bruce NGS-A, together with the estimated costs of three nuclear stations

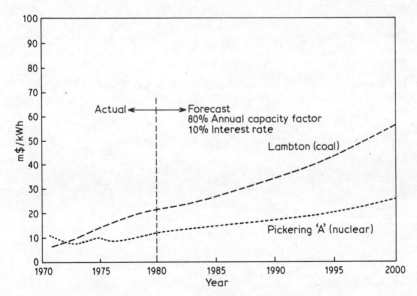

Fig. 10.3 Total unit energy cost for Pickering and Lambton. Basis: straight line depreciation escalation forecast of October 1978.

under construction – Pickering NGS-B, Bruce NGS-B and Darlington NGS-A.

10.5 ONTARIO HYDRO COST PROJECTIONS

The CANDU-PHW at Pickering NGS-A has demonstrated major cost advantages for base-loaded application in Ontario Hydro in the 1970s. The TUEC has been projected for CANDU-PHW and coal-fired stations for the period from 1980 to 2000. These projections exclude the possible retrofit of SO_2 scrubbers in coal-fired stations and exclude possible major retrofits in nuclear stations to meet new requirements.

Figure 10.3 compares actual TUEC for Pickering NGS-A with Lambton TGS (assuming base-load application) for the period up to 1980. It also compares the forecast TUEC for these stations for the period from 1980 to 2000, assuming Ontario Hydro escalation forecasts of labour, materials and fuels.

Figure 10.4 displays forecast TUEC for base-load application of five stations currently in service:

Coal-fired
1. Lambton TGS (4×495 MW)
2. Nanticoke TGS (8×490 MW)

CANDU-PHW
3. Pickering NGS-A (4×515 MW)
4. Bruce NGS-A (4×740 MW)

Oil-fired
5. Lennox TGS (4×495 MW)

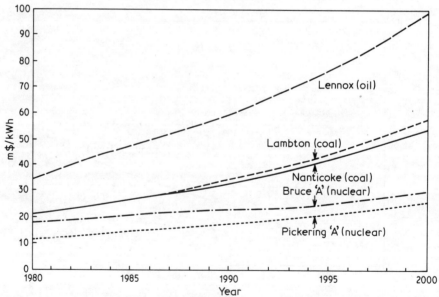

Fig. 10.4 Projected total unit energy cost for major operating thermal stations. Basis: straight line depreciation escalation forecast of October 1978, 80% annual capacity factor and 10% interest rate.

The economics of the CANDU reactor

These projections indicate:

1. That the base-load advantage of CANDU-PHW is expected to continue.
2. That the base-load advantage of CANDU-PHW is expected to increase with time.
3. The 'inflation-proof' characteristic of CANDU-PHW.

10.6 FACTORS TO BE CONSIDERED FOR INTER-UTILITY AND CONCEPT COST COMPARISONS

Relevant and meaningful comparisons of cost experience using data from two or more utilities demand a very objective and rigorous analysis. The following are some of the factors which should be considered:

Unit size

1. Larger units will tend to have lower specific capital costs ($/kW) and lower specific OM&A costs ($/kW).
2. Larger units will tend to have lower capacity factor performance for the same vintage and technology.

Units per station

1. Increased number of identical units will tend to have lower specific capital costs and lower specific OM&A costs.
2. Increased number of identical units will tend to have higher capacity factor performance.

Schedule upsets

Generating stations that suffer schedule delay because of programme changes or because of major problems in executing the project, will tend to have higher specific capital costs due to higher interest during construction and rescheduling costs.

Concept maturity

A well-developed and proven concept will tend to enjoy lower costs and higher performance. Promising new concepts will require tolerance while maturing.

Utility maturity

Independent of concept maturity, each utility adopting the concept must go through a learning process during which costs and performance will suffer.

Designer maturity

Independent of concept maturity and utility maturity, the cost and performance will depend upon the maturity of the designers.

Supply industry maturity

The cost and performance will depend, in part, on the composite capability of the many suppliers of station components.

Supply industry volume

The cost of a station will depend, in part, on the throughput volume of supply industries.

Project management efficiency

Nuclear station costs depend not only on the maturity of the individual project members, but also on the overall efficiency of the project members working as a team. New teams will tend to have higher costs.

Industry management efficiency

The cost of designing and manufacturing components in a given nation will depend, in part, on the general industrial capability and labour stability of that nation.

Labour costs

Nations with lower labour costs will tend to have lower costs assuming similar industrial efficiency.

Operating staff maturity

A rapidly expanding nuclear programme increases costs and reduces performance due to less than optimum operating experience.

Operations management efficiency

Some utilities will enjoy better management systems to reduce problems and respond to problems more effectively.

Research and development efficiency

The ready availability of research and development capability will enhance design and facilitate problem solving during design and operation.

Regulatory efficiency

A regulatory authority with good judgement and ability will enhance the achievement of objectives including schedule and cost. Immature or incompetent regulatory authorities will cause schedule delays with attendant cost penalties.

General society behaviour

Where a concept has been generally endorsed by society, the cost of building and operating a nuclear station will tend to be reduced.

Foreign exchange rates

Comparisons between utilities in different countries require conversion from one currency to another. The relationship between currencies can change dramatically over a period of a few years.

Supply policies

Costs and performance will depend, in part, on supply policies. For example, policy to utilize domestic sources may increase cost.

In view of the above factors, precise conclusions from comparisons between nuclear concepts are difficult, if not impossible, to make. Nevertheless, the authors have attempted to recognize these factors in the following comparison applicable to Ontario. The conclusions could be quite different for alternative assumptions and conditions in another location in Canada or a different country.

10.7 ONTARIO HYDRO COST COMPARISON – CANDU-PHW VERSUS LIGHT WATER REACTORS (LWR)

The Ontario Hydro nuclear programme to date has been limited to experience with CANDU-PHW units. Ontario Hydro has exchanged cost information, and operating performance with other utilities in the USA, Europe and Asia. In particular, this information applied to alternative nuclear types – light water reactors (LWR) and gas-cooled reactors (GCR). Ontario Hydro is continuing to observe the world progress on fast breeder reactors (FBR). At the present time the LWR is the only viable nuclear alternative to CANDU-PHW in Ontario Hydro. The LWR has two basic options – the pressurized water reactor (PWR) and the boiling water reactor (BWR).

Inasmuch as Ontario Hydro has had no design and operating experience with LWR, the cost comparisons between CANDU-PHW and LWR must be based upon the following:

1. Comparison of costs reported by other utilities for LWR with Ontario Hydro costs for CANDU-PHW.
2. Estimates of CANDU-PHW and LWR assumed to be built in Ontario under Canadian licensing requirements.

The judgements expressed below are those of the authors and are based upon the following:

1. The detailed insight Ontario Hydro possesses on CANDU-PHW with regard to cost.
2. The detailed insight Ontario Hydro possesses on CANDU-PHW with regard to performance.
3. Capital cost information on LWR units built in the USA, and extensive discussions with sister utilities in the USA.
4. Detailed performance information on LWR units throughout the world.
5. Interpolative judgement of the authors regarding expected LWR costs and performance of LWR in Ontario.

10.7.1 CANDU-PHW costs – Ontario Hydro

The actual cost data and projected cost data for CANDU-PHW units in Ontario have been presented above.

10.7.2 CANDU-PHW operating performance – Ontario Hydro

The CANDU operating performance has been documented by the authors and published in a companion paper [1]. The eight commercial CANDU units have demonstrated a net capability factor of 77% (from first electricity production) and 79% from the in-service dates. The authors use demonstrated capability for CANDU/LWR comparisons.

10.7.3 Capital cost information – LWR, USA

The actual or estimated initial specific dry capital costs for LWR units of 500 MW and greater built in the USA are shown in Fig. 10.5, based on information provided by a number of USA utilities, in US dollars. The actual or estimated initial specific dry capital costs for CANDU-PHW units in Ontario Hydro are also shown in Fig. 10.5 in Canadian dollars.

The following observations may be made:

1. There is a wide scatter in the specific capital cost data.
2. The CANDU-PHW units have a similar cost and cost trend compared to the average LWR.

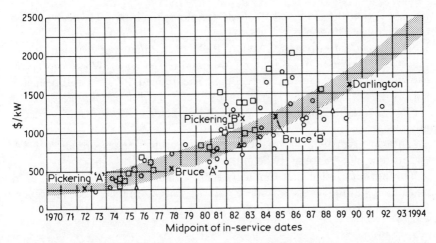

Fig. 10.5 Interutility comparison – nuclear projects, specific dry capital costs. Canadian stations in Canadian dollars, other stations in US dollars. □ is single unit plant, O is two unit plant, △ is multiple unit plant and X is Ontario Hydro plant.

10.7.4 LWR performance – actual

The actual LWR lifetime performance has been reviewed and documented by Ontario Hydro and published in document form [2].

The lifetime average Gross Capacity Factor of PWR and BWR has been 57% and 55% respectively. In the following comparisons, we have used the PWR performance of 57%.

10.7.5 Authors' judgements

Through examination, discussion, and study of available data, the following judgements have been made by the authors:

Performance

There is a wide range of PWR performance (capacity factor) with a world average of 57%. The PWR capacity factor performance is expected to improve further. The available data is based upon stations which usually have only one or two units per station. Four-unit stations (Ontario Hydro practice) are expected to have better performance due to improved diversity (spare parts, technical support, etc.). The authors also feel that Ontario Hydro enjoys better than average project management and operator training.

The authors judge that if Ontario Hydro had an extensive PWR programme (same number of units in service, same number of units per station and in-service dates similar to the actual CANDU-PHW programme) that the expected PWR performance in Ontario Hydro would be 67%. This is to be compared with a typical coal-fired capability of 74% in the USA and a demonstrated CANDU-PHW performance of 77% (since first electricity production). Both coal-fired and CANDU-PHW stations enjoy the advantages of on-power fuelling. The judged 10% capability factor superiority of CANDU-PHW assumes a judged 6% capability factor credit for on-power fuelling and a 4% capability factor credit for other concept advantages.

Ontario Hydro's performance for coal-fired stations is similar to USA experience at 74% for 500 MW units.

The authors judge that Ontario Hydro could achieve 77% in coal-fired stations if staffing levels and spare part diversity were increased. However, this is not economically justified for peak load application.

Capital cost

The above comparison of Ontario Hydro and US utility data (Fig. 10.5) indicates no major specific dry capital cost difference between CANDU-PHW units built in Ontario and LWR units built in the USA.

Examination of the design requirements of the two concepts suggests there should be no major dry capital cost differences for most facilities such as site, turbine-generator, cooling systems, instrumentation and controls, buildings and containment.

In the opinion of the authors, assuming identical supply capability and manufacturing volume, the CANDU reactor with on-power fuelling should be less expensive due to the absence of enriched fuel, very demanding pressure vessel specifications as compared with pressure tubes, the need for in-core high pressure regulating and shut-down devices and the like. However, in this chapter the cost comparisons which follow assume the Dry Capital Cost for CANDU-PHW and LWR in Ontario Hydro to be identical.

OM&A costs

The capacity factor achieved by Ontario Hydro has depended, in part, on maintaining round-the-clock maintenance staff at four-unit stations. This high staff level is economically warranted because of the high cost of burning coal whenever a CANDU unit is shut down. For example, in the Bruce NGS-A, 1% capacity factor is equivalent to the wages of about 100 people.

Since PWR units also have a low fuelling cost compared with coal, round-the-clock maintenance would also be justified. In the opinion of the authors, there is no significant difference in opimum staff levels for four-unit CANDU-PHW and four-unit PWR in Ontario Hydro. Similarly, the total OM&A cost is expected to be the same.

Fuel

Examination of available USA data suggests that, for a large PWR programme in Ontario, the Fuelling Unit Energy Cost for PWR would be 5.46 m$/kWh or higher in 1980. This evaluation assumes the mining and refining costs of natural uranium are identical for CANDU and PWR. Enrichment cost is peculiar to the PWR units. Fabrication costs are particular to each design.

10.7.6 Comparison of CANDU-PHW versus PWR – Ontario Hydro

Table 10.4 is a cost comparison of CANDU-PHW and PWR in Ontario Hydro. It is based on actual cost and performance experience of CANDU units in Ontario and the estimated performance and cost of PWR as outlined above.

Table 10.4 Ontario Hydro 1980. CANDU-PHW versus PWR costs (all UEC data in m$/kWh (1980$))

	CANDU-PHW	PWR	
	PNGS-A	High NCF	Average NCF
Station size (MWe net)	2060	2060	2060
Net capacity factor (NCF %)	77	67	57
Capital UEC			
dry capital	4.93	5.67	6.66
commissioning	0.17	0.20	0.23
fuel	0.07	0.35	0.41
heavy water	1.33	—	—
Capital UEC	6.50	6.22	7.30
OM&A UEC	4.24	4.87	5.73
Fuelling UEC	2.33	5.46	5.46
Heavy water upkeep UEC	0.46	—	—
Total UEC	13.53	16.55	18.49

Two sets of estimates are given for the PWR, corresponding to the world average capacity factor of 57% and the authors' judgement of 67% for a major PWR programme in Ontario Hydro. This comparison indicates that:

1. The TUEC for PWR operating in Ontario at the world average capacity factor (57%) would be approximately 37% higher than the TUEC of CANDU-PHW actually experienced.
2. The TUEC for PWR operating in Ontario at the authors' judgement of 67% capacity factor would be approximately 26% higher than the TUEC of CANDU-PHW actually experienced.

3. The costs of capital heavy water plus heavy water upkeep for CANDU are more than offset by the higher fuelling UEC of the enriched PWR fuel.

10.8 SUMMARY

1. Nuclear-electric and coal-electric generating stations are the primary options in Ontario for new generating requirements for the period from 1980 to 2000.
2. Coal-electric generating stations are the best choice to meet peaking requirements.
3. Nuclear-electric generating stations are the best choice to meet base-load requirements.
4. For base-load applications, the CANDU-PHW has a proven lower cost than coal-electric. In 1980, the total unit energy cost for Pickering NGS-A was 12.77 m$/kWh, compared with the same size, same vintage Lambton coal-fired station with a corresponding cost of 21.18 m$/kWh.
5. CANDU-PHW has inflation-proof characteristics due to its very low fuelling costs.
6. CANDU-PHW with on-power fuelling is expected to continue to enjoy high capacity factor performance.
7. The actual performance of CANDU-PHW units in Ontario is 77% capacity factor since first electricity production and 79% since the in-service dates. The actual average lifetime performance of PWR in the world is 57% since first electricity production. The authors judge that 67% PWR capacity factor would be achievable in Ontario for a mature PWR programme.
8. For Ontario conditions and requirements, the estimated total unit energy cost of PWR units as compared to CANDU-PHW experienced costs indicates the PWR to be 22% higher, assuming a 67% PWR performance and a 77% CANDU performance.

REFERENCES

[1] *CANDU Operating Experience* (NGD-9-1980).
[2] *Comparison of Ontario Hydro with World Power Reactors* (1979) Ontario Hydro, 700 University Avenue, Toronto, Ontario.

NUCLEAR POWER IN OPERATION

11

G.R. Corey
THE COMPARATIVE COSTS OF NUCLEAR AND FOSSIL FUELLED POWER PLANTS IN AN AMERICAN ELECTRICITY UTILITY

11.1 INTRODUCTION

The relative operating performance of different systems of electricity generation is customarily measured in terms of the average availability rates and capacity factors at which each system typically functions, recent trends in such availabilities and capacity factors and the consequent outlook for future changes therein. The relative economics of the various systems are measured by comparing the average cost of electricity delivered to the station bus (the 'bus-bar costs') prior to transformation, transmission and distribution to ultimate consumers. In view of the changing social and regulatory constraints, an attempt should be made to evaluate the likely effect of such constraints upon future availability rates, capacity factors and bus-bar costs.

This chapter compares the current and historic operating performances of twelve large nuclear and coal-fired units now operated by Commonwealth Edison Company (CECo.), and provides specific comparison of bus-bar costs of electricity generated by those units in recent years. It also provides cost comparisons for future nuclear and coal-fired units and attempts to deal realistically with the effect of future inflation upon these comparisons.

Finally, the chapter attempts to deal responsibly with the problem of uncertainty – how present-day comparisons may be affected by future developments and how my own published comparisons have varied over the past four or five years.

The chapter closes with the conclusion that, given the uncertain world in which we live, no electric power supplier can afford to put all its eggs in one basket. Utility managers have a strong incentive to diversify their sources of power generation, and society as a whole would do well to encourage such diversification.

11.2 RELATIVE PERFORMANCE OF EXISTING GENERATING UNITS

All of the following comparisons are for base-load steam electricity generating systems. There are other types of electricity generation – coal and oil-fired cycling and peaking capacity, hydro and pumped-hydro facilities, and the renewable energy sources (solar power, wind power, ocean-thermal and bio-mass). Large coal-fired and nuclear base-load steam electric units are not well adapted for cycling or peaking usage, because their higher investment cost results in a significant carrying charge penalty, when operating at the low capacity factors typical of cycling and peaking operation. Also, today's nuclear reactors are not capable of meeting widely-fluctuating loads with as much flexibility as units which are specifically designed for cycling or peaking service.

Because of the complexities involved in making economic comparisons between base-load and cycling or peaking service, the latter forms of generation are not considered in this chapter.

Table 11.1 shows recent performance data for all 12 of the large base-load nuclear and coal-fired steam electricity generating units which CECo. has placed in service since 1964. During the last three years, our six big nuclear units have been available for service 79.5% of the time, compared with 67.6% for the coal-fired units, and the 'nukes' have operated at an average capacity factor of 63.5% compared with 45.0% for the big coal units. This nuclear capacity factor is probably higher than can be expected year-in and year-out, since CECo.'s overall system capacity factor is only about 45%. Thus, when the nuclear machines are operating at 65% level, the other

Table 11.1 Commonwealth Edison Company performance of all twelve large base-load generating units completed since 1964*

	Availability rates		Capacity factors	
	6 Nuclear	6 Coal	6 Nuclear	6 Coal
1974	59.5%	72.6%	51.1%	58.0%
1975	64.4	67.3	50.7	53.5
1976	71.4	61.8	57.3	44.7
1977	79.9	66.7	60.7	46.2
1978	83.6	72.0	70.0	45.6
1979	74.9	64.0	59.8	43.2
Three-year average (1977–1979)	79.5%	67.6%	63.5%	45.0%

*The nuclear units are Dresden 2 and 3, Quad Cities 1 and 2, and Zion 1 and 2. The coal-fired units are Joliet 7 and 8, Kincaid 1 and 2, and Powerton 5 and 6. Data shown above represent the latest computations and have, in a few cases, been changed from previously published data.

two-thirds of the system generators cannot run at more than 35% to 40%, and the large base-load coal-fired units may as a consequence have their own operating rates unduly depressed.

11.3 RELATIVE ECONOMICS OF EXISTING UNITS

Table 11.2 compares Commonwealth Edison's nuclear and coal-fired bus-bar costs for 1979. The 'system average' data are for all nuclear and coal-fired steam electric units on the system. The 'six big units', both nuclear and coal, represent all of the large base-load nuclear and coal-fired units which have been placed in service since 1964, as listed in Table 11.1. Costs are shown separately for Powerton units 5 and 6 because these are our two newest and largest coal fired-units, both of the 800-MWe-class. They were placed in service in 1972 and 1975 respectively.

Table 11.2 Commonwealth Edison Company 1979 bus-bar costs per books* (mills per kWh)

	Fuel	Other O&M expense	Carrying charges	Total	Adjusted to 60% capacity factor
Nuclear					
System average	5.2	3.8	8.3	17.3	16.8
Six big units	5.2	3.7	8.0	16.9	16.8
Coal					
System average	19.3	3.7	9.6	32.6	29.6
Six big units	18.0	3.2	9.0	30.2	27.3
Powerton 5 and 6	17.5	3.1	13.3	33.9	28.5

*The six nuclear and six coal-fired units are designated in Table 11.1. Fuel costs per kWh, as recorded on the books, have been increased by one mill to reflect an allowance of two mills for spent fuel disposal. Carrying charges have been computed by applying an annual rate of 20% to gross plant investment costs for the units involved.

Fuel costs shown in Table 11.2 represent actual 1979 fuel expenses as recorded on the books, plus an allowance for carrying charges on the investment in both nuclear and coal fuel inventories and nuclear fuel in the reactors. Nuclear fuel costs also reflect an allowance for spent-fuel disposal of two mills per kWh (one-mill higher than recorded on the books in 1979). The two-mill figure represents the maximum likely cost of spent fuel disposition, expressed in 1980 dollars. For further discussion of this matter, see Section 11.5.

Carrying charges represent insurance, property taxes, depreciation, return on investment (both interest on borrowed money and return on equity) and income taxes. These were computed at an annual rate of 20%,

Comparative costs in an American utility 277

which is roughly consistent with the factors used in making investment and replacement decisions in today's money market. This comports closely with the detailed capital cost and tax assumptions used for evaluating the economics of future units, as set forth in Table 11.5.

In 1979, Commonwealth Edison's nuclear generating costs were substantially below any of the coal-fired combinations shown in Table 11.2. This was due in part to the unusually low capacity factors for several of our large coal-fired units. However, even after adjustment to reflect a 60% capacity factor, the nuclear costs were still substantially lower than coal.

11.3.1 Recent cost trends

Commonwealth Edison have been operating nuclear generating facilities for over twenty years. All of our large nuclear units have been in service for six full calendar years (1974 to 1979), and we have made economic comparisons of the kind shown in Table 11.2 every year since 1974. During that period, the bus-bar costs of coal have increased substantially faster than nuclear, due primarily to escalating fuel costs, as shown in Table 11.3. As a result, in 1979, the total bus-bar cost of all coal-fired generation was nearly double that of all nuclear – 32.6 versus 17.3 mills per kWh.

11.4 COMPARISONS OF PERFORMANCE AND ECONOMICS OF FUTURE INSTALLATIONS

The cost trends in Table 11.3 show how resistant nuclear power costs are to inflation, once the facilities have been placed in service. However, regulatory changes and uncertainties are continually increasing the construction cost, delaying the service dates and jeopardizing the availability of new facilities, both nuclear and coal. This requires continual evaluation of both performance and economics. The results of such recent evaluation of nuclear and coal-fired base-load units which might be ordered today for service in 1991 and 1992 are discussed below.

11.4.1 Performance comparisons

In the light of additional regulatory constraints resulting from the March 1979 accident at Three Mile Island (TMI), who can say what future nuclear plant availability may be. The matter seems a subject more appropriate for metaphysics than scientific appraisal. Though none were injured at TMI, we are today importing between 50 and 100 million barrels of oil a year (at an annual cost of between $1.5 and $3 billion) as a result of TMI-induced restrictions imposed upon nuclear units already completed and ready to operate. Nevertheless, we have assumed herein (for the future) that availability rates and capacity factors for new base-load nuclear and coal-

Table 11.3 Commonwealth Edison Company; historic comparisons of bus-bar costs for twelve big units (1974 to 1979)*
(mills per kWh)

	Fuel	Total	Adjusted to 60% capacity factor
1974			
Nuclear	3.9	15.3	
Coal	5.1	12.3	
Nuclear advantage	1.2	−3.0	not applicable
1975			
Nuclear	4.0	17.2	
Coal	6.8	15.6	
Nuclear advantage	2.8	−1.6	not applicable
1976			
Nuclear	5.0	15.7	
Coal	8.0	18.0	
Nuclear advantage	3.0	2.3	not applicable
1977			
Nuclear	4.5	14.1	14.2
Coal	10.1	20.9	19.0
Nuclear advantage	5.6	6.8	4.8
1978			
Nuclear	4.7	13.6	15.1
Coal	14.0	25.3	22.7
Nuclear advantage	9.3	11.7	7.6
1979			
Nuclear	5.2	16.9	16.8
Coal	18.0	30.2	27.3
Nuclear advantage	12.8	13.3	10.5

*Comparisons are for the twelve nuclear and coal-fired units listed in Table 11.1, except for 1974 and 1975, which exclude coal-fired Powerton Unit 5 which did not come in service until December 1975. Prior years' data have been revised from earlier published data to reflect fully the 2 mill allowance for spent fuel disposal.

fired generation facilities will be in roughly the same range. Specifically, we have assumed a 60% capacity factor rate for both nuclear and coal, although recent trends shown in Table 11.1 clearly indicate such an assumption may be on the high side for coal.

11.4.2 Economic comparisons

The data shown in Tables 11.4 to 11.8 compare the estimated costs of building and operating a new nuclear station consisting of two 1120 MW

Comparative costs in an American utility

electric (MWe) pressurized water reactors (PWRs), with a new coal-fired station consisting of four 557 MWe generating units. In each case, it is assumed that approximately 1100 MWe are scheduled for service in time to provide firm capacity for the summer of 1991 with an additional 1100 MWe scheduled for the summer of 1992. The cost estimates include the costs of meeting all current licensing, environmental and safety requirements, including flue-gas desulphurization equipment for coal-fired units and all nuclear back-fitting requirements growing out of Three Mile Island. Data for the coal-fired alternative are based upon low-sulphur coal, because this provides the lowest bus-bar cost. As mentioned above, both alternatives assume a 60% capacity factor although this seems less certain for coal than nuclear.

The effect of inflation

In order to make valid economic comparisons for generating units scheduled for installation a decade or more hence, whose service lives may extend to 2020 or 2030, one must either express all costs in constant dollars and use carrying charge rates which exclude the inflation component of money costs (perhaps as low as 5% a year, which seems unrealistic in the real world) or else one must project the estimated future course of inflation, expressing all costs in current dollars and making sure that the options compared have comparable service dates and lives.

We have chosen the latter course, because it is the procedure generally used in making business decisions. Since we live in a society of changing dollar values, managers had better deal in those values if they are to select the best options for their shareholders.

Accordingly, all economic comparisons shown in Tables 11.4 to 11.8 are expressed in current dollars. An array of assumptions are made as to future inflation. If future inflation rates are higher than those shown, the likely effect upon the economic comparisons can be derived from the data provided herein. As will become apparent later on, the economic advantage of nuclear plant increases significantly if inflation assumptions are increased.

Construction costs. Over the past decade and a half, nuclear plant construction costs have increased from the $150 to $160 per kW range for the four Dresden and Quad Cities units, which CECo. initially contracted for in 1965 and placed in service in the early 1970s, to estimates of well over $1000 per kW for the Byron and Braidwood units, which will go in service in the mid-1980s. At the same time, the costs of coal-fired plants have increased from under $120 per kW for Joliet Units 7 and 8, which were placed in service in 1965 and 1966, to $800 or $900 per kW for coal-fired units ordered today.

These upward trends, approximating 15% per year in each case, have resulted from general construction-cost increases, coupled with increasingly stringent licensing, environmental and safety requirements. These are nationwide phenomena affecting both nuclear and coal-fired options.

Similar upward trends in construction-costs can be expected to continue. However, we have assumed, with some optimism, that the rate of inflation may slacken somewhat in the future. In Table 11.4, we have provided an array of construction-cost estimates based upon future escalation rates of 6%, 7.5% and 10%. Surprisingly, if future inflation proceeds at a higher rate than assumed, installed costs will rise faster for coal than nuclear because construction expenditures for a coal-fired station are made later than for a nuclear station with the same scheduled service date. However, this may be offset to a degree by the likelihood that higher costs of construction funds (and, hence, higher accumulations of capitalized cost of construction, which is generally referred to as 'allowance for funds used during construction' or AFUDC) would be expected under more inflationary conditions.

Carrying charges. Carrying charges on new plant investment are estimated to be significantly higher for nuclear than coal because of the higher construction-costs shown in Table 11.4. Applying a total annual carrying charge rate of approximately 20%, including property taxes and insurance, and assuming a 60% capacity factor (equivalent to 5256 hours use per year) for both nuclear and coal – a procedure roughly comparable to that

Table 11.4 Estimated construction costs of future units*

	Installed cost/kW ($)
6% annual escalation	
Nuclear	1816
Coal	1458
Nuclear higher	358
7.5% annual escalation	
Nuclear	2035
Coal	1695
Nuclear higher	340
10% annual escalation	
Nuclear	2458
Coal	2172
Nuclear higher	286

*The above estimates are expressed in current dollars for units scheduled for service in 1991 and 1992. They include all overheads. Land and transmission terminal costs are excluded since they tend to be the same for coal and nuclear. For further assumptions, see the text.

Comparative costs in an American utility

used in developing the 1979 carrying charges shown in Table 11.2 – result in the array of carrying charges shown in Table 11.5. The carrying charges for nuclear are 18 or 19 mills per kWh higher than for coal, both for the first ten years of service and for the full service life.

Table 11.5 Estimated carrying charges for future units*

	Mills per kWh	
	First 10 years	Full service life
6% annual escalation		
Nuclear	78	68
Coal	59	49
Nuclear higher	19	19
7.5% annual escalation		
Nuclear	87	76
Coal	68	57
Nuclear higher	19	19
10% annual escalation		
Nuclear	105	91
Coal	87	73
Nuclear higher	18	18

*Based upon Table 11.4 construction costs which assume 1991 and 1992 service dates. Carrying charge factors include (1) annual money cost estimates of 10% for debt, 10.5% for preferred stock and 18% for common equity; (2) a composite corporate income tax rate of 49.456%; (3) an Illinois invested capital tax rate of 0.8%, and (4) a resulting present-value discount rate of 10.8066% (assuming a capital structure of approximately 50% debt, 15% preferred and 35% common equity).

Fuel costs. Fuel cost estimates are based upon the current market assumptions listed in Table 11.6. Interestingly, coal costs have increased six or seven times in the past decade and a half while nuclear fuel costs have only trebled, despite a five-fold increase in yellowcake prices and a four-fold increase in charges for enrichment services. This is because nuclear fuel burnup, that is the amount of energy produced by 1 kg of uranium, has more than doubled, going from 12 000 to 15 000 MW days per tonne (MWd/T) fifteen years ago to 36 000 MWd/T for nuclear fuel ordered today.

Table 11.7 shows a substantial fuel-cost advantage for nuclear, varying all the way from 32 mills per kWh for the first ten years with a mere 6% annual rate of inflation to 138 mills per kWh for the full service life with a 10%

Table 11.6 Fuel cost assumptions (1980 dollars)

Nuclear fuel
- $40 per pound for yellowcake (U_3O_8).*
- $2.20 per pound of uranium for conversion service.†
- $99 per separative work unit (SWU) for enrichment.‡
- Fabrication costs for future time periods are based upon current Westinghouse proposals which are generally less than $150 per kg of uranium, in 1980 dollars.
- Burnup – 36 000 megawatt days per metric ton (MWd/T) of uranium.†
- Spent fuel disposal – roughly equivalent to one mill per kWh (based upon DOE/ET-0055 estimates). (Possible increases in these costs are discussed in Section 11.5.)

Coal
- $1.35 per mill Btu for high-sulphur Illinois coal, 10 000 Btus per pound.*
- $1.50 per million Btu for low-sulphur Powder River Basin coal, 8200 Btus per pound.*

Inventories
- Fuel-cost estimates include carrying charges on fuel inventories as well as on nuclear fuel in the reactor.

* Based on current market.
† Based on current contracts.
‡ Current DOE price.

annual rate of inflation. In either case, the nuclear carrying charge disadvantage is far outweighed by the projected fuel savings. This is because nuclear fuel starts from a much lower base cost, which can be illustrated by assuming that each of the current fuel-cost figures shown in Table 11.2

Table 11.7 Comparative fuel costs – future units*

	First 10 years	Full service life
6% annual escalation		
Nuclear	22	30
Coal	54	80
Nuclear lower	32	50
7.5% annual escalation		
Nuclear	28	42
Coal	68	114
Nuclear lower	40	72
10% annual escalation		
Nuclear	39	75
Coal	99	213
Nuclear lower	60	138

* These are level premium averages (expressed in current dollars) for the first ten years of service as well as for the anticipated full service lives of the facilities in question. Money cost and present value discount assumptions are set forth in Table 11.5.

Comparative costs in an American utility

(roughly five mills for nuclear and 18 mills for coal) were to double. In that case, the fuel-cost differential in favour of nuclear would be 26 mills per kWh, not 13 mills. If future fuel prices escalate at only 6% a year, they will double in just twelve years, roughly the service date for new nuclear capacity ordered today. (It should be noted that the cost comparisons shown in Table 11.7 include miscellaneous operating costs – often referred to as 'O and M' – which generally fall in the two to four mill per kWh range, expressed in 1980 dollars and excluding flue-gas desulphurization equipment.)

Backfitting costs. Recent years have seen a proliferation of governmental regulations requiring repeated back-fitting of all types of energy supply facilities. Realistic appraisals of future costs must necessarily include provisions for additional requirements of this nature. Consequently, a separate allowance for backfitting equivalent 1% of the initial investment during

Table 11.8 Estimated total bus-bar costs – future units* (mills per kWh)

	First 10 years		Full service life	
	Nuclear	Coal	Nuclear	Coal
6% annual escalation				
Carrying charges	78	59	68	49
Fuel and miscellaneous	22	54	30	80
Cleaning and decommissioning	2	—	2	—
Backfitting	3	3	9	7
Total	105	116	109	136
Nuclear advantage	11 (9%)		27 (20%)	
7.5% annual escalation				
Carrying charges	87	68	76	57
Fuel and miscellaneous	28	68	42	114
Cleaning and decommissioning	2	—	2	—
Backfitting	4	3	13	11
Total	121	139	133	182
Nuclear advantage	18 (13%)		49 (27%)	
10% annual escalation				
Carrying charges	105	87	91	73
Fuel and miscellaneous	39	99	75	213
Cleaning and decommissioning	5	—	5	—
Backfitting	5	5	23	21
Total	154	191	195	307
Nuclear advantage	37 (19%)		112 (36%)	

*These are level-premium averages expressed in current dollars for the periods indicated. Money cost and present-value discount rate assumptions are set forth in Table 11.5.

every year of the service life of the facility in question has been included in the total bus-bar cost estimates for both nuclear and coal-fired plants. This allowance is not included in the plant costs or carrying charges shown in Tables 11.4 and 11.5 instead, it is listed as a separate cost factor in Table 11.3.

Interim cleaning and decommissioning. Chemical cleaning of the nuclear units is assumed to be required after 15 years of operation and again after 25 years. The 1978 dollar costs have been estimated by Commonwealth Research Corporation at $12 million for investment and $10 million for expense outlays for the first cleaning and $7 million additional expense for the second cleaning.

Decommissioning costs at end of life are estimated at $75 million per unit in 1980 dollars – double the estimate derived from the latest government publication [1].

Total bus-bar costs. Total estimated bus-bar costs in mills per kWh, as shown in Table 11.8, indicate a nuclear advantage varying from 9% to 36% and increasing as the annual rate of future inflation increases.

The authors are inclined to put the greatest weight upon the first ten year analysis shown in Table 11.8 and to assume a relatively high rate of future inflation. This implies that the most likely cost comparisons shown in Table 11.8 are in the 15 to 20% range, roughly the same as I have reported in earlier papers (see Table 11.9).

Table 11.9 Historic bus-bar cost comparison*

	Nuclear advantage
Three years ago at MIT (31 October 1977)	15%
Ill. C. C. testimony (14 April 1978)	17
California testimony (18 September 1978)	14
Two years ago at MIT (19 October 1978)	16
Today – assuming 10% annual inflation	19†

* All comparisons except the first one assume low-Btu low-sulphur coal. All earlier comparisons (October 1978 and prior) were expressed in constant rather than current dollars because the persistence of inflation was not then so apparent as it is today.
† If an extra four mill per kWh allowance were made for disposal of nuclear spent fuel, as described in Section 11.6.3 below, the nuclear advantage would be on the order of 17%.

11.6 SENSITIVITY TO VARYING ASSUMPTIONS

Our studies have included a wide variety of assumptions. The effects of varying some of these assumptions are reported below.

11.6.1 Type of plant

The foregoing data compare a two-unit PWR nuclear station with a four-unit coal-fired station designed to burn low-sulphur, low-Btu coal from the Powder River Basin and equipped with flue-gas desulphurization equipment. We estimate that the use of smaller PWR nuclear units or boiling water reactors (BWRs) would result in somewhat higher costs, but we have not made specific cost estimates for these alternatives.

We have estimated the costs of coal-fired units designed to burn high-sulphur, high-Btu Illinois coal and find that per kilowatt construction costs of such units would run about 10% more and that bus-bar costs would be about 8% higher than the costs shown for the coal-fired alternative.

We have not estimated the costs of coal-fired units designed to burn low-sulphur, high-Btu coal because such fuel has limited availability and most of it is of a more expensive metallurgical grade. Nor have we estimated the cost of oil-fired steam electric units because these do not appear to be a viable option for the future.

We have estimated the costs of installing four coal-fired units at two station sites instead of one because of the possibility that a four-unit installation might, in some instances, be prohibited by environmental regulations. Under this assumption construction costs would generally be expected to run 6 or 7% higher than for a single four-unit station and total bus-bar costs would accordingly be 2 or 3% more.

11.6.2 Escalation rates

There are four categories of bus-bar costs shown in Table 11.8: (1) carrying charges (including property taxes and insurance), (2) fuel and miscellaneous operating expenses, (3) backfitting costs, and (4) the costs of periodic nuclear plant cleaning and end of life decommissioning. Cost components within each category were assumed to respond to a common escalation rate. In turn, it was assumed that this rate might be different from rates for other categories. For example, construction costs might conceivably escalate at 6% a year while fuel costs escalate at 7.5% or 10% a year (or vice versa).

Following these assumptions, no matter how various escalation rates are applied to the various cost categories, the levelized bus-bar cost of a nuclear kilowatthour is always lower than that of coal, provided the escalation rate applied to a given category is common to both nuclear and coal. This is because nuclear fuel is so much cheaper than coal.

The nuclear advantage increases with an increase in the assumed future level of escalation. This is to be expected since a larger portion of coal-fired production costs is subject to escalation.

11.6.3 Nuclear spent fuel disposal

Considerable uncertainty surrounds the question of the ultimate cost of spent fuel disposal. We have reflected an allowance of roughly 1 mill per kWh (expressed in 1980 dollars) in the foregoing comparisons. When escalated over the future service lives, this comports roughly to the 2 mills per kWh allowance reflected in Tables 11.2 and 11.3, even assuming that the two mill allowance per books may be adjusted upwards from time to time if inflation continues, because that allowance represents the bookkeeping provision now deemed adequate to accumulate a reserve for expenditures which will not be made for a decade or more.

We have also estimated the costs of providing so-called away-from-reactor (AFR) storage pools adequate to hold the entire fuel discharges for the full service life of our nuclear stations. These cost estimates fall in the general area of 2 mills per kWh, level-premium, over the service lives of the stations and tend to reinforce the reasonableness of the foregoing.

However, even doubling the allowance reflected in Tables 11.6, 11.7 and 11.8 to provide for unforeseen cost escalation and contingencies, results in an addition of only about 4 mills per kWh to the first 10-year costs, far less than the nuclear advantage shown in Table 11.9.

11.7 THE IMPORTANCE OF DIVERSIFICATION

Whenever presenting economic comparisons of nuclear and coal-fired generation, one should emphasize the difficulty of predicting precisely what future costs will be. It is conceivable that one cost factor or another may change in such a way as to wipe out the economic advantage which nuclear power now enjoys. While I see no such trend developing, there are sufficient uncertainties to suggest that we should not, as individual suppliers or as a nation, select only one energy supply option. An economic advantage of 15 or 20% for either option is insufficient, in my opinion, to cause us to elect that option to the exclusion of the other. We need to preserve all available choices. And so, we must pursue both the nuclear and coal-fired alternatives with vigour.

11.8 CONCLUSION

The foregoing cost estimates indicate that nuclear power and coal are generally competitive, and that neither has such a distinct economic advantage as to rule the other out. In some parts of the country nuclear would be the clear preference. In others, like Montana, South Dakota or Wyoming, coal might well be the choice on economic grounds.

However, today's regulatory uncertainties are so great that, for the time being at least, nuclear power is not a viable alternative for new generating capacity ordered today. Uncertainty is a critical element. Today, no one can predict, with certainty, what the schedule for constructing and licensing a new nuclear plant will be. Unforeseen licensing delays can be extremely costly. These coupled with the possibility of improper regulatory response to nuclear interruptions have made the nuclear option too risky to take on.

In a highly regulated industry, investments can only be made if there is confidence in future regulation. The nuclear utilities and the rest of the nuclear industry have lost that confidence. They feel that decisions on nuclear power have been effectively removed from the marketplace by politicization of the nuclear regulatory process. Hence, we can be almost certain that no more nuclear units will be ordered until that confidence is regained.

ACKNOWLEDGEMENTS

This paper first appeared in the *Annual Review of Energy*, published by Annual Reviews Inc., Palo Alto Ca, whose permission to reprint is gratefully acknowledged.

REFERENCES

[1] NUREG/CR-0130, *Technology, safety and costs of decommissioning a reference Pressurized Water Reactor;* government publication.

12

D. Schmitt and H. Junk
THE COMPARATIVE COSTS OF NUCLEAR AND COAL-FIRED POWER STATIONS IN WEST GERMANY

12.1 OVERVIEW

Nuclear energy is the cheapest source for additional base- and medium-load electricity within the foreseeable future in West Germany. Although additional safety requirements, delays (caused by court decisions) and inflation have added another 40 to 50% to the construction costs of nuclear power stations since 1977, and although costs for reprocessing and final storage have quadrupled, the cost advantage of nuclear energy relative to coal-based electricity generation has even increased during the last few years. This is because the cost increases for electricity generating in coal-fired stations have been even steeper. The main reason is that prices for indigenous coal increased from 150 to 220 DM/tce and that prices for imported coal more than doubled at the same time and are expected to increase further (world market prices for coal steeply increased because of sharp demand rises, bottlenecks in the transport system and shortfall of exports from main suppliers like Poland. These effects have been intensified by the devaluation of the Deutsche Mark against the dollar).

Though water power and lignite are cheaper than nuclear energy in West Germany, the possibilities of increasing electricity production on the basis of these sources are very limited, mainly by environmental restraints. Heavy fuel oil (and, to a certain degree, natural gas too) has been exiled step by step from the German electricity market by energy policy since the end of the 1960s, in order to protect indigenous coal and to reduce supply risks in the electricity sector (the result is that the oil share in electricity generation in West Germany today amounts to only 5% as compared to more than 50% in Japan or Italy). Today oil and natural gas are 50% more expensive than German coal, which is itself one of the most expensive coal resources in the world. So these fuels are suitable for peak-load shaving only – using gas turbines whose low capital costs and technical characteristics fit them for this role.

It would, nevertheless, be totally misleading to assume that public utilities in West Germany henceforth will build only nuclear power stations for medium- and base-load. Utilities in Germany have voluntarily agreed to use increasing amounts of indigenous coal (until 1995 up to 57 Mtce/annum). This requires additional coal-fired stations, even if growth rates of electricity consumption fall short of the rates which were expected when the coal deal was agreed upon.* This will accordingly reduce, at least within the next fifteen years, the necessity to increase capacity on the basis of other kinds of energy.

The economic advantages of nuclear power against coal are impressive, but without any doubt, it is not pure economics that determines the future development of nuclear power in West Germany. The crucial barrier is public acceptance. The expansion of nuclear energy thus becomes a problem of political decision-making.

12.2 BACKGROUND

In spring 1981 the nuclear energy debate seemed more open than ever. Uncertainties about the risks of accidents, dangers of plutonium misuse and necessity and benefits of nuclear power, split the public, political parties and scientists.

Two issues are of paramount importance for the development of nuclear power in West Germany which is characterized by a *de facto* moratorium over the last few years. Regulatory bodies ask for the incorporation in nuclear plant of the most recent standard of technological development and scientific knowledge, according to the German nuclear energy law (Atomgesetz). This has resulted, for example, in the request for retrofitting already-commissioned plants, and even parts of nuclear power stations under construction. The delays caused by the requirements of regulatory bodies and court decisions must be seen as one of the most important reasons for increased construction costs. The second issue is the vague energy policy with regard to nuclear energy for which the term 'Restbedarfsphilosophie' was coined. It argues that nuclear energy should supply only energy needs otherwise remaining uncovered, regardless of the merits of following this course.

In addition, the debate about the economic advantages of nuclear power in West Germany has continued unabated. Part of the criticism has been raised by the lobbyists (antinuclear, environmentalists and supporters of competing energy sources), part has been stimulated by construction cost escalations, and some because aspects like decommissioning, reprocessing or final storage have not yet been sufficiently demonstrated, at least not in large-scale applications.

With respect to these reservations about the economic advantages of

* After growth rates of more than 7% per annum during the 1960s and 1970s we expect, for the next one to two decades, growth rates of only 2 to 3% per annum.

nuclear power we have re-investigated the cost differences between nuclear power plants and the most important alternative technology to generate electricity in West Germany (i.e. coal-fired stations), on the basis of most recently available data. The main results are presented in this chapter.

12.3 METHODOLOGY AND DATA BASE

Cost comparisons between existing power plants have limited significance, because they reflect only historical, nominal costs, i.e. not necessarily actual investment costs,* technological safety standards or requirements, and not actual variable costs. Such comparisons also take no account of possible future increases for main cost components. They form no adequate basis for the strategic decision of how to minimize the costs of supplying additional electricity demand in the future.

For this purpose it is necessary to estimate the costs of generating electricity in plants to be built in the future reflecting the latest available standard of technology and scientific knowledge. In order to get comparable results and not to compare apples and pears, costs have to be investigated for comparable conditions. Therefore cost comparisons have to be based on either base- or medium-load power stations, or similar load factors, start up, lifetime, size, etc. Such exercises are somewhat limited. They produce electricity generating costs for certain types of power plants, going into operation at a determined date, running for x years under similar hypotheses about load factor, fuel prices, etc. If the costs of a total generation system are to be minimized the decision about fuel selection is more complex because existing capacities must be taken into account.

The costs for each alternative were computed with a dynamic investment approach. As the output characteristic of the two alternatives is similar (production profile over running period, availability and time of economic life are assumed to be equal) revenues can be neglected. The appropriate methodology for our purpose is, therefore, the so-called RR (requirement revenue) method, which compares the cost-covering revenue requirements of different projects with a dynamic investment planning approach and thus delivers costs to be compared. Figure 12.1 shows the output of this method.

According to this method, total costs during the time of construction and operation for the plants to be compared are estimated and discounted to the start-up time of the plant. Discounting is necessary to allow for different profiles of costs over time (of construction and operation) for the various types of power station.

Actual quantities and prices for plants, labour, insurance, fuel, etc. form

---

* In the initial phase of nuclear power, industry offered plants at prices which made nuclear power competitive. On the other hand, coal-fired stations have been heavily subsidized by government in the past.

the basis for the estimate of costs. Assumptions are then made about the development of these costs in the future.

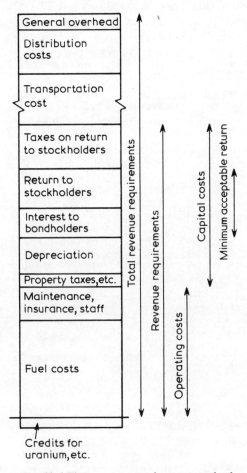

Fig. 12.1 The revenue requirement method.

For capital costs, we can base our calculations on actual tenders to public utilities from March 1981 (data from industry have been confirmed by a Delphi study with big utilities). Actual fuel prices are published. Other data, like decommissioning, investor's additional capital expenses (for sites, planning, recruitment and training of staff), time of construction and operation, number of staff, costs per man per year or specific fuel consumption have been derived from the literature, discussed and confirmed by industry and utilities. For some figures we had to make our own estimates. This holds especially for factors like interest rate or rate of inflation, value of the currency (Deutsche Mark against the dollar) and the escalation factors for labour, material and fuel prices.

The main assumptions may be found in Table 12.1, the results in Tables 12.2 and 12.3, and Figs 12.2 and 12.3. Calculations were made with the aid of a model which allows for sensitivity tests to check the impact of variations of important factors on the results. The results of these tests are presented in Figs 12.4 and 12.5.

Table 12.1 Assumptions supporting the cost comparison between coal-fired and nuclear power stations

Subject		Nuclear power station (a)	Coal-fired power station (b)	Escalation rate of cost item (a) (b)	
Generating plant					
Capacity	(MW$_{net}$)	1255	2×675		
Load factor		0.75	0.75		
Start-up	(year)	1989	1989		
Construction time	(a)	6	5		
Station life	(a)	20	20		
Interest rate	(%)	9	9		
Construction costs (including owner costs without escalation of interest and taxes during construction)	(DM/kW$_{net}$)	2550	1310[a]	Before start of construction: 6.0 6.0 During construction: 5.5 5.2	
Decommissioning	(10^6 DM)	340	—	6.5	—
Operating costs					
Maintenance, insurance and staff	(DM/kW)	62	55.90	6	6
Variable costs other other than fuel	(Pf/kWh)	0.05	0.10	4.5	4.5
Coal price (including shipping to plant)				Price profile A	B
indigenous coal			220	5.5	7.5
imported coal[d]			130	6.5	9.0
Heat rate	(kcal/kWh)	—	2340		
Fuel cycle costs					
heat rate	(kcal/kWh)	2650	—		
price of uranium[b,d]	($/lb U$_3O_8$)	30	—	6.5	
enrichment[d]	($/kg SPW)	100	—	4.5	
fuel fabrication	(DM/kg fuel)	500	—	5.5	
reprocessing[c]	(DM/kg fuel)	2200	—	6.0	

[a] Including 80% stack gas desulphurization.
[b] Including 3 $/lb for conversion U$_3O_8$ to UF$_6$.
[c] Including storage, treatment, credit for uranium and plutonium and waste disposal.
[d] DM/$ = 1.80.

Comparative costs of power stations in West Germany

12.4 ASSUMPTIONS

12.4.1 Capital costs

As mentioned above, we have to compare the costs of electricity generation in proposed new power stations. Therefore the assumed start-up date of these stations had to be set far enough into the future to accommodate plants with the longest construction time. With a construction time of about six years for nuclear power stations (four years for coal-fired stations) and date of order 1981 to 1982, the analysis had to be confined to plants starting operation towards the end of the decade. In our comparison, plants were assumed to start electricity generation on 1 January 1989, and to be typical of plants in the electricity sector of West Germany at the end of the eighties. The nuclear power station (PWR) was taken to have an installed capacity of 1255 MW (net). Units of this type are offered today by Kraftwerk Union, the leading supplier of nuclear power plants in Germany. It is intended to build a series of five to seven of these plants by 1990.* For coal-fired stations we based costs on actual offers by German industry.

The cost computation was based on two units with an installed capacity of 2×675 MW (net), 100% stack gas desulphurization using either German or imported hard coal on sites 150 to 250 km distant from minehead and coast.† Both types of power stations were taken to use wet cooling towers.

For these units the total investment up to the start of power generation was calculated. Determinants of expenses were: actual construction costs, adjustment to cost increases, until the start of construction and during construction, interest and taxes during construction and additional expenses of the utilities for planning, sites, training of staff, etc. For the nuclear power station decommissioning costs were assumed to arise 10 years after shutdown. In the case of coal-fired stations shutdown is assumed not to raise additional costs.

For nuclear power stations, as described above, the actual construction costs (March 1981) amount to some 2550 DM/kW (net). Total costs (including adjustments to cost increases during the planning and building period, interest and taxes during construction and costs for site, planning and staff) up to the assumed start of operation (1 January 1989) sum to 4455 DM/kW in nominal terms. For coal-fired stations these figures amount to 1310 DM/kW (net) and 2240 DM/kW (net) 1989. Costs for decommissioning (including a

---

* Konvoi-type, i.e. with a standardized regulation procedure which enables reciprocal admission of examinations by different regulatory bodies.
† The coal-fired station is commissioned for medium-load operation, because there exist no actual offers for base-load plants. Investment for base-load plants is said to be 10% higher and fuel consumption 5 to 7% lower than in medium-load plants. If we include these data in our calculations, electricity generation costs in coal-fired stations are only slightly lowered for operation at 6500 h/a (see Table 12.2).

ten-year custody) are estimated to range between 300 and 350 million DM (in real terms) for a nuclear power unit of 1255 MW (net).

Operating time for both plants was assumed to be 20 years;* the interest rate 9%. Inflation rate was assumed to decrease to 4.5% per annum from today's 5%. Escalation factors for prices and cost data are heavily determined by the general price level each year. The results are expressed in then-current Deutsche Mark.

12.4.2 Fuel costs

One of the most important cost factors (according to plant type and fuel prices) is that of fuel costs. Fuel has to be ordered early enough to start operating, as assumed, in January 1989 and to run the plant during the 20-year period of operation with the assumed load factors. Thus, fuel-price profiles for the next three decades have to be estimated. For the complex nuclear fuel preparation process the first quantities of yellowcake, separative work and fuel fabrication have to be ordered, and the first payments for these services are necessary more than three years in advance of use. Interest for this period has also to be taken into account, as has reprocessing and final storage of radioactive waste, because utilities have to provide for the reprocessing of burned-up nuclear fuel and the storage of radioactive waste, as planned, in a salt mine. (According to the new 'Entsorgungsgrundsätze' it will be investigated whether final storage of spent fuel without reprocessing is feasible.) These activities are assumed to take place after a cooling time of about ten years and possibly intermediate storage.

Reprocessing and facilities for final storage were assumed to be delayed by at least five to ten years in comparison with earlier plans. Originally an integrated concept with a reprocessing plant (capacity: 1400 t metal/annum) and final storage in a salt dome at Gorleben (Lower Saxony) had been developed. It was cancelled owing to political objections. It is now planned that a small-scale demo-plant (capacity: 350 t/annum) will start operation elsewhere in 1993 and that another 700 t/annum unit will probably be in use by the late 1990s.

Requirements for reprocessing should be met until the end of the century by purchases of reprocessing services from the French factory at La Hague (2300 t until 1985) and some intermediate storage of burned fuel elements. Final storage will be provided for by the government in West Germany. Cost estimates for reprocessing and final storage (as well as decommissioning) have been the reason for heavy criticism of the economics of nuclear energy. We have estimated them as follows:

---

* The actual operation time for both types of plants may be longer. This could promote capital intensive systems like nuclear energy. Experts agree that there is no case for adopting a longer time of operation, a higher availability or a greater flexibility for coal-fired stations as compared to nuclear power stations.

We took contract prices for reprocessing services from the French factory at La Hague, plus expected costs for one 350 t/annum and one 700 t/annum plant, plus final storage as estimated today and escalated to the time of expected operation. As a result we assume costs for reprocessing and final storage at 2200 DM/kg (in real terms). This assumption may be regarded as rather high at first glance (three times higher than estimated in 1978), but it is a realistic view for the changed conditions in West Germany.

By far the most important question is the future development of fuel prices (and services as indicated above). The uncertainties which surround every attempt to forecast prices need not be emphasized. For West Germany, assumptions on future price paths for different kinds of energy are based on the following hypotheses:

1. The energy sector will remain, for decades to come, highly dependent on energy imports, even if strong conservation efforts succeed, and indigenous sources are developed as far as economically useful and ecologically tolerable. The import dependence for hard coal and uranium will increase because the expected additional coal demand could be met from German mines only at high costs and because the indigenous uranium resources developed up to now are negligible.
2. Therefore, world market conditions will determine the supply conditions for new power stations in the future, last but not least because prices for indigenous coal will be highly influenced by world market prices.
3. With world energy demand still increasing (by another 50 to 100% by 2000) and oil production still at a plateau (or even decreasing), the bulk of the additional energy demand (as well as the substitution for oil) has to be supplied from non-oil resources. Reserves are sufficient. The question is how fast will they be developed and marketed and what will be the conditions that consumer countries like Germany will have to accept.
4. Under the prevailing conditions it seems highly optimistic to assume that prices for coal will increase only with long-run marginal costs, i.e. possibly 1 to 2% per annum (in real terms). On the other hand, the expected oil-price development may define an upper line of future coal and uranium prices. In this connection we have to take into account that it is not the crude-oil price, but the price for heavy fuel-oil which determines (with a reduction for handling disadvantages and/or higher capital costs) the long-run upper price limit for coal and uranium. Crude-oil prices are expected to increase (with some level of world GDP growth in real terms) by 3.5 to 4.5% per annum, but may increasingly diverge from the general price trend on heating markets as oil progressively withdraws from these markets, in favour of higher earnings in premium markets like transport and petrochemicals. We assume that prices for gasoline, diesel and naphtha will increase at higher rates than crude oil (owing to lower price and higher income elasticities). The price for these products and the

conversion costs (in hydrocrackers today, about 350 DM/t) will then determine the long-run floor price for heavy fuel-oil, and only competition between coal, uranium and natural gas (and conservation) will lead to lower price increases than 3 to 4% (in real terms).

5. Prices for German hard coal up to now have been calculated according to a cost formula for coal mining (Schwantag-Formel) based on wages, productivity and cost increases for material. It is questionable if this formula will be valid beyond the next decade and only very optimistic assumptions on productivity gains would lead to prices according to Case A (see below, point 6). Much more realistic are prices that are linked to the market conditions.

6. Given these uncertainties, we have decided to calculate electricity generation costs in coal-fired stations on the basis of two different sets of price assumptions: Case A which reflects an optimistic view (escalation for indigenous coal 1% per annum and imported coal 2% per annum in real terms) and Case B with a steeper price profile (indigenous coal increasing at 3%/annum and imported coal at 4.5%/annum in real terms) based on the assumptions made concerning future oil prices. The base price of 220 DM/tce for German hard coal (site 150 to 200 km from minehead and coast) is valid for the beginning of 1981 (neglecting the price increase of 6.5% from the spring). The assumed base price for imported coal (130 DM/tce) does not reflect the actual prices on this market which for long-term contracts have reached 170 to 200 DM/tce. We believe that this price is artificially high and that some factors (like the low rate of exchange of the Deutsche Mark against the dollar, high transport costs resulting from bottlenecks in harbour facilities, deficits from Polish exports and sharp increases in demand) will not necessarily hold in the long term. On the other hand, we have to recognize that coal prices could rise by another 40 to 50% without making coal uncompetitive with heavy fuel oil and natural gas.

7. Prices for uranium have decreased sharply recently. The reason is that production capacity in mines had been based on a faster development of nuclear energy than has actually occurred. It is expected that overcapacity will endure at least until the 1990s. As base-price we take $30 per pound U_3O_8 (including conversion), escalating at 2% (in real terms) over the next three decades. The price for enrichment is set politically and it is only at the margin that it is influenced by costs. Competition from non-USA plants (in France and Russia) and construction of production facilities in the Netherlands, the UK and West Germany may give reason to believe that prices for separate work will remain constant (in real terms). Fuel fabrication which costs 500 DM/kg today may increase at 1% per annum. For reprocessing and final storage we have assumed 2200 DM/kg (in real terms) escalating at 6% per annum. The price of plutonium for reprocessing is derived from its value as a substitute for uranium

in light water reactors. The price for the remaining uranium in spent fuel is mainly determined by reference to the cost of the separative work saved. All other cost components are of minor importance. They may be taken from Table 12.1.

12.5 RESULTS

The results of our calculations are shown in Tables 12.2 and 12.3 and Figs 12.2 and 12.3. Nuclear energy is the cheapest option for electricity production from prospective new base-load power stations, even if very optimistic assumptions on the movement of future coal prices are adopted. The break-even point with the second-best alternative, i.e. imported coal with low-price escalation, is only reached at a load factor of 45%.* The point of intersection with the cost curve for power stations fired by indigenous hard coal occurs at a load factor lower than 35% even if the calculation is based on the assumption of low coal-price escalation.

Table 12.2 Costs of electricity generation in coal-fired and nuclear power stations – start-up 1989 (in DPf/kWh)

	Nuclear power	Indigenous hard coal		Imported hard coal	
		A	B	A	B
6500 h/a	16.61	25.06	33.13	19.58	26.76
		(24.97)[a]	(32.67)[a]	(19.74)[a]	(26.59)[a]
5000	20.25	27.25	35.35	21.72	28.93
			(35.10)[a]		
4000	24.21	26.68	37.84	24.07	31.34
3000	30.80	33.71	41.96	27.99	35.33

a Base-load plant.

Costs for electricity generation in (base-load) nuclear power stations amount to 16.6 DPf/kWh and range in coal-fired plants from 19.6 to 33.1 DPf/kWh depending on the assumptions on coal prices and coal-price escalation. It has to be borne in mind that: (1) these figures refer to the value of money in 1989 (at 1981 money values, costs amount to 11.62 DPf/kWh and 13.7 to 23.17 DPf/kWh) and (2) these costs are levelized unit costs for plants which start up in 1989. The levelized unit costs must not be confused with costs calculated for the year of start up.

The costs in the first year of operation are 13.5 DPf for nuclear power and 14.0 DPf/kWh for the best coal case. The differences between these cost categories can be explained by Fig. 12.3. As can be seen the costs (and the

* If a currency rate of 2.00 DM/$US is considered instead of 1.80 DM/$US (2.15 DM/$US), the break-even point decreases to a load factor of less than 40%.

Table 12.3 Cost structure in coal-fired and nuclear base-load power stations – levelized unit costs (DPf/kWh)

	Nuclear power station		1A		Coal-fired power station 1B		2A		2B	
	(Pf/kWh)	(%)	(Pf/kWh)	(%)	(Pf/kWh)	(%)	(Pf/kWh)	(%)	(Pf/kWh)	(%)
Capital charges	9.11	54.8	4.57	18.2	4.57	13.8	4.57	23.3	4.57	17.1
Decommissioning	0.3	1.8	—	—	—	—	—	—	—	—
Coal stocks	—	—	0.15	0.6	0.11	0.3	0.07	0.4	0.04	0.2
Other operating costs	2.62	15.8	2.4	9.9	2.4	7.5	2.47	12.6	2.47	9.2
Fuel (cycle) costs	4.59	27.6	17.87	71.3	25.98	8.4	12.46	63.7	19.68	73.5
Total generation costs	16.62	100.0	25.06	100.0	33.13	100.0	19.57	100.0	26.76	100.0

1A/B = indigenous coal.
2A/B = imported coal.

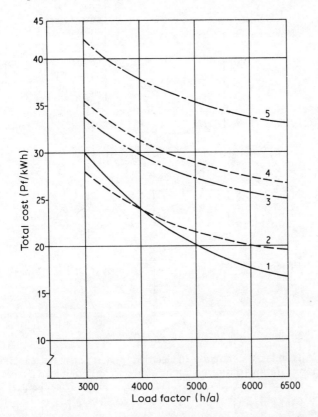

Fig. 12.2 Total costs of electricity generation in nuclear and coal-fired power plants as a function of the load factor. 1. Nuclear power station, coal-fired power station; 2. imported coal, price profile A (low case); 3. imported coal, price profile B (high case); 4. German coal, price profile A (low case); 5. German coal, price profile B (high case).

cost differences between nuclear and coal-fired plants) for the initial year of electricity generation are far lower than the costs at the end of the operating time. The reasons are the differences in the cost structure between nuclear-powered and coal-fired plants and the different assumptions about price increases during the time of operation. Only the levelized unit costs represent the costs and cost increases during the total life cycle of the plants.

12.6 SENSITIVITY ANALYSIS

In order to test how sensitive the results are to variations of the main parameters we have investigated the impacts of differing assumptions. The results have been drawn in Figs 12.4 and 12.5 which show how far the

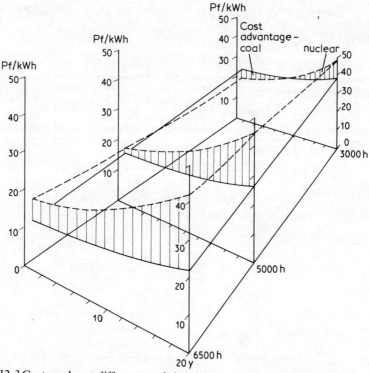

Fig. 12.3 Costs and cost differences of electricity generation in nuclear and coal-fired power stations during operating period for different load factors.

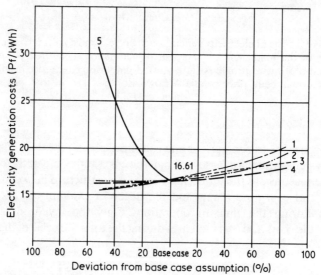

Fig. 12.4 Sensitivity analysis for electricity generation costs in nuclear power plants. Note: levelized unit costs in 1989 prices, escalation rates for 1. reprocessing costs, 2. decommissioning costs, 3. construction costs, 4. price of uranium and 5. hours per year.

Comparative costs of power stations in West Germany 301

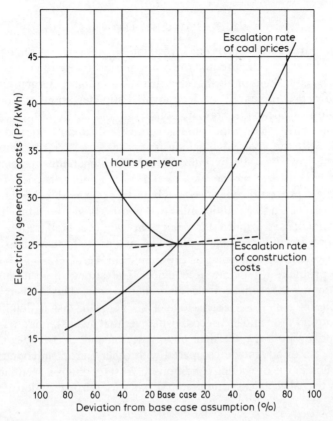

Fig. 12.5 Sensitivity analysis for electricitry generation costs in coal-fired power plants. Note: levelized unit costs in 1989 prices, German coal, fuel price profile A.

levelized unit costs increase (or decrease) with isolated variations of given parameters.

The result must be interpreted with care, because strong variations of single parameters do not normally occur in isolation because of interdependencies between markets for different goods and services. If nuclear and coal-fired stations are compared it must be assumed that variations of parameters like capital cost or fuel-cost escalation, apply to both plant categories but that the impact on cost differences is related to the differences in the cost structures. Nevertheless, it can be stated that only very low coal price assumptions or a somewhat improbable combination of assumptions to the disadvantage of nuclear power stations would absorb the cost advantages of nuclear energy in prospective new base-load power stations in West Germany.

12.7 CONCLUSIONS

If we compare the results of our investigation with plans of the electricity sector we have to recognize that the economic advantage of nuclear energy does not by any means imply that electricity generation in Germany in future will take a predominantly nuclear form. Estimates about the mid-term penetration of nuclear energy are conservative.* Public utilities last year have 'voluntarily', i.e. with some governmental support, agreed upon a contract with the coal mining industry to increase the consumption of indigenous coal by 1995 by more than 50%. Under this agreement the German government has decided to reduce the import barriers (existing since 1959 to protect the German coal industry) step by step.† Therefore, it has to be expected that utilities must first honour their commitments under this agreement, and install as far as necessary new coal-fired stations, even if nuclear power would be cheaper. In addition it has to be taken into account that an acceleration of the nuclear programme in West Germany would raise difficult problems of public acceptance. The strong opposition against nuclear energy in recent years has continued to grow. This is reflected in an official energy policy that assigns to nuclear energy the task of filling the gap left by conservation and use of coal (whatever that means). It is obvious that such a policy puts the German energy sector and electricity-consuming industry at a disadvantage compared with other European countries like France. Under the prevailing conditions in West Germany it seems to be the only way to realize at least a minimum strategy for the penetration of nuclear energy.

---

* Installed capacity in nuclear power stations today amounts to almost 10 000 MW, with another 7000 to 8000 MW under construction (to go into operation before 1985). Only under rather optimistic assumptions could this capacity double again by 1995. Within the next 15 years, capacity in coal-fired power stations is expected to increase by another 50%.
† Utilities will be allowed to import in addition 20/40/60 million tce 1981–85/85–90/90–95 if 200% of these quantities up to 1987 (and 100% beyond) will have been ordered in addition from German mines.

13

W.H. Lunning

THE ECONOMICS OF FINAL DECOMMISSIONING AND DISPOSAL OF NUCLEAR POWER PLANT

13.1 INTRODUCTION

Decommissioning requires that nuclear facilities permanently withdrawn from operational service be put into a safe condition to protect man and the environment, the ideal ultimate objective being complete removal and disposal. The UKAEA in conjunction with other organizations has, for the past five years, been examining the development issues and practical aspects of dealing with nuclear facilities no longer required in the UK. The objective is to include all types of nuclear installations but attention has been directed initially to nuclear power stations. Close liaison has been maintained with the Generating Boards.

At the present time world experience of decommissioning nuclear reactors is comparatively limited. A number of early, low power units have been closed down and taken to various stages of decommissioning. The largest reactor to be completely dismantled to date was the Elk River boiling water reactor (22.5 MWe) in the USA. There is, however, a wealth of practical experience of working under active conditions during plant maintenance, modification and adaptation which is directly relevant to the type of work involved in decommissioning.

International liaison is maintained through a technical committee of the International Atomic Energy Agency, which has concluded that there are no unsurmountable technical problems to decommissioning. Within the European Community a Commission proposal for a collaborative R & D programme on topics specific to decommissioning is currently under consideration.

Within the UK the UKAEA reactors in support of the power development programme together with the currently operating 26 Magnox reactors in eleven stations totalling some 5 GW will probably be retired by the end of the century. The timing of withdrawal from service will be dictated by development-programme requirements in the case of UKAEA reactors and by economic and technical considerations in the case of commercial reactors. Decommissioning aspects were not a primary concern in the design of these facilities, but future designs will take this factor into account to reduce, where practical, the complexity of decommissioning problems.

The UKAEA selected the Windscale advanced gas-cooled reactor (WAGR) as the initial reactor for decommissioning studies and a similar study is being undertaken by CEGB of a typical steel pressure vessel Magnox station.

13.2 DECOMMISSIONING OPTIONS

Three generally accepted stages of decommissioning have been identified from national and international studies. For the current classes of UK reactors these have been interpreted as:

Stage 1. Shut down, remove fuel, remove coolant and make safe. Maintain under surveillance.

Stage 2. Reduce installation to the minimum practical size without penetrating into those parts which have high levels of induced radioactivity. Ensure the integrity of the reactor primary containment and biological shield to prevent personnel and environmental hazard. Maintain under surveillance.

Stage 3. Complete removal of the reactor and all other plant and waste off-site followed by the return of the site for redevelopment or general use by the public. No further requirement for surveillance.

These stages, each of which establishes a safe condition, define the status of the reactor in terms of its physical state and required degree of surveillance. Accepting that complete removal of the facility is the ideal ultimate objective two main options are open: to proceed to Stage 1 or 2 and to delay Stage 3 operations to allow radioactive decay which will ease dismantling, or to proceed continuously from reactor closure to Stage 3. The decision as to which option to adopt will be influenced *inter alia* by the dose commitment to persons during dismantling operations, the economic attraction of reusing all or part of an existing site, and environmental considerations.

13.3 RADIOACTIVE INVENTORY

Decommissioning of a nuclear power station compared with conventional types of industrial installations is unique, due to problems associated with radioactivity. It should, however, be recognized that radioactivity is limited to specific areas and that a large proportion of a nuclear power station has no associated activity and can be decommissioned and disposed of using conventional methods.

To identify the development issues and practical aspects of decommissioning requires, among other data, a knowledge of the total radioactive

Final decommissioning and disposal of nuclear power plant 305

inventory and its decay, together with its distribution within the system. The radioactive inventory includes:

1. Neutron-induced activity in the fixed structure of the plant.
2. Neutron-induced activity of removable components remaining in the reactor after de-fuelling, e.g. control rods.
3. Contamination around the primary cooling circuit arising from activated corrosion products of burst fuel.
4. Contaminated/activated operational waste arising during the life of the reactor and stored in designated facilities.

Estimates for (1) and (2) can be made by calculation but the accuracy which can be achieved is dependent upon the assumed chemical composition, in particular the abundance of trace elements which become radioactive, of the construction materials. For UK reactors currently being studied these are essentially mild steel, stainless steel, concrete and graphite. The composition of the last is well defined due to the 'nuclear' specification required for its use and analytical control. The specifications for WAGR steels were nominal and the decision was taken to extend them, after consultation with UKAEA metallurgists, to include reasonable quantities of inevitable trace elements. The abundance of trace elements in concrete is controlled principally by the aggregate which in turn is dictated by the geographical source. The concrete composition adopted is based on a nominal specification modified by measurements carried out on samples from the WAGR biological shield.

The calculation of the inventory for WAGR was based on a mean flux in the moderator of 5.7×10^{13} n/cm^2/s at a nuclear load factor of 0.7 for a period of 15 years. Although a calculated inventory is considered adequate as a basis for technical judgements it should be validated and corrected if necessary by physical measurements of samples wherever possible from within the reactor. Item (3) is dependent upon the operational history of the reactor and can only be estimated on the basis of sampling. Item (4) should be identified from records. In the case of WAGR, operational waste is disposed of as it arises to general facilities on the Windscale site.

13.4 WAGR ACTIVATION AND CONTAMINATION

Figure 13.1 is a diagram of WAGR indicating the main features of the reactor and its ancillary plant. The major neutron-induced activity occurs in the steel pressure vessel and the steel structure within it, which have collectively a mass of approximately 600 tonnes of mild steel and 40 tonnes of stainless steel. Also within the pressure vessel is some 300 tonnes of graphite forming the core moderator and reflectors, together with the neutron shield situated above the latter. The degree of activation varies within the pressure vessel due to neutron attenuation by internal components. The induced

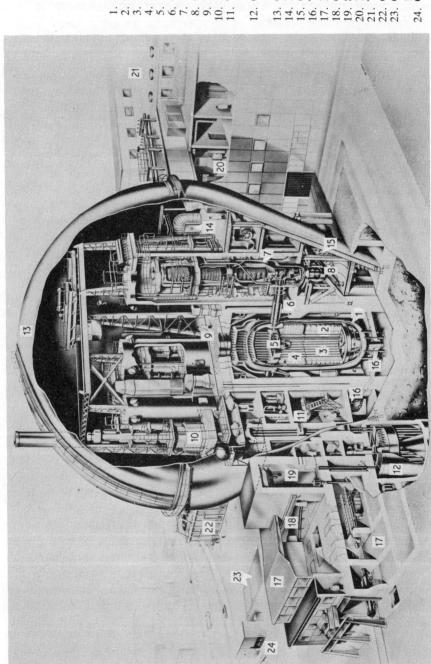

1. Reactor vessel
2. Graphite moderator
3. Fuel element channels
4. Neutron shield
5. Hot box
6. Concentric gas duct
7. Heat exchanger
8. Gas circulator
9. Refuelling floor
10. Refuelling machine
11. Transit station for irradiated fuel
12. Carousel for storage of irradiated fuel
13. Containment building
14. Personnel air lock
15. Goods air lock
16. Test loops
17. Fuel element building
18. Clean fuel preparation room
19. Stringer breakdown cave
20. Reactor control room
21. Turbine hall
22. Cooling towers
23. Gas discharge treatment plant
24. CO_2 and CO plant.

Fig. 13.1 The Advanced Gas-cooled Reactor at Windscale.

activity is concentrated in the steelwork, which incorporates the bulk of the stainless steel, in the immediate proximity of the core. The significance of the stainless steel is that it has a higher proportion of cobalt and nickel than mild steel and the overall inventory is influenced by the radioactive isotopes of these two elements. The neutron-induced activity and its decay with time, of the pressure vessel and its internal structure, is shown in Fig. 13.2. The initial decay over the first 40 to 50 years is dominated by ^{55}Fe (half-life 2.6 years) and ^{60}Co (half-life 5.27 years) which are then superceded by ^{63}Ni (half-life 92 years) as the principal isotope. The exponential decay of the system over this second phase therefore is much reduced. The radionuclides resulting from neutron irradiation are exclusively $\beta\gamma$ and no α active nuclides are produced. From Fig. 13.2 it can be seen that the β decay follows the total curie decay but the γ activity stabilizes at a virtually constant value after about 100 years.

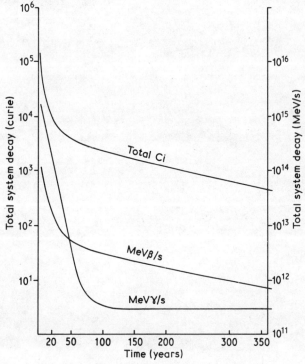

Fig. 13.2 Total system decay – curie and MeV/s.

The degree of activation of the main concrete biological shield and its mild steel reinforcement which surrounds and supports the pressure vessel on internal corbels will vary with location. The maximum depth of activation of the concrete measured from the internal face is approximately 1 m. The

Fig. 13.3 Aerial views of a part of the Windscale site showing the Advanced Gas-cooled Reactor in the left foreground; (a) as it is today; (b) as it might appear at a late stage in the decommissioning of the reactor. Conceptually, the whole area now occupied by the reactor might eventually be returned to 'green field' condition.

mass of the biological shield is approximately 4000 tonnes of concrete containing some 200 tonnes of mild steel reinforcement. It is calculated that after seven years' decay following shutdown the active portion will consist of around 750 tonnes of concrete and 90 tonnes of the inner reinforcing steel.

The four heat exchangers, each 20 m high by 7.3 m diameter and weighing 150 tonnes, are external to the main reactor biological shield. They are not exposed to direct neutron irradiation but are, however, contaminated internally with ^{60}Co, ^{137}Cs and ^{134}Cs. The degree and distribution of this contamination is monitored on a routine basis.

13.5 DECOMMISSIONING PRACTICE

The prelude to any decommissioning of nuclear reactors is the removal of the fuel and its contained fission products. This significantly reduces the radioactive content of the system leaving the bulk of the residual activity as neutron-induced and therefore in a safer form within the central region of the reactor protected by the massive concrete biological shield structure.

Stage 1. Decommissioning to this stage consists essentially in sealing the pressure vessels with plugs in the fuel element channels after fuel removal, and securing the integrity of ancillary circuits. This in effect renders the reactor safe and substantially intact on a 'care and maintenance' basis, backed by the appropriate degree of monitoring. This stage is the most economic to achieve but will attract high maintenance costs particularly if this stage is of long duration.

Stage 2. The decommissioning of WAGR to this stage would include the removal of all plant external to the reactor proper, together with the reactor containment building and all plant and equipment within it, outside the reactor biological shield but including the heat exchangers. The residual structure, which is the reactor biological shield containing active components of the reactor within the sealed pressure vessel, would be a 15 m diameter by 15 m-high right cylinder. This would occupy about one-fiftieth of the present WAGR site area and would also reduce the visual impact by a factor of 6 based upon the present vertical area compared with the 41 m-diameter containment building which currently dominates the WAGR complex.

The most economic Stage 2 situation would be to leave the plant with both the pressure vessel and the interspace between the latter and the biological shield filled with air. No external ancillary operating plant would be required except for monitoring purposes. The following aspects relating to a long-term Stage 2 condition have been considered:

1. Temperature. The residual heating in the system associated with activity after two years' shutdown is less than 1 kW. Assuming no forced cooling

of the graphite and applying a simplified model based on pessimistic assumptions it was calculated that the graphite temperature would not exceed 40°C. This temperature is well below the minimum graphite operating temperature (230°C) and hence there is no possibility of spontaneous energy release from the graphite, or of its combusion. Graphite temperature monitoring would be maintained as a safety measure.
2. Structural integrity. This is basically dependent upon the corrosion of steel. Since the site is coastal a pessimistic corrosion rate of 0.075 mm per year on each exposed surface (i.e. 0.15 mm total thickness) was assumed. Ignoring corrosion retardant factors such as temperature due to residual activity, major component-failure periods have been estimated. It is concluded that the integrity of the reactor pressure vessel and its supports would be satisfactory for at least 100 years.

The sealing of penetrations through the biological shield resulting, for example, from cutting through gas ducts to release the heat exchangers, is essential. The top of the biological shield would be capped with concrete with an access provided to enter the reactor top void space for inspection.

The biological shield is a large reinforced concrete structure and judging from available experience of this material no significant problems of deterioration should arise for at least 50 years if it were left exposed to weather. This period could be substantially extended by the construction of a relatively light weight structure around and braced to the biological shield to afford protection.
3. Radiological aspects. No radiation hazard should exist at the external face of the biological shield after fuel removal and all penetrations have been sealed and checked.

Throughout the Stage 2 condition, corrosion, both within the pressure vessel and external to it, could produce loose particulate activity. This would be entirely contained within the sealed biological shield the penetration of which by particulates to create an environmental hazard is discounted. The possibility of radiolytic chemical reactions within the pressure vessel between the air constituents and materials of construction, such as graphite, cannot be discounted absolutely. They are unlikely to occur to any significant degree since the radiation fields are relatively low; the reactor vessel and the interspace between the pressure vessel and its biological shield would, however, be equipped with sampling points for routine atmosphere monitoring.

The barrier to any contamination reaching the surrounding ground water is the steel diaphragm floor beneath the reactor, which would require to be made weather-tight at ground level with provision made for access, maintenance, and monitoring.

Access to within the biological shield of the reactor under Stage 2 conditions would only be available through facilities engineered to permit monitoring or inspection. These entrances would be secured and only used by authorized persons. A boundary fence would be erected around the structure.

The engineering requirements to establish Stage 2 have been examined and no major problems have been identified except for the dismantling of the contaminated heat exchangers. These will require to be handled under special shielding and contamination control conditions.

The establishment of a Stage 2 situation will attract a higher initial cost than for Stage 1 but to offset this the long-term cost of maintenance and surveillance will be considerably reduced.

Stage 3. The two factors which dominate the technical approach to Stage 3 decommissioning are the radiological hazards which will require to be countered during dismantling operations, and the availability of suitable disposal facilities for the dismantled components. During the period between reactor closure and dismantling of active components, radioactive decay will occur and so reduce the radiological problems.

A feasibility study has been carried out of the Stage 3 decommissioning of WAGR as a continuing process from reactor closure. The study took account of engineering requirements, radiological aspects and waste management and concluded that on technical grounds such an operation could be undertaken safely and efficiently. The study proposed that a demolition plan should be prepared on the basis of engineering logic. The plan should then be examined against the known or assessed magnitude of radiation/contamination problems which will arise at the various demolition stages; and that consideration then be given to their solution by methods – such as remote handling, shielded working or controlled access – which do not entail modification to the engineering logic. Only if at particular stages such methods prove impracticable will there be a departure from the strict engineering logic, and the overall plan amended accordingly.

In broad outline the plan proposes the initial removal of inactive components, other than those associated with services which must be retained to a later date. With the same qualification, the active components external to the pressure circuit would than be removed, followed by removal of the internal structure within the pressure vessel, and of the pressure vessel. The concrete biological shield would be demolished in a manner which would segregate the active and inactive sections. The final operations would be to dismantle the steel reactor containment building, clear the site and back fill the reactor foundations.

A team is now developing a detailed decommissioning plan on the basis of the feasibility study, which includes conceptual engineering studies for remote handling equipment and the modification of existing facilities for dismantling and waste management. It is relevant to comment that demolition will not need research into any new technology; but existing techniques will require development to adapt them to meet special dismantling problems.

It is important to appreciate that the engineering logic differs between leaving decommissioning at Stage 1 or 2 for an unspecified period, and continuing to Stage 3, particularly in the retention and adaptation of existing plant facilities. Hence if the policy relating to the fate of the reactor can be declared well in advance of retirement it should be possible to select the optimum plan for decommissioning. The cost of direct Stage 3 decommissioning must exceed those of Stages 1 and 2 but no continuing costs are involved.

13.6 DECOMMISSIONING WASTES AND DISPOSAL

Effort will be applied during decommissioning to salvage the maximum quantities of materials suitable for recycling or reuse from all areas of the site. Such materials will be subject to rigorous monitoring before release. There will however be large quantities of materials which due to their radioactive content cannot be released and will require controlled disposal. The routes currently available are disposal to land and to sea, but at this juncture no firm statement can be made of overall UK policy. This topic is under consideration in the current review of the Government White Paper *Control of Radioactive Wastes* (Cmnd 884), – which is an advisory document and forms the basis of UK practices. The recommendations of the review cannot be anticipated but work has been carried out to assess the practical application of the options.

13.7 COSTS AND TIMESCALES

The removal of fuel from a reactor, which is the initial operation leading to a defined decommissioning stage, will in the case of WAGR extend over a period of about 3 years. The time required beyond this period to complete Stages 1 and 2 will be of the order of a further 1 and 3 years respectively, and in the case of continuing progression to Stage 3 from reactor closure the corresponding extension is about 5 years. Indicative costs, excluding the cost of defuelling (which is an operational charge) and with no allowance made for the value of recovered plant and scrap, have been assessed. For Stages 2 and 3 these costs represent less than 10 and 15% respectively of the current replacement cost for WAGR at around £70m.

13.8 CONCLUSION

This article has concentrated on WAGR, which differs in design and size to commercial stations. Although the detail and scale of operations will differ, the general principles which have been discussed are applicable.

Decommissioning has not been a primary consideration in the past, but more attention is now being given to both the design and specification of materials of reactors to ease the problems of dismantling, and also to power station layouts to optimize land re-utilization.

From the studies summarized in this article and those carried out in other nations there are no technical reasons to suggest that nuclear power stations withdrawn from service cannot be rendered safe and ultimately removed.

ACKNOWLEDGEMENTS

This paper is reprinted from *Atom,* whose permission to publish here is gratefully acknowledged.

NUCLEAR POWER AND ELECTRICITY TARIFFS

14

J.M.W. Rhys
THE IMPLICATIONS OF NUCLEAR ENERGY FOR ELECTRICITY TARIFFS

14.1 INTRODUCTION

Analysis of costs and prices is fundamental to the examination of any economic activity. Nuclear power, if developed on a large scale, is bound to have a major impact on the cost structure of electricity production, on electricity tariffs, and hence on electricity use. This chapter attempts to give a broad indication of the consequences that might be expected to follow from a system of electricity supply based predominantly on nuclear power, in terms of the commercial implications for electricity tariffs, and the effects of these on the actual uses to which electricity is put.

In this context, the analysis that follows does not deal directly with the comparison of economic or other benefits between nuclear and conventional methods of electricity generation. In the long run it may be assumed that the extent and pace of the development of nuclear power as a source of electrical energy will be conditioned by the extent to which it is accepted as a safe and economic means of electricity generation. The economic basis for nuclear power must rest primarily on assessments of power station construction costs, fuel-cycle costs, and the costs of its alternatives, notably fossil fuel. Nevertheless the structure of electricity costs as a whole, and its intelligent translation into tariffs, can be shown to be an important factor in determining the proportion of total energy demand met by electricity and the proportion of electricity production economically met by nuclear power. It will therefore be relevant in determining the extent to which nuclear power can satisfy future energy needs. The extent to which nuclear energy can provide an increased share of total primary energy supply will be conditioned by the extent to which electricity can provide an economic means of satisfying the requirements of final consumers. The latter will depend on both the level and the structure of electricity tariffs, in relation to other fuels.

It is not possible in a short chapter to do complete justice to either the elaborate and well-developed theoretical basis for electricity tariffs, or to the full range of practical approaches to pricing policies that are taken by electricity supply undertakings across the world. The former already has a substantial literature (see [1] for example). The latter reflects the very different forms of public or private ownership, of regulatory control, and of financing arrangements, that apply in different countries. The analysis that is set out below is based on a broad view of general principles of economic analysis that are likely to be applicable, with modifications for particular circumstances, to most electricity supply systems where a substantial component of nuclear power is developed. These principles provide a general basis for tariff policies within an electricity utility, and also serve to spell out the economic logic which will dictate the actual impact of nuclear power.

14.2 PRINCIPLES OF ECONOMIC PRICING

Economic theory attaches importance to price, as the fundamental mechanism by which messages pass between producers and consumers. The price of any good has to be high enough to cover the producer's cost and persuade him to sell, but low enough to persuade the consumer that it is worth his while to purchase. If prices are based on costs, therefore, the price mechanism should ensure the best possible allocation of resources by establishing a link between the value to the consumer and the cost to the producer.

This principle can be stated more explicitly and more precisely. When a purchaser causes an additional unit of any good to be produced, he should pay a price that covers the additional cost involved in the production of that additional unit. The principle is therefore put forward, in terms of economic theory, that prices do, and should, tend to equate to marginal costs. The production of many goods is associated with heavy capital investment, on which the producer has to earn a return. The above principles must therefore be interpreted, for both practical and theoretical reasons, so that the costs of production are defined to include the costs of this capital provision. The concept of marginal cost must therefore be related to a sufficient period of time to allow this inclusion. Prices based on long-run marginal cost should then cover the full costs of production, and provide the producer with the necessary incentive to invest.

One can view the application of this economic principle in two contexts. Firstly it can be seen as a statement about how prices are likely to be formed in a free market with perfect competition. Secondly it can be taken as a basis on which prices ought to be formed where no real competition exists, as in the case of monopoly, for example. Both views can be criticized, and there is a large body of economic literature that deals with the subject of marginal cost pricing, and its variants, in the context of public utilities. Practical and theoretical debate surrounds the extent to which pricing policies can or should take account of long-run marginal costs. The issues relate to wider aspects of political economy, such as taxation and income distribution, and to practical difficulties such as the correct identification of resource costs, and the identification and interpretation of what is 'long-run' and 'marginal'. Nevertheless, in the absence of any theoretically perfect solution, the concept of long-run marginal cost pricing provides a useful practical indicator in looking at the influence of economic events, such as the development of new means of production, or as a basis for pricing policies appropriate to particular situations.

Electricity can be considered in terms of these ideas. Electricity supply is typically the responsibility of a publicly- or privately-owned monopoly subject to regulation. In either case there is a need to formulate rules, preferably with a sound theoretical underpinning, to guide and regulate the setting of electricity tariffs. Electricity has a central role in the energy economy, both as regards the supply of electrical energy to consumers, and as a major consumer itself of primary energy inputs, whether these come from oil, coal, uranium, or alternative energy sources. As energy pricing is a fundamental aspect of energy policy, this role gives a particular importance to electricity tariffs. It is also clear that the pricing of electricity must, in principle, be examined in conjunction with the pricing of all fuels, whether as inputs to electricity generation, or for supply to consumers. If one fuel is overpriced and another underpriced, in relation to true costs, this may lead to excessive demand for, and over investment in, the second fuel–and under utilization of the first.

14.3 COST STRUCTURES IN ELECTRICITY SUPPLY

Two particular features of electricity supply are: (1) long lead-times for investment in productive capacity, and (2) electricity cannot be stored but has to be produced at the moment of consumption. Differential pricing, according to the time of use, has to be employed in order to reflect varying costs of production and the cost of providing capacity to meet peak-loads. It is not enough, in other words, to look only at the prices of separate fuels in relation to one another. The efficient distribution of resources requires that 'off-peak' and 'on-peak' electricity should also be properly priced. The costs of supplying electricity at different times of day or year should be separately reflected in prices in all cases where differences can sensibly be identified, and provided the benefits of a differential pricing system outweigh the costs of operating it.

It is helpful to begin with an examination of the components of electricity costs. Costs of supplying the consumer include costs of metering and invoicing, costs of localized distribution systems, costs of higher voltage transmission systems, and costs of electricity generation. It is typically the latter that account for the greater proportion of total costs, and define the particular qualities of electricity pricing structures. It is also the latter that will be affected most directly by the development of nuclear power and by

The implications of nuclear energy for electricity tariffs

changes in methods of production. This article therefore concentrates on the relationship of generation cost structures to electricity prices.

Analysis must commence by identifying and describing the fundamental task of electricity production. It is to meet a consumer demand that varies continuously over the day, and which has a daily pattern that varies according to the day of the week, the season of the year, and the weather. The description of this requirement is itself potentially complex and requires some simplification. It is a convenient convention for many purposes to identify the electrical energy (kWh) consumed in half-hour periods as a basis for the description of electrical power demand (kW). (To avoid confusion with other uses of the word 'demand', the latter will be referred to in this article mainly as load.) Thus the peak-load (kW) over (say) an annual period is the average actual electric power demand (kW) for the particular half-hour with the maximum kWh consumption of electrical energy.

Having set out a practical definition of power demand (kW) or load, it is helpful in analysing the structure of production costs, to provide a summary description of the continual variations that occur within the system of electricity supply. A very useful description is given by the load duration

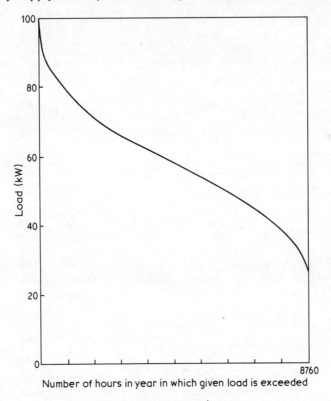

Fig. 14.1 Load duration curve.

curve, which shows the proportion of time during which load exceeds a given level. A typical load duration curve would have characteristics similar to those of the curve shown in Fig. 14.1.

Although the load duration curve in Fig. 14.1 tells us nothing about the daily load shape, hourly variations in load, or seasonal variations in load, it does describe key features of load variation very concisely. Thus Fig. 14.1, as an actual load duration curve, would enable us to deduce, for example:

1. Load never falls below about 25% of its maximum level.
2. Load is at 80% of its maximum level, or above, for only about 10% of the time.

Since the vertical axis measures kW, and the horizontal axis time, the area under the load duration curve corresponds to electrical energy. By drawing a horizontal line at 50 kW, and measuring the areas under the curve above and below this line, as in Fig. 14.2, one can see that the additional electrical energy consumed by an excess of load or power demand (kW) above 50% of the maximum load accounts for only about 20% of total electrical energy (kW).

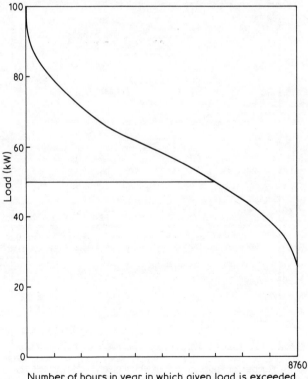

Fig. 14.2 Load duration curve.

14.3.1 Short-run marginal costs

Any given pattern of demand will be met by a corresponding pattern of production, aimed at achieving a lowest overall cost of production. The means of production available for use consists of the variety of generating plant currently in operation, in many systems based on a variety of sources of primary energy (e.g. coal, oil or hydro-power), with varying efficiencies and operating costs. The operation of the available generating plant will be aimed at minimizing the short-run variable costs of supply. In order to achieve this, generating plant can be ranked in a merit order. High-merit plant, typically modern plant based on the cheaper fossil fuels, or hydro or nuclear plant whose operating costs intrinsically tend to be lower, provide the first choices for base-load operation. This is followed by the use of less efficient plant, lower in the merit order, with progressively higher operating costs. Finally, actual peaks may be covered by use of the available plant from the bottom of the merit order, which typically operates for a very small part of the year. Such plant may be older plant, or more modern plant, perhaps gas turbines, installed with the planned intention of operating for relatively short periods. Plant installed for the prime purpose of meeting peak-loads will be likely to have lower capital but higher running costs.

A hypothetical application of the merit-order principle is shown in relation to the load duration curve in Fig. 14.3 below. Ignoring, for the purposes of simplification, any consideration of plant availability, or of maintenance scheduling, Fig. 14.3 shows the use of different types of plant, against the load duration curve, for a system with a small amount of nuclear plant, under conditions where coal is generally cheaper than oil. In Fig. 14.3, it is evident that nuclear plant is run as 'first choice' base-load, and only as base-load. Some coal stations are run as base-load, but coal stations are the 'marginal' stations for 20% of the year, while oil-fired stations are 'marginal' for most of the remainder. Obviously, many variations on this particular combination are likely to occur in practice.

It should be evident from the example of Fig. 14.3 that the cost of electricity sent out during the peak periods may vary from the very low levels associated with high-merit base-load plant, to much higher figures associated with the last few kWs of total demand – the cost of supplying electricity at the margin. The economic principles outlined earlier indicate that it is the cost of the marginal units that should be used as the basis for pricing calculations. The same principle applies to all periods of the day. As Fig. 14.3 makes clear, there will be a schedule of marginal running rates, according to the marginal plant in operation at different times. These constitute the short run marginal costs of supply. In broad terms, however, a distinction may be drawn between off-peak costs (lower) and on-peak costs (higher), and one should expect that these distinctions will be reflected in the actual tariffs offered to consumers. The actual scope for differential pricing

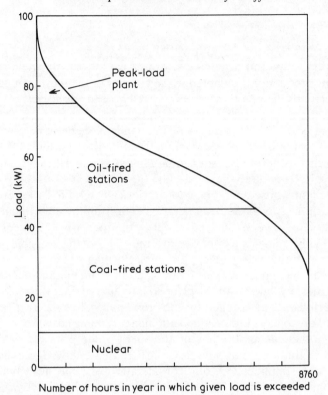

Fig. 14.3 Load duration curve.

will be limited by practical considerations. Tariffs are set in advance, and differential pricing will normally focus on those cost differences that can reasonably be expected to be significant and persistent.

14.3.2 Long-run marginal costs

Marginal changes in variable costs, assuming existing capacity as fixed, are described by short-run marginal costs. However, in the longer run, almost all costs, including the capital costs of provision for a level of capacity, may be regarded as variable. Electricity supply is a capital-intensive industry, and these costs form a major part of its overall cost structure. As argued earlier, therefore, they also need to be reflected in pricing structures. These costs are therefore included in any analysis of long-run marginal costs; the latter reflect the incremental cost of capacity, plus the short-run marginal costs described above. Considerations applicable to the calculation of long-run marginal cost are described below.

In the context of short-run marginal costs, the analysis related to the optimization (cost minimization) of an electricity supply system with existing plant. For future investment, an electrical utility is faced with choices between investment in a variety of types of generation, including stations with relatively high capital but low running costs, and those with low capital but high running costs. Investment appraisal, under these circumstances, will normally be based on comparison of the total costs of operating the system, over a period, with or without the addition of different types of plant. This appraisal requires forecasts of several factors as a basis for the comparisons. These factors include growth (positive or negative) in consumer demand for electricity, the future pattern of load, relative fuel and capital costs, and other technical factors. In the long run, therefore, cost structures are defined by the opportunities for optimization through choice of new plant. The choice of new plant determines the capital costs actually incurred for the purposes of electricity generation. It also determines the mixture of future capacity available for short-term optimization measures, and therefore exerts a powerful influence on the future structure of short-run marginal costs.

The amount of new capacity to be installed is based primarily on expectations of the total capacity required to meet the anticipated maximum load on the system. However the type or types of generating plant chosen to meet the capacity need will depend on the way in which the system is expected to operate, and, in particular, the load factors* that will apply to stations at different points in the merit order. Questions of plant choice are therefore very closely related to the mixture of load (kW) and energy (kWh) requirements, whose chief characteristics are defined by the load-duration curve.

Choice of an ideal combination of plant to meet the requirements described by an actual load-duration curve involves trading, one against another, the relative advantages of different types of plant. Thus gas turbine plant enjoys low capital but high running costs, and provides the ideal plant to operate at a low load factor. Base-load stations have higher capital but lower running costs, and are the most economical choice of plant provided they can be run at sufficiently high load factors. In a system with an ideal combination of plant, these relative advantages are balanced against each other, and, in principle, different types of plant may be allocated to different modes of operation, according to load factors, in the manner shown in Fig. 14.4. At certain levels of the load factor for the utilization of plant, one type of plant may cease to be economic, while another becomes economic.

Incremental consumption of electricity, whose costs should, in principle, be matched by the price charged, normally consists of load growth which occurs not only at times of maximum load on the system, but at other times of the day and year. This growth can be met by additional gas turbine plant,

---

* In this context, load factor refers to the proportion of the year for which a station is expected to operate. It is defined as the number of units supplied as a percentage of the plant capacity multiplied by the number of hours in the year.

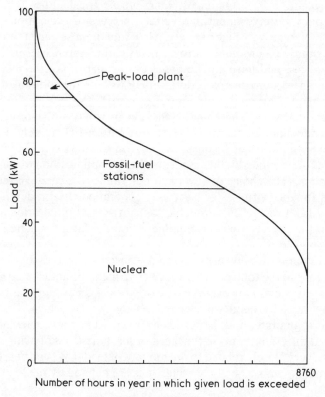

Fig. 14.4 Load duration curve.

to cope with extra peak load, and increased use of existing, older (lower merit), plant at other times. Alternatively the growth may be met by installing a new base-load station, with higher capital cost but savings on running costs. In a system where the existing combination of plant approximates fairly closely to the theoretical optimum, one may well be indifferent as regards the choice between these two options. This is because the differences in capital costs are offset by gains in running costs attributable to use of more efficient base-load plant. The net costs of providing capacity by either method may then be the same.

It may be deduced that the incremental cost of capacity can be related to the relatively low capital cost of plant used to meet peak loads, even though actual investment takes place in base-load plant with higher capital cost. One can envisage that the units produced by base-load plant, valued at the short-run marginal cost levels associated with actual merit order operation, yield a 'profit' to base-load plant which represents a return on the additional capital employed. One should not, therefore, seek to create an identity between the incremental costs of capacity to meet peak loads and the actual capital costs incurred in the construction of new plant. Some part of this capital investment may, normally, be regarded properly as investment in maintaining an economic means of generating electrical energy (kWh), and not merely as the provision of capacity to meet maximum load (kW).

It may be noted that an existing combination of plant is not always ideally balanced with the particular load requirements it has to serve. This may happen when conditions change quickly or unexpectedly, as, for example, when the price of oil changes dramatically in relation to the price of other fuels, or when new methods of production emerge. It may then be economic to construct new plant even though supplies could be maintained with existing plant, and construction is not required to meet growth in load. Once the new plant has been installed, it becomes subject to the same optimization routines as existing plant, and runs according to its position in the merit order. In displacing older, less efficient plant, it yields operating savings measured by the differences in short-run marginal costs. It will be evident that, if oil becomes very expensive, the much lower fuel costs of coal or nuclear stations may give rise to savings in operating costs that are large enough in themselves to justify investment in new plant.

In summary, minimization of the costs of electricity production follows a logical but complex set of rules, both for short-term optimization based on the use of existing plant, and for long-run optimization which includes the choice of new plant. This process defines a cost structure for electricity production, and allows consideration to be given to the long-run marginal cost of increments of load. The main elements in the calculation of long-run marginal costs are the costs of providing incremental capacity, together with whatever are the current short-run marginal costs. The former reflect the capital costs associated with plant planned to run for a small proportion of the year, and to meet peak loads. These different elements, taken together, form a structure from which one can calculate the incremental costs associated with any particular type of load growth.

14.3.3 Prices based on long-run marginal cost

The translation of complex generation cost structures into retail tariffs for particular groups of consumers, or for particular categories of load, requires an assessment of the contribution from those groups or categories to the overall pattern of load. The marginal costs imposed by the different incremental loads can be calculated from the overall cost structure, and appropriate retail tariffs, based on long-run marginal cost, can be formulated.

The purpose of a pricing structure is to influence consumer behaviour. Consumers should respond to the price 'messages' they are given by deciding on their own patterns of electricity use that will best satisfy their own preferences, given the costs they will incur for themselves. An initial pattern of consumer load may be significantly altered by carefully constructed tariffs, with differential pricing based on the principles of long-run marginal cost. Changes in the pattern of consumer behaviour may in turn provoke re-examination of the different elements in the cost structure of electricity supply, and lead to a continual monitoring of the validity of the elctricity tariffs on offer to the consumer. Long-run marginal cost pricing should not therefore be viewed purely as a static reflection of production cost structures. It is rather the means by which the latter are brought into harmony with the preferred patterns of consumption of electricity consumers. In so far as principles of economic pricing can achieve an improved allocation of economic resources, this must be done by a simultaneous optimization in the choices made in both production and consumption.

In the above analysis of long-run marginal cost pricing for electricity supply, the emphasis has been on the development of a basis for differential pricing, and on the economic costs most appropriately associated with production at different times. It is also important to realize that the application of economic methods of pricing (based on long-run marginal cost) also establishes a general relationship between the overall level of prices and the overall level of costs of extra production. Some elements of production, such as a limited and fixed quantity of hydroelectric power, can be excluded from consideration of marginal costs, but provide an endowment of lower cost production. Normally, however, the economic principles oulined above give a formulation of prices that corresponds to recovery of the total costs of production. This is obviously of great practical importance for the financing of electricity supply. It also provides a guide to the effect of introducing cheaper means of production into a system of electricity supply, in terms of the effect on the general level of electricity prices, expressed for example as average revenue per kWh sold. In the long run, the overall level of electricity prices will be determined by changes in the total costs of production.

14.4 THE IMPACT OF NUCLEAR POWER

It is against this general theoretical background that the effect of a large-scale development of nuclear power for electricity generation can be examined. Nuclear power does not add any new dimensions of principle to the theory of electricity pricing. In this context its chief characteristic is the balance between capital and running costs, which seems likely to be very different for nuclear stations, compared with fossil-fuel stations. On cost, the general expectation of nuclear capital costs is that they will be higher than those of fossil-fuel generation, although they may fall as experience is gained. The expectation of running costs, where fuel costs predominate, is that these will be much lower than those for coal or oil stations, even though the price of uranium or the cost of the fuel cycle may rise. This latter

The implications of nuclear energy for electricity tariffs

expectation has, of course, been greatly influenced by the large rises in the real price of fossil fuels in recent years.

In Fig. 14.3, nuclear plant provides only a relatively small quantity of base-load plant. If we assume this quantity is fixed, equating the situation of nuclear power to what is often the case in practice for hydroelectric power, then it is obvious that nuclear electricity, even if produced at zero cost, has no effect on marginal costs or on prices*. If, however, the balance of cost advantage is substantially in favour of nuclear power for base-load generation, then in order to provide an optimum mix of generating capacity to produce at lowest possible overall cost, the quantity of nuclear generation will rise until nuclear stations are at the margin for some part of the year.

Nuclear stations, with higher capital but lower running costs, will be the most economic form of generation provided they operate for sufficiently long periods during the year. The relative capital and running costs of nuclear and conventional fossil-fuel plant will determine a 'break-even' level of operation, at which it is not economic to install more nuclear plant. Figure 14.4 shows a revised structure of production costs. If we ignore important complicating factors, such as plant availability and maintenance schedules, Fig. 14.4 may be interpreted as the representation of a 'break-even' point in which the last nuclear station operates for 70% of the year. On this diagram the capacity of the nuclear stations accounts for 50% of maximum load, but these stations produce about 80% of the total electrical energy. Nuclear operating costs determine short-run marginal costs for 30% of the time.

One may therefore make certain general deductions about the impact of nuclear power in circumstances where it provides a substantially cheaper means of base-load production than fossil-fuel based alternatives. Firstly, nuclear power is likely to account for a substantial proportion of capacity, and a significantly higher proportion of production. In overall terms, therefore, the price of electricity is likely to be dominated by the costs of nuclear power, even though fossil-fuel stations remain an important determinant of short-run marginal costs. If the price of fossil fuels continues to rise, then the overall competitiveness of electricity, relative to other forms of energy, is likely to improve. Secondly, one may simultaneously see a greater differentiation in pricing according to time of day or year. This differentiation will itself play an important role in establishing the competitive position of electricity in the energy market.

14.4.1 The impact on electricity consumption

Analysis of the overall impact of nuclear power on energy and fuel use therefore requires attention to be given to its impact, through tariffs, on the

---

*This statement requires some qualification. It is true where all prices have settled down at levels that fully reflect long-run marginal costs. The utility profits implied by a slice of low-cost production are, however, sometimes used to subsidize consumer prices generally (pricing at an average cost), or to subsidize a particular group of consumers, e.g. one particular industry.

consumer – the final user of energy. The economic function of a pricing policy is not just the reflection of costs as a means of fair and equitable charging, but the provision of messages about the use of resources, that will actually influence consumer behaviour. If the consumer is given the correct market messages, and then chooses to dispose of his requirements in the ways that best suit his needs, then the best possible overall allocation of resources will, hopefully, have been achieved.

It is evident that some types of electricity usage will not be strongly influenced by pricing structures that emphasize differences between time of day or between seasons, at least within the context of most likely ranges of price variation. This is obviously true, for example, of use of electricity for lighting, where the need for use cannot easily be postponed. It is also true, in the case of many miscellaneous domestic and commercial uses where convenience is of paramount importance, that energy costs are relatively insignificant, and the technical possibilities for manipulating load are limited. However there are many other areas where there is much greater scope for manipulating the incidence of electricity usage, and where an economic incentive based on differential pricing may be very important.

Space- and water-heating load provides an opportunity for storage of energy. Even though the consumer's requirements for water-heating may relate to a particular time, hot water can be stored in well-insulated tanks, heated during off-peak hours to take advantage of lower-price electricity. Use of differential pricing by time of day is well established in many countries for domestic consumers. Space- and water-heating happen to be applications where the choice between alternative fuels is most closely influenced by considerations of relative cost. It follows that if differential pricing allows the consumers' requirements to be satisfied with a lower total electricity bill, then there will be a greater incentive for consumers to use electricity in preference to fossil fuels. In the context of nuclear power, therefore, the development of appropriate cost-reflecting tariffs, which emphasize the potentially much lower prices available for electricity consumed at times outside peak periods, will tend to increase the proportion of total energy requirements met by electricity, and at the same time increase the proportion of nuclear energy as a share of total primary energy.

Similar manipulation of electrical load can be applied in the case of large industrial loads. It is already a fairly common practice to offer substantial tariff incentives to industrial consumers to manipulate their load requirements. For some large consumers these incentives may even be sufficient to change the annual pattern of plant maintenance and so affect the seasonal incidence of load. The possible accentuation of differential pricing with the advent of nuclear power will probably tend to reinforce incentives for load manipulation.

Many types of large industrial loads share properties which, in principle, allow the industrial consumer to respond to a given pattern of costs by manipulating load demand in order to achieve a reduction in electricity costs. Typically, energy costs and, specifically, electricity costs may be among several other important cost components, which have to be considered in an industrial process. A typical example of a large energy usage in an industrial process is that of metal-melting furnaces. This example is used by Langman [2] in a paper on the control of large industrial loads, but other examples include cold-storage plant, induction heaters, pumping equipment, drying ovens, and battery charging. Also of potential importance is the use of bivalent systems, in which different fuel inputs are used at different times, according to changes in relative cost. The industrial process, in the case of the metal-melting part of a steel works, constitutes a major problem of control, in which half a dozen or more separate factors may need balancing at any one time. Obviously, the development of an overall control strategy includes determination of energy demand patterns and energy costs. The operation of the processes requires a sophisticated series of choices about the sequence and timing of different operations, each of which may have different energy requirements. Processes may be advanced or delayed, to make use of a cheaper energy input, using energy storage within the process. The use of sophisticated control systems, which are increasingly associated with modern industry, allows more attention to be paid to cost optimization. Clearly, the greater the tariff incentive to control the timing of electrical load, the greater the number of options that become open to the industrial operator. These options may be apparent not only at the operating stage, but also at the design stage. In the case of furnaces, plant may be designed to allow the incorporation of melting and heating furnaces, or of melting furnaces only. The selection of plant will therefore be closely related to consideration of operating costs, and development of electricity tariffs in particular.

The availability of cost-reflecting tariffs once again assists in achieving a better overall use of economic resources, including energy resources. In this case the attractions of electricity to industrial consumers are enhanced by a tariff in which the cost advantages for electricity production to meet particular patterns of electricity use are reflected in price messages to the industrial consumer which encourage that form of use.

For both heating and process uses, and for both domestic and industrial consumers, the possibilities for load manipulation are likely to be improved by technical progress in the development of control devices and systems, and of the means of monitoring and recording energy use. Developments in metering may assist in the further practical development of cost-reflecting tariffs, and the conveying of price messages to the consumer, while developments in the field of control assist the consumer to respond to the messages he receives.

The nature of potential effects on the shape of the load-duration curve is shown in Fig. 14.5. It may be noted that the load gains obtained from

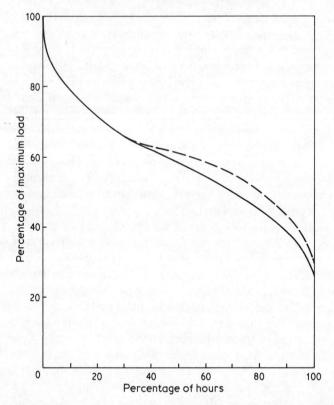

Fig. 14.5 Load duration curve, where ———— is the development of the shape of the normalized load duration curve with more differentiation in pricing by time of day or year.

differential pricing need not be confined to the lowest price periods. There will inevitably be spillover effects if entire processes are converted to electricity.

14.5 CONCLUSIONS

Much of the above analysis is inevitably general and hypothetical. However, it does serve to show some of the commercial implications of the development of nuclear power as an economic means of electricity production. In the long run, provided nuclear power can be established as the cheapest means, it is likely to be the dominant factor in determining the overall level of electricity prices, and to account for the major proportion of production. The extent to which nuclear power can satisfy total energy requirements will be linked to the success of electricity in meeting consumer needs. Improvements in the price competitiveness of electricity will be important in allowing the substitution of electricity for other fuels, and the development of different types of electrical load. Also important in this regard will be the accurate reflection in electricity tariffs of the changes in cost structures implied by nuclear development.

REFERENCES

[1] Turvey, R. and Anderson, D. (1977) *Electricity Economics*, World Bank Research Publication, The John Hopkins University Press, Baltimore.
[2] Langman, R.D. (1981) *The Control of Large Industrial Electrical Loads*, Watt Committee on Energy/Council of Europe Seminar, Reading University, England, September.

NUCLEAR POWER IN LESS-DEVELOPED COUNTRIES

15

L. Kemeny

THE PRACTICAL PROBLEMS AND ISSUES PERTAINING TO NUCLEAR ENERGY IN LESS-DEVELOPED COUNTRIES

15.1 INTRODUCTION

Early enthusiasm for the transfer of nuclear technology to developing countries during the decades of the 1950s and 1960s has, in recent years, become an object of scornful scepticism on the part of some experts and ill-founded disbelief and fear on the part of the general public, especially in the Western World.

The above attitudes relate, in particular, to the sharing of technology in the specialized areas of the nuclear fuel cycle and nuclear power engineering. The subjects, related to the production and uses of radioisotopes in agriculture, medicine and engineering are not considered to represent the same degree of complexity and hazard and, in fact, have gained a large measure of public acceptance.

Surprisingly, the deteriorating environment of public opinion in which the nuclear industry finds itself in the Western World – especially in the United States – is not manifested to anywhere near the same extent in the development attitudes and programmes of the developing and newly industrialized countries, which in the context of this chapter includes countries of the COMECON group. This apparent paradox is largely due to:

1. An understanding by nuclear experts in the Third World that the present situation in the West can, to a great extent, be attributed to media orchestration and political posturing.
2. The scenario of risk proliferation, environmental desecration and the potential of risk currently painted by nuclear opponents is a pseudo-scientific fulfilment of a successful campaign by self-styled 'environmentalists, ecologists and consumer protectionists'. It is a state of affairs which, to future generations, will probably appear as a disastrous aberration. To the countries of the developing world it is apparent that these false perceptions represent far from informed realism and that the demands and expectations of an exploding world population present no

option but to further increase the utilization of uranium and coal as base-load energy supplies.

It is the purpose of this chapter to examine the circumstances which must prevail before a reasonable technical, administrative and sociological case can be made to justify the introduction of nuclear power technology to a developing country. Consideration is also given to the global advantages which would accrue to developing countries if the industralized countries of the world accelerated their nuclear power programmes, thereby making greater contribution to the expansion of fossil-fuel usage amongst their developing neighbours, hopefully at a mutually agreeable price level.

15.2 ROLE OF THE INTERNATIONAL ATOMIC ENERGY AGENCY

According to Sigvard Eklund – when Director General of the International Atomic Energy Agency – the IAEA activities [1] in the nuclear field are determined primarily by the requests it receives from the Member States for assistance in establishing the scientific, technical, legal and organizational infrastructure which is necessary for the practical application of various nuclear technologies. The requests may include the planning, construction and operation of nuclear facilities and, of course, may well originate from Member States classified as 'developing countries' and listed in the Appendix.

Key aspects of the work of the Agency as summarized by Eklund include:

1. Assistance in establishing and developing Member States' own capabilities in basic and applied nuclear chemistry and nuclear physics, including theoretical nuclear physics.
2. Coordination of applied research and development, particularly in areas of relevance to developing countries. Through coordinated research programmes, work in specific fields, carried out in research establishments in developing countries, is brought together with corresponding work in leading institutes of developed or scientifically advanced countries. These research programmes usually run over a period of three to five years. Progress is reviewed at periodic meetings in which leaders of research from participating institutes take part. Leading world experts are often invited to contribute to these meetings, where plans for follow-up are also made.

A typical example of the important role played by the Agency in these areas is support for the establishment of a Nuclear Agriculture Research Centre on the campus of the Agricultural University of Bangladesh at Mymensingh. Further assistance to the same country has included, over a number of years, feasibility studies of the construction of a small nuclear

power station at Ruppor, some 100 miles south-east of Dacca. Some aspects of this project will be considered later in this chapter.
3. Reviews of progress in selected fields of nuclear techniques through symposia, advisory groups and technical committees. In selecting these topics, attention is paid to the needs of developing countries.
4. Administration and scientific and technical guidance of large and small-scale technical assistance programmes and projects and the provision within these projects of experts, advisory services and equipment.
5. Training of scientists, engineers and technicians through the organization of training courses, seminars, study tours, the preparation of training manuals and the award of fellowships.
6. In collaboration with leading experts provided by Member States the specification and elaboration of internationally agreed recommendations, codes of practice and guides, providing a basis for countries wishing to establish indigenous regulations in health protection, plant safety and the transportation of nuclear materials.
7. Guidance to the Member States in developing and improving their nuclear materials accounting systems. Increasingly this aspect of the Agency's work will involve it in recommending, and possibly monitoring, the physical protection of nuclear materials. This will include general recommendations, training courses and advice by experts. It may also incorporate periodic visits by Agency inspectorate to Member States to ensure that all aspects of the *Nuclear Non-proliferation Treaty* are being honoured.

The foregoing discussion on the role of the International Atomic Energy Agency in responding to the needs of developing countries is central to the subsequent expositions and considerations developed in this chapter. It seems most unlikely that nuclear energy will play a major role in the economy and technology of developing countries without the full-scale involvement of the Agency. Whilst, on the industrial and commercial side, developing countries may well form bilateral links with industrialized trading partners, past precedents and future projections associated with possible expansion of the nuclear industry indicate that at all times the Agency will play a role in nuclear power development as consultant, assessor, staffing instructor and safeguards monitor.

15.3 INFRASTRUCTURE AND TRAINING

The definition of what constitutes a developing country is frequently ambiguous. The author himself is puzzled by the inclusion, within such a context, of a number of COMECON countries with a highly advanced economy based on secondary industry and with a vigorous and well-advanced nuclear power programme, based on a highly organized and

The practical problems and issues

skilled infrastructure. On the other hand, Australia is regarded as a 'developed' country, despite the fact that she may not have operating nuclear power plant for another two decades. For this reason this country will also be considered in the context of this chapter. Some case studies of nuclear energy progress in other countries will be considered towards the end of the chapter.

In a preliminary document [2] prepared for the International Nuclear Fuel Cycle Evaluation project by the IAEA, the following special problems of developing countries were noted as being of special interest:

1. The long lead-times involved in programme establishment and project realization.
2. The highly advanced technology involved in design, construction and operation of nuclear plants, which is without equivalent in other industrial fields.
3. The strict safety requirements, with corresponding requirements for a special regulatory structure and strict standards for design, manufacturing, construction, operation and maintenance.
4. The need for highly qualified and specially trained manpower at all levels.
5. The requirements for non-proliferation measures in connection with the transfer of technology, equipment and materials, and the provision of specialized services.

The initiation of a nuclear power programme in a developing country can be considered only if the following key factors are clearly kept in the forefront of the decision making process.

1. The availability of capital obtained on terms which favour the local economic environment of the developing country.
2. The supply of trained and qualified personnel.
3. The existing industrial infrastructure of the developing country.
4. The *status quo* regarding the base-load energy supply which is available already within the developing country under consideration.
5. The indigenous fuels and nuclear fuels which may or may not be available to determine the selection of the fuel cycle.
6. The state of both public opinion and informed, professional, technical opinion concerning nuclear power within the country, the stability of its government and commitment to and motivation for the successful completion of a reasonable programme.
7. The availability of a reasonable level of basic technical expertise at the plant operator, technical and workshop levels to ensure that both routine maintenance and unexpected malfunction can be adequately handled.

15.4 FUEL CYCLE SUPPLY

A number of different avenues of approach are open to a developing country when it decides to become involved in nuclear fuel-cycle programmes. In the initial stages an assessment is usually made of the possibility of a domestic supply of nuclear raw material.

It is believed, at the present time, that fourteen developing countries have operations in uranium mining and milling, geophysical prospecting, ore extraction and mines under development. At least one of these countries is also involved in the mining of thorium. Some 308 000 tonnes of uranium are contained in these ventures. This represents a 'reasonably assured resource' recoverable at a cost of less than $80 US per kg and an annual capacity at present for mining and milling of about 29 000 tonnes of uranium.

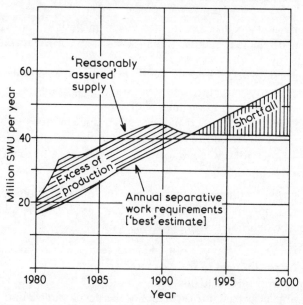

Fig. 15.1 Western world enrichment – comparison of supply and requirements. (Australian Atomic Energy Commission 1979/1980 Annual Report.)

In respect of the front end of the fuel cycle, one developing country has plans for enrichment by means of the jet nozzle technique, with a pilot plant under construction and the development of a small fuel-fabrication facility*. Another has a uranium conversion/uranium oxide fuel fabrication facility in operation with a capacity of 120 tonnes per year and commercial plants producing heavy water for domestic requirements. Yet another has a pilot uranium oxide facility.

Most developing countries are fully satisfied with the possession of technology related only to the construction, commissioning and operation of base-load nuclear power stations. This is understandable, as the basic criterion for the adoption of a nuclear fuel cycle will always be the convenience,

* The need for further enrichment capacity in the future is brought out in Fig. 15.1.

economics, safety and environmental desirability of utilizing the energy content of nuclear fuels. At the present time the IAEA listing of nuclear plant in operation [3] and projected for the future is detailed in the Appendix.

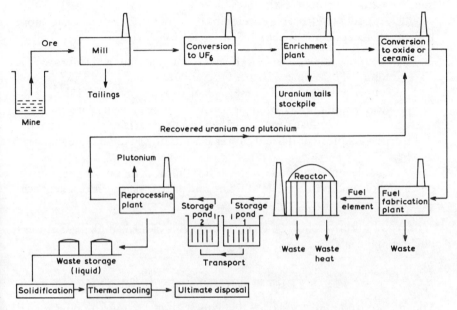

Fig. 15.2 The light water reactor nuclear fuel cycle.

For many of these countries, the 'leasing' of nuclear fuel – as already practised by the COMECON block – appears to have attractive possibilities for the future. This is a scheme whereby a single country or a consortium of countries, with full access to the 'front end' as well as the 'back end' of fuel-cycle technology (see Fig. 15.2), contracts to supply to a developing country, on a leasing basis, the fuel rods required to operate the recipients' power stations. Such contracts would usually require the return of the spent fuel to the lessor who might also undertake the fuel loading, cycling and management of the developing countries' nuclear installations. Considerable financial benefits could accrue to the recipient country under such an arrangement, ranging from tax benefits to electric utilities to a more attractive cash flow situation as long as lead-time project minimizes its capital outlay before it receives its first return on investment. Equally important benefits which could arise from a leasing arrangement include closer controls on the supply, accounting and safeguards associated with fuel transport and a strict supervision over weapons proliferation risks.

On a long-term basis, a developing country might consider a national or multinational participation in the 'back end' of the fuel-cycle activities. At present one developing country is planning a pilot facility with a reprocessing capacity of 10 tonnes per year for oxide elements. Another has a reprocessing facility with a capacity of 125 tonnes per year, with a similar plant under construction and a pilot facility for reprocessing 30 tonnes per year for natural uranium fuel.

The key issues confronting developing countries in their decision to adopt a nuclear fuel cycle have been summarized as follows [2]:

1. The adequacy and efficiency of bilateral and multinational assistance for development of a nuclear power programme.
2. The supply availability of suitable plant sizes and types and their technological, economic and ecological viability. The present problems of the nuclear power industry create an increasing uncertainty concerning the future availability of nuclear power plant in the small and medium size range.
3. Reasonable and attainable targets for the nuclear power programme being formulated. Long lead-times and heavy capital investment will necessitate careful decision making.

Consideration will now be given to problems of nuclear plant safety and nuclear materials' safeguards.

15.5 SAFETY AND SAFEGUARDS

The nature and functions of a regulatory body for nuclear power in a developing country have been reviewed by Rosen [3]. Once again it is obvious that the IAEA will have a significant role to play in achieving the training of suitable staff, and in organizing some form of indigenous regulatory commission.

A Safety Code of Practice on Governmental Organization for the Regulation of Nuclear Power Plants has recently been published by the IAEA. This includes the statement: 'It is regarded as essential that the Government of a Member State embarking on or implementing a nuclear power programme establish a regulatory body.' This regulatory body, under the supervision of a Director would require:

1. Administrative and legal support.
2. Consultants and advisory bodies.
3. Assessment and licensing.
4. Regulatory inspection.

A brief salutory reminder of the responsibilities inherent in transferring the appropriate nuclear technology to a developing country is to consider the expertise [3] required for the regulation of nuclear power plants. No fewer than 26 different scientific and technical disciplines are involved.

As in developed countries, the key issues in this context include:

The practical problems and issues

1. Early planning and progressive development of manpower capability.
2. The need to develop internationally-agreed safety assurance criteria leading to uniform inspection practices.
3. The necessity to provide functional independence for the regulating authority.

All Member States of the International Atomic Energy Agency undertake the acceptance of systematic expert inspection of their nuclear facilities by specialists nominated by the Agency. These inspections may range from radiation safety to the complex procedures associated with accounting for fissile materials inventories at any stage of the nuclear fuel cycle.

Developing countries gradually acquiring nuclear expertise and facilities are frequent recipients of expert advice from professional scientists and technologists, usually stationed *in situ* under the sponsorship of the Agency. It is not unreasonable to expect that, as the peaceful nuclear fuel cycle proliferates through five continents, the Agency's nuclear inspectorate will grow and be granted a clear mandate for international surveillance of the production, transport, utilization, reprocessing and disposal of nuclear materials. The thoughtful reader might well at this stage rightly reflect on the effectiveness of any form of inspection in the present world climate of arms escalation and military posturing. Questions in this context frequently arise in respect of small, under-developed nations precipitately acquiring nuclear technology for the sake of international prestige and, it is alleged, military advantages.

Exaggerated dangers associated with the above scenario can be shown to be unrealistic, both qualitatively and quantitatively for the following reasons:

1. The capital investment in nuclear power plant is so great that it is most unlikely that a developing country would contemplate complex weapons technology based on the peaceful nuclear fuel cycle. The acquisition of a nuclear arsenal is, from a technical point of view, completely divergent from the supply of base-load energy for industry and agriculture.
2. On the other hand, to withhold peaceful nuclear technology from a developing country which has established a legitimate need for it, is tantamount to driving its government into an impasse which may motivate it to undertake covert action.
3. The energy needs of developing countries, based on their population increase over the next five decades and on their growing industrialization will have a two-fold effect. The developed countries of the world will rely increasingly on a fuel mix of uranium and coal to supply their base-load energy requirements. At appropriate times a developed 'donor' country may, dependent upon agree political, technical and economic ground rules, inject viable nuclear technology to a recipient 'client' country, in return for raw materials and other commodities which the industrialized

country may require. This strategy may assist the developed country not only in providing technical 'know-how' but also in alleviating the recipient country's very considerable economic problems, by providing the capital needed to initiate a nuclear power programme.

15.6 DEVELOPING COUNTRIES IN THE NUCLEAR FUTURE

Conservative projections indicate that within approximately 40 years the world's population will be more than twice the present 4 billion and the total energy demand will have also more than doubled.

Even if the Western democracies practise every measure of energy conservation, this will be more than offset by the legitimate and increasing energy demands of the developing countries. Over this period of time most countries will not have the option of deciding whether additional resources should be developed. They will have to act quickly and decisively to provide the additional energy which will be required. Any credible scenario for the future must include full use of uranium as a fuel.

On a world-wide basis, the pattern for the increasing use of the uranium fuel cycle could take the sequences shown in Table 15.1.

Table 15.1 Introduction of uranium fuel cycle

1980–1990	Increasing use of nuclear energy for electricity generation and marine propulsion. Phasing out of oil and natural gas as energy-producing fuels for base-load power stations.
1990–2000	Introduction of nuclear energy for industrial process heating, water desalination, steel making and coal gasification.
2000–2040	Introduction of commercial fast breeder reactors and thermal breeders based on thorium. Reassessment of the uranium and plutonium fuel cycles based on the 'state of the art' of thermonuclear fusion, hybrid reactors and solar technologies.

Taking account of maturity of technology, safety, economics and environmental impact, most European countries logically choose uranium as the best available fuel to replace depleting hydrocarbons. That developing countries should not be denied a similar solution was recognized in the final communique of the International Nuclear Fuel Cycle Evaluation (INFCE) working party, initiated by the United States in 1976, and made up by many of the member states of the International Atomic Energy Agency. This communique, issued from Vienna in early 1980, concluded as follows:

'Finally, the Conference wishes to state that the findings of INFCE have strengthened the view:

- that nuclear energy is expected to increase its role in meeting the world's energy needs and can and should be widely available to that end;

The practical problems and issues

- that effective measures can and should be taken to meet the specific needs of developing countries in the peaceful uses of nuclear energy; and
- that effective measures can and should be taken to minimize the danger of proliferation of nuclear weapons without jeopardizing energy supplies for peaceful purposes.

The Conference recognized that the objectives mentioned above can only be achieved through continued international cooperation and the participants are determined to preserve the climate of mutual understanding and cooperation in the international nuclear energy field that is one of the major achievements of INFCE.'

The economics of nuclear energy will continue to provide the stimulus. Every country with a nuclear power programme – and the EEC industry is expanding at the rate of some 30% per annum despite the frustrations of the American utilities – finds that the uranium fuel cycle is around 40% less expensive (capital and operating charges included) than coal or oil. Consider figures (1980) from *Electricité de France* in a country which expects to be around 60% nuclear by the year 1990.

Cost per kWh of:	
Nuclear generated electricity	13.52 centimes (2.1 cents Aust.)
Coal generated electricity	24.49 centimes (3.8 cents Aust.)
Oil generated electricity	36.32 centimes (5.6 cents Aust.)

Table 15.2 shows an updated list to 30 June 1980 – prepared by the Australian Atomic Energy Commission – of all nuclear electricity generating plants of capacity greater than 30 megawatts. It can be seen that some developing countries already make significant use of nuclear energy.

15.7 A ROLE FOR AUSTRALIA

It is with some trepidation that the author will now include Australia as a 'developing' country, in the firm belief that it will operate nuclear power stations within two decades. The fact that the International Atomic Energy Agency regards Australia as 'developed' is probably based on the country's industrialized background and its small but highly trained corps of nuclear scientists and technologists.

On 15 April 1953, the Royal Assent was given to the *Atomic Energy Act*. The formation of the Atomic Energy Commission followed, and some of Australia's top engineers and scientists were sent overseas to receive specialized training in many facets of nuclear technology. In particular, it was intended that many of these would return with expertise in the field of

Table 15.2 Nuclear generating plant greater than 30 MW, June 1980

Country	In commercial operation	Under construction	On firm order	Total no.	Megawatts
Argentina	1	1	1	3	1 617
Belgium	3	4	0	7	5 505
Brazil	0	3	0	3	3 116
Bulgaria	2	2	0	4	1 640
Canada	10	10	4	24	14 875
Cuba	0	1	0	1	410
Czechoslovakia	3	8	0	11	4 210
Finland	2	2	1	5	3 113
France	15	31	8	54	48 322
Germany (East)	5	8	0	13	5 995
Germany (West)	11	11	4	26	25 183
Hungary	0	4	0	4	1 640
India	3	5	0	8	1 664
Italy	4	3	2	9	5 356
Japan	22	7	2	31	21 385
Korea RO	1	4	2	7	5 398
Mexico	0	2	0	2	1 308
Netherlands	2	0	0	2	502
Pakistan	1	0	0	1	125
Philippines	0	1	0	0	600
Poland	0	1	0	1	410
Romania	0	0	2	2	820
South Africa	0	2	0	2	1 844
Spain	3	8	4	15	12 155
Sweden	6	5	1	12	9 485
Switzerland	4	2	0	6	3 793
Taiwan	2	4	0	6	4 960
UK	33	6	0	39	11 794
USA	71	74	26	171	162 462
USSR	26	22	3	51	34 930
Yugoslavia	0	1	0	1	632
TOTAL	230	232	60	522	394 249

nuclear power plant design, safety analysis, construction and commissioning.

In retrospect, despite the undoubted competence of our nuclear technologists and the early enthusiasm for nuclear power development shown by both politicians and the scientific administration, the historical perspective appears bleak and fruitless. On the credit side there are the benefits to the country from the production of radioisotopes and the training of professional staff. To this must be added the more intangible and long-term returns of fundamental research carried out in the fields of physics, chemical technology, metallurgy, engineering and applied nucleonics and systems analysis. On the debit side we see the spectre of the aborted Jervis Bay nuclear power station project and the very limited involvement of this country, thus far, in any aspect of nuclear fuel cycle technology.

Has Australia a nuclear future? The author believes that the inevitable answer to this question is 'yes'. It would appear that a second opportunity now exists to frame a nuclear energy policy for the nation based on a logical sequence of nuclear energy development involving:

1. Uranium mining and export including possible new methods for uranium marketing to minimize the risks of nuclear weapons proliferation.
2. Nuclear fuel cycle technology from enrichment through fuel fabrication to reprocessing and high level radioactive waste disposal.
3. Nuclear power applications in the areas of electricity generation, process heat production and marine propulsion and consultative services in these areas to developing countries.

Exploration for uranium in Australia began in earnest two years after the end of the Second World War. By 1971, production from minor discoveries at Radium Hill in South Australia, Mary Kathleen in Queensland and Rum Jungle and South Alligator River in the Northern Territory had amounted to about 9100 tonnes of yellowcake (U_3O_8), of which 7200 tonnes were exported and the balance stockpiled by the Australian Atomic Energy Commission. Production then ceased because existing contracts had been fulfilled and there seemed to be no further market for Australian uranium.

In the meantime, major ore-bodies had been discovered in the Nabarlek, Ranger, Koongarra and Jabiluka regions of the Northern Territory, in the Mt. Painter-Lake Frome area of South Australia and Yeelirre in Western Australia. However, Australian reserves of currently marketable uranium are more modest than is generally believed. The total recoverable, at a cost of $27.25 (Aust.) per pound (454 g) of uranium oxide or less, is 380000 tonnes – or only 9.9% of the non-communist world's resources in this acceptable cost category. However, unlike most other uranium-producing countries, Australia has substantial production capacity uncommitted.

Furthermore, new finds are almost certain and could increase uranium reserves by five or even tenfold. Because substantial Australian production capacity is, as yet, uncommitted, the country, in some three years time, could become a major supplier of uranium to a world hastening into the nuclear age.

According to a recent survey in *Australian Business* [4], Australia's uranium producers have little to fear from either the present depressed state of the market or long-term prospects or fluctuations. By 1990, according to a report by the Massachusetts Institute of Technology (MIT), Australia will be the non-communist world's biggest producer, contributing up to 23000 tonnes. The Nuclear Exchange Corporation expects Canada to be next, on

12 230 tonnes, followed by South Africa (9500 tonnes), the USA (7000 tonnes) and Niger and Gabon (6800 tonnes).

The MIT forecast takes into account the huge potential of Roxby Downs, which is expected to be one of the biggest uranium mines in the world. The processing plant, if it is ever built, would be likely to take its raw material from the mine, but it could also take uranium from other South Australian prospects, such as Beverley and Honeymoon, the Northern Territory mines yet to be developed, and prospects in Western Australia, such as WMC's Yeelirrie Project.

There seems little doubt [5] that Australia is well situated and equipped to contemplate becoming involved in every operation connected with the nuclear fuel cycle. However, world energy demands, markets, domestic demands, workforce availability and national development and energy-policy constraints will govern the timing of such projects.

15.8 SOME NATIONAL PROGRAMMES

In the ensuing discussion, the case studies considered are almost a 'random sample' of developing countries with nuclear aspirations. The common denominator in the choice of national programmes being considered is the fact that in most cases the author has had the opportunity to lecture in the designated country and to discuss with some of their top experts the problems and goals which guide and motivate their energy strategies in general and nuclear power in particular.

An interesting pattern emerges from these technical discussions and clarifies and enhances the rationality of the logic behind nuclear aspirations in even the most unlikely communities desiring access to nuclear technology. In brief summary, senior decision makers and technical experts in developing countries expect the following benefits to accrue from the introduction of nuclear power into their country:

1. The liberation of conventional, indigenous fuel resources, when and where available, for expert purposes and, for a limited and usually very inefficient national usage in remote areas with a very low population density which, at the present time cannot be supplied with energy from a national grid system.
2. The systematic acquisition of technological skills and the development of new industrial centres and peripheral manufacturing complexes as a direct concomitant of the technology transfer associated with the provision of nuclear facilities.
3. Somewhat surprisingly, following rigorous assessment by both national energy planning institutions and with support from the International Atomic Energy Agency, the realization that even small nuclear plant within the 100 MWe to 600 MWe range can become competitive with other energy options based on indigenous or imported fuels. This is a

generalization but has been shown to be true in many locations despite high capital costs.

In connection with the last of the above contentions, Japan, Germany and France are three of a number of countries investigating the design and construction of small nuclear reactors.

In Japan, the Ministry of International Trade and Industry (MITI) has decided to investigate the development of a small to medium-sized light water reactor with a capacity of 100–200 MWe. The initial feasibility studies are being carried out by the Japan Atomic Industrial Forum and the Comprehensive Engineering Research Institute for Energy.

In West Germany, Kraftwerk Union (KWU) are developing a concept for a reactor in the 200–400 MWe range which would be economic for countries with a small grid system. The design is for a simplified boiling water reactor which will use tested components, however it is not a scaled down version of the 1000–1300 MWe reactors being built in industrialized countries today. The Company believes that many developing nations will be unable to afford an energy supply which relies on oil or coal and that in the area of electricity generation they will have to substitute nuclear energy for oil.

On the basis of a current crude oil price of $34US per barrel, KWU estimate the fuel cost for a 200 MWe oil-fired plant operating at full load for 6000 hours would be approximately DM130 million. By comparison the fuel costs for a 200 MWe nuclear reactor operating on the same basis would be about DM24 million. The difference in annual fuel costs alone is sufficient, they believe, to offset the higher capital costs for a nuclear reactor. Over a 20-year period, assuming a 9% annual increase in the cost of crude oil and natural uranium, the oil-fired plant would cost around DM4 billion more than a nuclear plant.

At the International Conference on Nuclear Power and Its Fuel Cycle, Salzburg (Austria) 2–13 May 1977, the listing of the many factors which might adversely influence the adoption of the new technology was given in numerous papers; however, reports from many countries which have already passed through a systematic and in-depth planning programme showed a high degree of maturity and a commitment to local involvement. It is therefore evident that technological, financial, environmental and sociological problems can be overcome through determined resolution and major effort. The present increase in the rate of escalation of fossil fuel prices should add incentive and economic motivation to the accelerated development of nuclear power programmes.

Typical national plans for nuclear energy in the developing countries include:

15.8.1 Indonesia

The introduction of nuclear power into Indonesia is likely to be hastened following the signing of an agreement in Jakarta in March 1980, by Italy's

Minister for Scientific Research and Technology and Indonesia's State Minister for Research and Technology. Continuation of an existing joint nuclear research project between the two countries could lead to the construction of a 600 MW heavy water nuclear unit and a 30–40 MW research reactor by the end of the decade.

The 600 MW facility, estimated to cost $1 billion, is likely to be located 100 km east of Semarang in central Java whilst it is intended that the research reactor will be located close to Jakarta. In addition the Indonesian Government plans to build a nuclear research station in Serpong (southwest of Jakarta) which will include a heavy water reactor and research facilities. Officials from Indonesia's National Atomic Energy Agency estimate the cost at $377.5 million.

At present Indonesia has a small heavy water reactor in Bandung used for research and the production of radioactive isotopes for medical, agricultural and industrial use. Indonesia has also begun a major campaign to develop other energy resources in order to meet the soaring domestic demand for energy and also allow the continued lucrative export of crude oil instead of having to divert it to domestic use.

A French exploration team has discovered rich uranium deposits near the Kalan River in West Kalimantan (Borneo) and are continuing exploration in Sumatra. Rapid development of uranium deposits is not expected however as, in the short term, officials have said Indonesia may turn to Australia for uranium to fuel its nuclear power plants.

15.8.2 Bangladesh

Despite the small generating capacity of the interconnected electricity generating system which exists at the present time (around 1200 MWe) studies have indicated a technological and economic justification for the the installation of a small 120 MWe nuclear power station at Rupoor, some 160 km southeast of Dacca. One or two nuclear generating units of this size could be justified as soon as the East–West H.V. electricity link across the Brahmaputra River is completed. The chairman of the Bangladesh Atomic Energy Commission has stated that the new technology could greatly assist in providing the motivation for building up skilled technical manpower in his country.

15.8.3 Egypt

The Egyptian nuclear programme was first planned in 1963, but has been delayed due to political events and economic factors. A recent government-ordered study has shown economic savings could be obtained in operating costs by the use of nuclear power plants as opposed to oil-fired plants. The study estimates these savings could equal the capital cost of a nuclear plant in less than ten years of operation.

Last year Egypt signed the *Nuclear Non-Proliferation Treaty* and announced its intention to sign a safeguards agreement with the International Atomic Energy Agency in the near future.

Egypt is now anxious to implement a nuclear programme and orders have been placed with France for two 900 MWe pressurized water reactors. Within the past few weeks an agreement was signed whereby the United States would supply two 1000 MWe nuclear reactors and sufficient enriched uranium for operation. Formal ratification by the US Congress and the Egyptian Parliament is expected soon. By the year 2000, Egypt plans to have eight nuclear power reactors in operation with a capacity of 6600 MWe.

Egypt has some uranium deposits and mining is expected to commence soon. Initially, production from a small mine, located in the Eastern desert will be 30.5 tonnes a year expanding to 100 tonnes a year at a later stage. Two other mines will open in the near future. These mines will not be sufficient to support the country's nuclear programme and Egypt might therefore be a prospective buyer for Australian uranium.

15.8.4 Mexico

In November (1979) Sweden's Asea-Atom was commissioned by the Federal Electricity Commission (CFE) of Mexico to conduct a feasibility study into the development of nuclear power in Mexico. Other feasibility studies currently being carried out are on heavy water systems by Atomic Energy of Canada Ltd and by France's Sofratome on pressurized water systems.

Sofratome notes Mexico's plans to construct two or three 900 MW units per year between 1990 and 2000, three units per year after 2000 and four units per year after 2010.

The studies concluded that Mexico is in a favourable position to proceed with a nuclear power programme and has the capability to build up its own domestic nuclear reactor manufacturing industry.

15.8.5 China

Writing in a Peking publication, the Vice Minister of the Second Machine Building Ministry (which includes the nuclear industry) noted, 'at the present rate of coal extraction China cannot keep up with electric power requirements'. He concluded 'that without nuclear power China would find it impossible to achieve its programme of industrial, military and agricultural development by the year 2000 target date.'

At China's first national conference on nuclear power, recently held in Peking, it was stated that the People's Republic is capable of building its own nuclear power stations if it cannot import technology at prices it can afford. In the 1980s, China plans to develop and complete two nuclear plants, a 300 MW pressurized water reactor and a 100–200 MW heavy water reactor. Last year China decided against the purchase of two 900 MW pressurized water reactors from Framatome.

The author's personal contact with electricity generating authorities in Hong Kong has elucidated a number of enquiries from Chinese sources and proposals from Hong Kong related to possible joint nuclear power projects. So far these are only at the discussion stage. However, it would appear that the first large 'hard currency' funded nuclear power station in the People's Republic will, in fact, be in Quandong Province – a joint project with the government of Hong Kong. The rationale for this decision is based on the difficult problems of transporting energy from coal and hydropower rich Northern China with little industry to an energy scarce southern region which contains most of the country's industrial complexes.

15.8.6 South Korea

The Seoul government has invited tenders from American, French, Swiss, Swedish, West German and Italian firms for Korea's eighth and ninth nuclear units. These units are scheduled for construction by 1987 and 1988, at Buku-Ri on the east coast about 140 miles south-east of Seoul.

Westinghouse has completed the one operating reactor in South Korea, is in the process of constructing four others and was successful in tendering for the sixth and seventh units. Framatome is a possible supplier for the next two 950 MW units.

15.8.7 COMECON countries

Although it is certainly not intended to classify the Soviet Union as an 'under-developed' country, a complete picture of the rapid growth of nuclear energy in the COMECON group would not be complete without reference to the achievements of the Soviet Union and her partners. Tables 15.3 and 15.4 [6] show a reasonably up-to-date status of the industry in seven COMECON countries.

The countries listed in Table 15.4, with the possible exception of one, all rely on Soviet type PWR technology.

A case study of Hungary [7] is typical of energy problems confronting Eastern European countries:

'At the present time, Hungary's electric power system is based on coal (52%), hydrocarbons (43%) and hydro (5%). The growth of electricity generation to the year 2000 is projected at 7% per year reaching 130–140×10^9 kWh in the year 2000. The capacity of the power generating system in that year is estimated at 25.5 to 27.5 GWe. It is desired to meet future electricity demands with reduced dependence on hydrocarbons and also to stabilize the contribution of coal at about 30% of electricity

Table 15.3 Nuclear power stations in the USSR

Name	Number	Type	MWe	Year
In operation				
Obninsk	1	PTR	5	54
Troitsk	6	PTR	100	58–62
Beloyarsk 1	1	PTR	100	64
Beloyarsk 2	1	PTR	200	67
Leningrad	2	PTR	1000	73–75
Bilibinsk	4	PTR	12	73–76
Chernobyl	1	PTR	1000	77
Kursk	1	PTR	1000	76
Novovoronezh 1	1	PWR	210	64
Novovoronezh 2	1	PWR	365	69
Novovoronezh 3, 4	2	PWR	440	71–72
Novovoronezh 5	1	PWR	1000	78
Kola	2	PWR	440	73
Armenia	1	PWR	405	76
Ulyanovsk	1	FR	12	62
Shevchenko	1	FR	*120	73
Under construction				
Leningrad	2	PTR	1000	
Kursk	2	PTR	1000	
Chernobyl	1	PTR	1000	
Smolensk	1	PTR	1000	
Ignalino	2	PTR	1500	
Kola	2	PWR	440	
Armenia	1	PWR	405	
Kalinin	1	PWR	1000	
S. Ukraine	1	PWR	1000	
Rovno	2	PWR	440	
Beloyarsk	1	FR	600	
Planned				
Kursk	1	PTR	1000	
Chernobyl	2	PTR	1000	
Smolensk	1	PTR	1000	
Kalinin	3	PWR	1000	
S. Ukraine	3	PWR	1000	
W. Ukraine	4	PWR	1000	

*This reactor also produces 50 000 cubic metres of distilled water per day.
Notes
1. There is a small experimental boiling water reactor of 50 MWe at Ulynovsk.
2. There is an experimental fast reactor of 5 MWe at Obininsk.
3. Total MWe in operation is 8825.
   Total MWe under construction is 13 765.
   Total MWe planned is 14 000.
   Total = 36 590 MWe.
4. PTR refers to pressure tube reactors at the RBMK type.
   PWR refers to pressurized water reactors of the VVER type.

Table 15.4 (a) Nuclear power stations in Eastern Europe

Country	Name	Number	Type	MWe	Total	Year
In operation						
Bulgaria	Koslodui	2	PWR	440	880	74–75
Czechoslovakia	Bohunice	1	—	144	144	72
Czechoslovakia	Bohunice	1	PWR	440	440	79
	Rheinsbert	1	PWR	70	70	66
East Germany	Lubmin	3	PWR	440	1320	74–79
Hungary	Paks	1	PWR	440	440	79
Under construction						
Czechoslovakia	Bohunice	3	PWR	440	1320	
Czechoslovakia	Dukovany	4	PWR	440	1760	
Czechoslovakia	Mochovice	4	PWR	440	1760	
East Germany	Lubmin	1	PWR	440	440	
Hungary	Paks	3	PWR	440	1320	
Planned						
Bulgaria	Kozlodui	2	PWR	440	880	
Czechoslovakia	Malovice	4	PWR	1000	4000	
East Germany	Magdeburg	2	PWR	440	880	
Poland	Zarnowiec	3	PWR	2 × 440 1 × 1000	1880	
Romania	Olt	1	PWR	440	440	
Romania	Cernovada	4	CANDU	600	2400	

(b) Totals by countries (MWe)

Country	In operation	Under construction	Planned	Total
Bulgaria	889	—	880	1760
Czechoslovakia	440	4840	4000	9280
East Germany	1390	440	880	2710
Hungary	440	1320	—	1760
Poland	—	—	1880	1880
Romania	—	—	2840	2840
Total	3150	6600	10480	20230

generation. Thus, nuclear power will play an important future in Hungary. The first 4 × 400 MWe stations will come into operation during the 1980–1984 period, followed by a second stage of 2 × 1000 MWE nuclear plants between 1986–1990. By the year 2000, Hungary expects to have 12–14 GWe of nuclear capacity in operation. At that time the composition of electricity generation might be coal (31.5%), hydrocarbons (13%), nuclear power (48%) and gas turbine/hydro (7.5%). Electricity would then represent 58% of total energy consumption.'

15.9 NUCLEAR POWER COSTS IN DEVELOPING COUNTRIES

An excellent study of nuclear energy cost trends in developing countries has been carried out by Lane [8]. He states that early analyses by the IAEA in fourteen developing countries indicated that capital costs would be less than that prevailing in the country supplying the plant and expertise. However, special training requirements for locally recruited workers, the lack of industrial infrastructure, the high cost of construction supervision and uncertainties associated with local raw material supplies more than outweigh these advantages in 1980.

Unfortunately, therefore, in most developing countries at the present time, the capital cost of nuclear power plants of given size are at least as high, if not greater than those in industrialized nations.

Table 15.5 Extrapolation of costs to developing countries

Plant type	MWe	$/kW*
PWR	600	1285
	900	1000
	1200	845
Low-sulphur oil	600	395
	900	355
High-sulphur oil	600	450
	900	405

* Excluding fuel and escalation during construction.

Another factor militating against nuclear power in this context relates to savings of scale. If the cost of, say a 1200 MWe PWR is only 30% greater than a 699 MWe PWR, the attractiveness of nuclear energy for a developing country with a small installed generating capacity is greatly decreased. Lane [8] summarizes these factors in Table 15.5. He also makes some interesting comparisons in Table 15.6 between the relative costs of nuclear and low-sulphur oil-fired plants in situations which might prevail in a typical developing country – in this instance Pakistan.

We finally return to the main question at issue. Will nuclear energy ever serve the needs of developing countries safely, economically and without undue environmental problems?

The author believes that the answer is 'Yes', provided that the following structures are carefully observed by the national decision-makers. It should be self-evident that problems unique to each developing country might necessitate deviations in the approach to the adoption of the new technology. The steps along the road to nuclearization might well be summarized as follows:

1. Over the next one or two decades developed countries should expand their own nuclear power programmes at an accelerated rate to free hydrocarbon fuels for use in developing countries.
2. By the mid-1990s the escalation in hydrocarbon fuel prices and transport costs will be such that nuclear plant – possibly even in smaller sizes and perhaps in a mobile configuration – will be economical in the most remote conditions of geographical isolation and for the use of diverse ethnic backgrounds and national energy requirements.
3. At this stage of the peaceful nuclear industry it is probable that through the work of organizations such as the IAEA many of the problems of proliferation, safeguards, staff training and the creation of an appropriate national infrastructue for nuclear plant construction will have been implemented by the end of this period.

Table 15.6 Power costs from nuclear and oil-fired plants in developing countries (m/kWh)

	Light water reactor		Low-sulphur oil-fired	
Plant capacity (MWe)	600	900	600	900
January 1974 basis Capital at 12%, 65% plant load-factor	9.0	7.7	4.6	4.1
Fuel	3.0	2.8	14.1	14.2
Operating/maintenance	0.7	0.5	0.4	0.3
TOTAL	12.7	11.0	19.1	18.6
January 1978 basis Capital at 12%, 65% plant load-factor	27.0	21.1	8.3	7.5
Fuel	7.3	7.1	20.8	20.8
Operating/maintenance	2.2	2.0	1.4	1.2
TOTAL	36.5	30.2	30.5	29.5
Reduction for capital at 10%, 70% plant load-factor	−6.1	−4.8	−1.9	−1.7
NET POWER COSTS	30.4	25.4	28.6	27.8

Like their Western counterparts, developing countries will, increasingly in the future, seek the implementation of nuclear power programmes as their economic and environmental advantages became apparent and the socio-political opposition crumbles.

15.10 SPECIALIZED TECHNOLOGIES

A remarkable fact of life, often forgotten in developed countries, is the expertise invested in nuclear energy research and development centres in the Third World. Unfortunately, for developing countries, the highly trained nuclear research and development staff at their disposal is frequently not matched by the numbers of skilled tradesmen, operation engineers, health physicists and plant safety and maintenance staff required to operate, and if necessary repair malfunctions, in newly installed nuclear generating units.

Vendors seeking to attract business from developing countries must take these very practical problems into account. If necessary, even small nuclear plant should be sold on the basis of staff secondment and training programmes to benefit the recipient country for possibly five years or longer after the completion of an initial project.

For many countries in the Middle East, the South-East Asian Region, Africa and the West Indies the prospect of nuclear power plant placed off-shore, or on-board ship, holds great appeal. In most instances these units are offered as dual cycle electricity generation and sea water distillation units. Some typical design studies which have been completed are reviewed below [9].

Floating nuclear power plants (FNP) have already been considered by the Philippines and South Korea. In the case of the Philippines the plants were only 125 MWe. Such very small plants would only be attractive economically in very unusual circumstances.

An even smaller floating nuclear power plant, the Liberty ship Sturgis, with a 10 MWe PWR aboard, was used to provide additional power for the Panama Canal. The reactor was operated by the US Army.

South Korea had discussions with Offshore Power Systems of Florida USA (OPS) concerning a 600 to 900 MWe floating platform nuclear plant in 1975. The discussions were discontinued because OPS did not have a licence from the US Nuclear Regulatory Commission (NRC) to manufacture and had not decided policy on licensing its design for manufacture overseas. Other candidates for large FNPs may be Bangladesh, Chile, Indonesia, Malaysia, Singapore and Thailand.

The 200 MWe, prefabricated, barge-mounted, Rolls Royce plant is based on ship propulsion PWR technology. It could be floated complete to anywhere in the world where it could be used floating, or beached, or moved to a land-based site via a loch. It is 120 m by 30 m, 10 m depth, 4.2 m draft and 15 000 tonnes displacement.

Rolls Royce say that larger plants up to 500 MWe would be constructed on several barges and connected up on site. Rolls Royce also offer smaller PWRs down to 28 MWe. From 45 to 67.5 MWe they are very similar to the reactor they offered for the Canadian nuclear icebreaker Rolls Royce

(1979). They also offer world-wide through-life maintenance and refuelling service.

The Offshore Power Systems have a design for an 1150 MWe barge mounted FNP. They have been trying for seven years to obtain a manufacturing licence from the US Nuclear Regulatory Commission for this design. Offshore Power Systems had an order from Public Service Electric and Gas Company of New Jersey. This company spent 189×10^6 (after allowing for tax credits) on feasibility and environmental studies. The reason for the cancellation was given as reduced load growth (US Nuclear Regulatory Commission 1979). OPS is still in existence, although staff have been much reduced.

There still seems to be considerable interest in a nuclear floating island in Japan, where the high population density (and possibly earthquake problems) are a special incentive. The Japanese design is a 1200 MWe PWR, approximately the same size as the OPS design, but it displaces 303 000 tonnes compared with 157 000 tonnes for the OPS design. The dimensions of the barge are 140 m × 140 m, compared with 122 × 116 for OPS. Also it would would be moored in such deep water (150 m) that it either has no breakwater, or a floating breakwater, which may be adequate for the Sea of Japan (tidal waves?) but not for the North Atlantic or North Pacific.

At very small sizes, nuclear plants are uneconomic for electricity generation. A nuclear plant of 100 MWe will require a staff of about 80 people while one of 1200 MWe will require only about 200 people. However, if capital is hard to raise and load growth is uncertain it may be better to build a 500 MWe plant every x years than a 1000 MWe plant every $2x$ years. In the case of a land-based plant (LBP) a coal-fired 500 MWe plant would often be a better proposition than a nuclear 500 MWe plant. However, in the case of offshore plant (even at only 500 MWe) a nuclear plant would probably be the best proposition, except perhaps in the case of a coal-slurry-pipe-line supplied plant. The 500 MWe coal-fired offshore Hawaiian proposal weighed 200 000 tonnes compared with the 1150 MWe OPS nuclear design at 157 000 tonnes and the 1200 MWe Japanese nuclear design at 303 000 tonnes.

15.11 CONCLUSIONS – ETHICAL AND SOCIETAL CONSIDERATIONS

Few intellectuals, scientists and technologists exercising informed realism, and with a wealth of experience centred on nuclear power development in the Western world, would doubt the eventual transfer of this technology to the developing ones.

Some – with a heritage vested in the Westminster system of parliamentary democracy – might well envy the speed and decision-making capacity of developing countries with totalitarian right or left wing governments in implementing major national projects of which the decision to 'go nuclear' might well be one. It may be argued that it is unlikely that a significant percentage of the population of a developing country has much understanding of the technology, safety, environmental impact or economics and reliability of existing energy systems, let alone a comprehension of the risks and benefits of nuclear energy. Having said this, one might well ask, how many well-educated laymen in our Western Society have done the necessary homework to arrive at their own conclusions? In such a complex situation is it not pertinent to claim that a little bit of pseudo-scientific secondhand information might be more dangerous than blissful ignorance?

Without entering into the depth and sheer magnitude of the 'nuclear debate' it might be fair to observe that, given the legitimate lifestyle demands of an expanding world population, the aspirations of the developing countries for energy and equity and the dangerous depletion of fossil fuels, coupled with increasing concern over their environmental impact, planet Earth will, over the next few decades, face an inevitable nuclear future.

The new technology will undoubtedly permeate the world's developing countries. Its impact on them could be so dramatic that there could well be role reversals in terms of industrialization, the availability of cheap energy and the acquisition of technological competence and efficiency.

ACKNOWLEDGEMENTS

The author is deeply grateful for useful discussion and the supply of information by his colleagues at the International Atomic Energy Agency in Vienna, the Australian Atomic Energy Commission and the University of New South Wales.

REFERENCES

[1] Eklund, S. (1977) *Experience in the Transfer of Nuclear Technology*, Transaction of the First Conference on Transfer of Nuclear Technology, Iran, 10–14 April, IAEA.
[2] INFCE Working Party WG3/21/Rev.1. *Evaluation and Definition within the Scope of INFCE of the Specific Conditions in and Needs of the Developing Countries*, IAEA, Vienna.
[3] Rosen, M. (1979) Developing Countries – The Transfer of Regulatory Capability. *IAEA Bulletin*, Vol. **21,** No. 2/3.
[4] *Australian Business,* 13 August 1981.
[5] IAEA (1977) Nuclear Power in Developing Countries; Edit Report of the International Conference on Nuclear Power and its Fuel Cycle, Salzburg, May 1977. *IAEA Bulletin*, Vol. **19,** No. 3.
[6] *Atom* **276,** October 1979.
[7] Ocsai, M. IAEA Conference Paper CN 36/240.
[8] Lane, J.A. (1978) *Latest Trends in the Economics of Nuclear Power,* Lectures given at Conference on Physics and Contemporary Needs, Pakistan, June.
[9] Rodd, J.J. (1980) *Offshore Power Plants for Developing Countries.* First International Conference on Technology for Development, Canberra, 24–28 November, pp. 154–160.

APPENDIX

Developing member states of the IAEA 1978

Afghanistan
Albania
Algeria
Argentina
Bangladesh
Bolivia
Brazil
Bulgaria
Burma
Chile
Columbia
Costa Rica
Cuba
Cyprus
Czechoslovakia
Democratic Kampuchea
Dominican Republic
Ecuador
Egypt
El Salvador
Ethiopia
Gabon
Ghana
Greece
Guatemala
Haiti
Hungary
Iceland
India
Indonesia
Iran
Iraq
Ivory Coast
Jamaica
Jordan
Kenya
Korea, Republic of
Kuwait
Lebanon
Liberia

Libyan Arab Republic
Madagascar
Malaysia
Mali
Mauritius
Mexico
Mongolia
Morocco
Niger
Nigeria
Pakistan
Panama
Paraguay
Peru
Philippines
Poland
Portugal
Qatar
Romania
Saudi Arabia
Senegal
Sierra Leone
Singapore
Socialist Republic of Viet Nam
Sri Lanka
Sudan
Syrian Arab Republic
Thailand
Tunisia
Turkey
Uganda
United Arab Emirates
United Republic of Cameroon
United Republic of Tanzania
Uruguay
Venezuela
Yugoslavia
Zaire
Zambia

Reactor units and net electrical power (MWe), in developing countries, 1978–84

Country	Operating in 1978	Planned for 1984
Argentina	1 (345)	2 (945)
Bulgaria	3 (1257)	4 (1677)
Brazil	—	3 (3116)
Czechoslovakia	2 (491)	9 (2971)
Cuba	—	2 (880)
Hungary	—	3 (1224)
India	3 (602)	8 (1689)
Iran	—	6 (6582)
Korea	1 (564)	4 (2698)
Mexico	—	2 (1308)
Philippines	—	1 (621)
Pakistan	1 (126)	1 (126)
Poland	—	1 (408)
Romania	—	1 (440)
Thailand	—	—
Turkey	—	1 (620)
Yugoslavia	—	1 (632)
TOTAL	11 (3385)	49 (25 937)

Nuclear and total electric growth in developing countries

	Installed capacities			Average growth rate*	
Year	Nuclear GW(e)	Total electric GW(e)	Nuclear (% of total)	Nuclear (% per annum)	Total electric (% per annum)
1980	6–11	270	2–4	24	7.4
1985	22–36	385	6–9	26	7.3
1990	72–121	550	13–22	15	7.0
1995	156–246	770	20–32	11	6.6
2000	281–419	1065	26–39		

*Average growth rate for peak demand: 12%.

Example of power system data

Year	Peak demand (GW)	Installed capacity (GW)	Nuclear capacity (GW)	Nuclear capacity (%)
1977	4.4	5.8	0.6	10
1985	12.0	15.8	1.8	11
1990	20.0	26.5	6.3	24
1993	26.6	37.3	15.9	43

THE MACROECONOMIC
ROLE OF NUCLEAR ENERGY

16

S.H. Schurr

ENERGY, ECONOMIC GROWTH, AND HUMAN WELFARE

A couple of propositions widely accepted until recent years are now subject to great controversy. The first asserts that there is a direct and strong connection between economic growth and the growth in human welfare; the second, in a parallel fashion, claims a strong link between energy growth and economic growth.

The controversy that surrounds these propositions is a mixture of ideology and facts and the interpretations placed on the facts. My purpose here is to try to sort things out – to see where ideological considerations end and facts begin and to determine what we can and cannot learn from the factual record. I will also deal with policy approaches that are appropriate in the light of existing knowledge on these subjects.

16.1 ECONOMIC GROWTH AND HUMAN WELFARE

Let us consider first the proposition that posits a strong connection between economic growth and human welfare. The controversy now surrounding this question appears to be largely a matter of ideology. There are those who point to relatively high levels of per capita income in advanced industrial countries and to a number of social indicators, such as health conditions, access to higher education, upward mobility of population, labour-saving appliances, and physical mobility, to support their belief that human welfare has increased as a result of economic growth. In direct opposition are those who question the significance of a higher Gross National Product (GNP) per capita and point to other indicators, such as pollution, crowding, crime, alienation, time lost in commuting, and declining durability of products, which are said to demonstrate a pervasive deterioration in the quality of life.

Unfortunately, there appears to be no objective way of comparing the contradictory sets of criteria that would be acceptable to those holding the opposing viewpoints. In the absence of an acceptable common denominator, the debate is bound to continue without resolution because it is the basis of the value judgements that are in conflict.

A more objective basis of judgement probably can be reached if one asks not about human welfare *per se* but, instead, about the conditions of economic growth that are most compatible with minimizing political and social conflict in today's world. In response to this question, I believe it is fair to begin with the observation that those who have less usually want more – a generalization that appears to hold among both people and nations. It is obviously far easier to provide more for everyone by distributing shares of an ever-growing economic pie than by reapportioning the shares of an unchanging one.

Economic growth also appears to be the solution to coping with many specific social and economic conditions urgently in need of attention. In the United States, unacceptably high levels of unemployment and price inflation are a case in point. How is reasonably full employment to be achieved if not through higher rates of growth? And how is full employment without higher inflation to be achieved except through growth that is accompanied by increased productivity? Solutions to a large number of national needs in the United States, such as housing and urban rehabilitation, would be similarly expedited by higher rates of economic growth.

16.2 WORLD-WIDE ECONOMIC GROWTH

The need for faster world-wide growth is equally compelling if conflict is to be reduced within the less-developed countries and between those countries and the industrialized world. To deal with their problems, the less-developed countries have adopted industrialization as a major goal. A declaration issued in 1975 by these countries set a goal for themselves of providing at least 25% of the world's industrial production by the year 2000. Subseqently, in May 1975 the United Nations Conference on Trade and Development (UNCTAD) declared that this goal would mean an 11% annual growth rate in manufacturing output for each of the intervening years. Serious questions can be raised about the feasibility of attaining this target. For our purposes, though, what is important are the aspirations and the need to move toward their satisfaction in the interests of international stability.

It is important to aim for a high economic growth rate in order to deal with potential sources of domestic and international conflict. The debate over whether such growth will finally serve to enhance human welfare probably cannot be resolved because of a clash in value judgements. However, questions concerning the quality of life deserve serious attention in mapping growth strategies, in order to guard against the undesirable consequences that social critics have brought to our attention forcefully in connection with past economic growth.

16.3 ECONOMIC GROWTH AND ENERGY CONSUMPTION

In respect to our second proposition, which asserts a strong connection between economic growth and growth in energy consumption, there is a detailed factual record for the United States and other countries to refer to. However, even with an extensive historical statistical base, serious questions are raised about the interpretation to be placed on the facts, particularly as they apply to the anticipated future circumstances of the United States and other industrialized countries.

First, let me offer a brief summary of what US history says about how economic growth has affected growth in energy consumption. Later I will consider the converse of the relationship – how energy supply has affected economic growth. This aspect is of critical importance, but it is frequently overlooked.

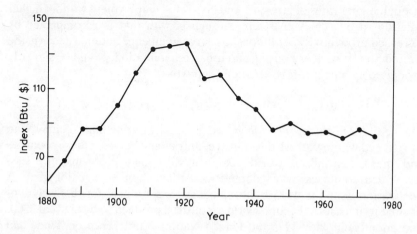

Fig. 16.1 This index of United States energy consumption per unit of Gross National Product (GNP) between 1880 and 1975 reveals three distinct periods, 1880–1920, when energy consumption rate grew faster than GNP, 1920–1945, during which growth in energy consumption rate was slower than growth in GNP and 1945–1975, when there were several short-term fluctuations with no persistent trend up or down, displaying a close relationship between energy consumption and GNP. Source data for the 1880–1950 period are from [1]. Data for the 1955–1975 period are from [2]. The Alterman data were adjusted by linking to series used for the 1880–1975 period.

The statistical record tracing the relationship between energy consumption and GNP in the United States is available from the latter part of the nineteenth century to the mid-1970s. When GNP (measured in constant dollars) and the Btu of total consumption of mineral fuels and hydropower are compared, this time span divides into three periods (see Fig. 16.1):

1. An early period – from the latter half of the nineteenth century to about the second decade of the twentieth century – in which energy consumption grew at a faster rate than GNP.
2. A middle period – from about the end of the First World War to mid-century – in which energy consumption grew at a slower rate than GNP.

3. The most recent period, in which there have been numerous short-term fluctuations, but no persistent secular trend either up or down.

The relative stability in energy/GNP ratio during the last period stands in sharp contrast to the two earlier long-term movements. This is the period usually referred to when a close relationship between GNP and energy consumption is said to be displayed in the historical record.

Two major points should be emphasized. The first is that despite the close relationship between GNP and energy consumption since the Second World War, the long historical record does not support the view that these factors have grown at essentially the same rate. Not only did they not grow at the same rate, but their comparative rates show divergence in different directions, depending on the period of US economic history being covered.

The second major point is that the changes over time in the relationship between energy and GNP have not been drastic. Even over the several decades when energy consumption declined persistently relative to GNP, the reduction in the ratio came to no more than about one-third. These were turbulent years for energy supply and use. There were fundamental changes in the composition of US energy output, including a phenomenal rise in the importance of energy in the form of electricity and liquids as opposed to the earlier heavy domination of coal. And there were sharp increases in thermal efficiency in such major areas of energy use as railroad transportation, electric power generation, and space heating. Yet the one-third decline in the ratio between energy consumption and GNP, though significant, was far less drastic than some of the forecasts now being made for the next 25–35 years.

Thus, although the recent historical record points to a strong relationship between economic growth and the demands placed upon energy inputs, earlier periods show substantial departures from the parallel movements of these two aggregate measures. The profound changes in energy supply and use technology that occurred, during the single sustained period of decline in energy consumption relative to GNP (1920–1945), place a strong burden of proof on forecasts that project a substantial decoupling of energy consumption and GNP in the future. To be credible, such forecasts should be accompanied by specifications of those changes in energy production and utilization technology that are supposed to produce such a decoupling and by plausible evidence of their technical and economic feasibility.

16.4 ASSESSING THE FUTURE

In drawing such a cautious conclusion, I do not mean to deny that there are numerous opportunities to utilize energy more efficiently without serious (if any) impairment of the services that energy yields. Examples that come readily to mind include smaller and more efficient automobiles, insulation that makes the same level of comfort available with less energy, and the utilization of waste heat generated in various processes.

Greater efficiency in energy use is likely to be the natural result of higher energy prices, of programmes being launched to provide consumers with information on energy conservation opportunities, and of incentives for the consumer to practice conservation. These effects will be enhanced by new energy-conserving technologies – in the home, in industry, and in commerce – that are certain to emerge and become cost-effective under higher energy prices. Such developments are now under way and are bound to accelerate in the future.

Here again, though, we run up against the lack of a firm fact base for predicting how large these impacts will be. It simply is not known how responsive energy demands will be in the long run to relative increases in energy prices. The available factual record was written in a period when energy prices were generally low, and so it cannot tell us what to expect in price-demand relationships during a period when relative energy prices are expected to be much higher.

Nor is it known to what extent the response of demand to price rises will be modified by demand responses to the higher incomes that also will prevail in the future. Also, just as the consumer products that emerged in the past and created new markets could not have been predicted, so we cannot know today what product developments will take place in the future. To be sure, with higher energy prices it is likely that such products individually will tend to be more energy-efficient than in the past, but what their aggregative impact on energy demands will be is impossible to say.

It is also to be expected that in the future the composition of national output will shift towards goods and services that are less energy-intensive. In general, service activities, which are playing an ever-larger role in the output of advanced economies, require less energy than the manufacture of most goods. For these reasons alone, GNP growth should require a lower rate of energy growth in the future.

16.5 CAUTION ADVISED

Yet a word of caution is called for. Services are highly heterogeneous and some of them may turn out to be quite energy-intensive. For example, consider leisure activities, which in the future will account for an increasingly larger percentage of the personal services consumers in advanced economies will demand. It is not unusual for people to travel great distances by airplane or automobile for a skiing weekend or to engage in other types of leisure activity that require substantial travel. This is obviously an energy-intensive form of service. There is also a growing trend in second homes – the future counterpart, perhaps, of the earlier phenomenon of second and third cars. The construction of such homes and the travel required to go from the city residence to the weekend house may both turn out to be comparatively energy-intensive. Thus, we should not fall into the trap of believing that the growth of non-industrial activities in the future will necessarily be associated with lower intensities of energy use.

In trying to evaluate how energy demands in the future will differ from those in the past, the absence of a dependable fact base must be recognized. A growing amount of valuable research is becoming available, designed to extract insight into future dynamics from the existing body of economic statistics, including international energy use comparisons. Statisticians and economists will be squeezing the data very hard with the aid of sophisticated methodologies and imaginative research approaches. Such research is vitally important, but expectations should be modest because of circumstances to be expected in the future that are vastly different from those of the past. It will be important, in my judgement, to supplement econometric approaches, which rely essentially on historical data, with research approaches from other social sciences that may yield improved behavioural insights applicable to the future. Greater attention should also be devoted to engineering data and comparative international practices, which serve to set forth the range of feasible technology alternatives for energy use that might become economically attractive as energy prices rise.

16.6 ENERGY SUPPLY AND ECONOMIC GROWTH

One of the most neglected areas in energy analysis concerns the effects on economic growth of changes in the conditions of energy supply. During the past century, as noted earlier, there have been several profound changes in the composition of energy supply, including the emergence of essentially new energy forms, such as electricity. Particularly noteworthy is how these changes in energy supply technology have served to remove constraints that otherwise would have severely impeded the rate and diversity of economic growth and development. Again, let me illustrate with a brief examination of US experience.

As late as 1870, about three-quarters of all the energy used in the United States was still coming from fuel wood, but the transition to coal was under way and coal soon became the dominant source. What was of primary significance in this transition was not that coal could substitute for wood in existing energy uses but rather that this was a change from a fuel resource severely limited in supply to another available in apparently endless amounts. The use of coal thus opened the way for the large-scale, unimpeded growth of iron and steel production. Ample supplies of iron and steel, in turn, made it possible to build and operate a railroad network that blanketed the country and to produce the machines required for the expansion of manufacturing. Once the fuel constraint was broken, one development led to another in a dynamic sequence that laid the foundation for modern industrial society.

In the twentieth century, as the composition of energy supply moved toward liquid fuels and electricity, other major constraints to economic growth were overcome. Electricity and the electric motor removed the limitations imposed on factory processes by the earlier mechanical energy systems, which used shafts and belting to transmit power from the in-house prime mover. Through the reorganization of factory production, which they made possible, electricity and the electric motor paved the way for large-scale productivity increases in manufacturing. Liquid fuels, the tractor, and energy-based fertilizers led to enormous increases in crop yields by removing the limits that had been imposed on agriculture by the availability of natural fertilizers and by animal draft power. And as agricultural productivity rose, farm workers became available for other sectors of production.

Geographic constraints also were removed through developments in energy supply. During the nineteenth century, railroad transportation and the mobility of coal removed the strict limits formerly imposed on industrial locations by waterways required for transportation and water wheels needed for mechanical power. In the twentieth century, the truck and the automobile broadened the availability of transportation routes, and the coming of liquid and gaseous fuels further increased the mobility of fuels, thereby removing the constraints on industrial locations previously imposed by railroads and coal. More recently, air conditioning and air transportation have removed other limitations to economic growth in many regions of the United States and the world.

The removal of all these contraints has resulted in national, regional, and local development and growth; increased productivity in industry, agriculture, and transportation; greater production of goods and services for human consumption; and marked improvements in personal living comforts and amenities. It is important to observe that in these adaptations, energy was not substituted marginally for other factors of production, such as capital and labour, which could have produced essentially the same final outcomes. Instead, energy supply and associated technologies made practical by developments in energy supply, together produced results that could not have been achieved in other ways.

16.7 SUPPLY AS CONSTRAINT

I want to draw particular attention to the relationship between developments in energy supply and associated developments in the capital equipment used in industry, transportation and agriculture. The reason for emphasizing this relationship is that most econometric energy models assume (contrary to what history teaches) that the conditions under which energy is supplied in the future will not impede future productivity growth. In most models, both capital formation and productivity are taken to be unrelated to energy development.

But if energy supply and associated capital equipment have fostered rising productivity in the past, can it be assumed that major changes in energy supply conditions – that is, considerably higher prices and constrained availability – will not seriously impair productivity growth in the future? This critical issue is usually 'assumed away' in econometric models of the relationship between energy and economic growth, which leads to conclusions that may seriously under-estimate the effects of energy supply constraints on economic growth in the future. My own reading of the evolution of energy-economic relationships leads me to conclude that supply effects are critically important and that strong policy efforts will be needed to prevent energy supply itself from becoming a constraint on economic growth. It is very worrisome to contemplate a future in which energy supply – the constraint-breaker *par excellence* in the past – becomes a constraint itself.

This is a tremendously important consideration because there is a pervasive mood of pessimism today concerning energy supply for the future. The mood reflects the conjuncture of a number of separate events – in particular, the Arab oil embargo of 1973; OPECs imposition of massive price increases on internationally traded oil; the emergence of widespread concern over the environmental impacts of energy processes; and, in the United States, a shortage in natural gas supplies. Many have jumped to the conclusion that these separate events point to a fundamental structural change in the underlying conditions of energy supply, a change with which the world will be forced to live forever. But we should be exceedingly cautious in accepting such a conclusion.

Two propositons are usually offered to support the view that a permanent structural change is in the making:

1. The world is running out of its mineral fuel resources, particularly those needed for the production of liquids and gases; and/or
2. The costs to the environment and human health and safety of continued expansion of energy supply and use that are based on mineral fuels will be too severe for society to bear.

16.8 SOUND POLICIES NEEDED

There is ample evidence that these propositions stand on shaky ground. The dilemma, however, is that the 'facts,' as they will be revealed by future developments, will be largely determined by the policies now put into effect. In other words, the factual *preconceptions* of policy actions may themselves be the most important determinants of eventual outcomes. This emphasizes the need for pursuing policies whose objective is to surmount supply and environmental constraints in an acceptable manner rather than to bow to their supposed inevitability.

To devise positive approaches to the simultaneous achievement of energy supply and environmental objectives is probably the most urgent task energy policy faces today. Unfortunately, it does not appear to be receiving the attention it deserves. We continue to be transfixed by the adversary aspects of the energy – environment conflict, while the needs of the future cry out for technical and institutional solutions that will permit forward movement on energy supply and environmental protection.

ACKNOWLEDGEMENTS

This paper is reprinted from the May 1978 *Journal of the Electric Power Research Institute,* whose permission to publish here is gratefully acknowledged.

REFERENCES

[1] *Energy in the American Economy, 1850–1875.* John Hopkins Press, Baltimore, 1960.
[2] Alterman, J. Bureau of Economic Analysis Staff Report, U.S. Dept. of Commerce.

17

L. Thiriet
NUCLEAR ENERGY AS AN INSTRUMENT OF ECONOMIC POLICY

17.1 INTRODUCTION

Energy policies generally encompass multiple objectives to which countries attribute varying importance, particularly depending on whether they have planned or liberal economic backgrounds.

One of the constant, traditional objectives of energy policies in most nations is the security of energy supplies, currently the object of considerable concern because of prevailing uncertain prices and procurement difficulties. It is nonetheless a relative objective, since a high degree of security at excessive cost would jeopardize the achievement of another fundamental objective: the maintenance of satisfactory industrial competitiveness, to prevent degradation of the balance of payments and subsequent restrictions on the potential for economic growth. As a consequence, efforts clearly should be made to obtain the most reasonable prices possible for energy, especially imported sources.

While all energy policy objectives are interrelated, some oppose the political goal on which they are founded, namely the concern for a large measure of economic independence, one of the cornerstones of national independence. Moreover, it is important to recall that the opening of borders and the development of international trade lead to increasing interdependence of national economies. This mutual relationship can be balanced by attempting to lessen domestic dependence.

There are thus two main conditions for achieving economic independence: first, preserve the competitiveness of industry indispensable for its growth, easy access to financing and even survival; second, provide the potential for technical progress necessary to maintain and, in the long term, control industrial competitiveness.

This chapter is a review of how nuclear power can help achieve energy policy objectives, illustrated with examples based on experience in France. It is preceded by a preliminary consideration of the global economic background for the development of nuclear power today.

17.2 WORLD-WIDE ECONOMIC ENVIRONMENT

The world economy is presently characterized by a breakdown of the main economic mechanisms which spurred exceptional expansion of international trade, economic growth and living standards over the past 25 years.

17.2.1 Inflation and deflation simultaneously accelerated by 1973 oil crisis

High inflation now coexists with serious deflation, so that the two are no longer antagonistic. Although the world-wide increase in inflation preceded the 1973 oil price rise, the oil crisis did have an accelerating effect, causing domestic prices to soar to varying extents in all countries throughout the world. Furthermore, it damaged the balance of payments and monetary parity of oil-importing nations, as well as the stability of their currencies.

In addition, the creation of an international currency surplus due to the massive oil price rises, without adequate consideration in goods and services, contributes substantially to inflation throughout the world. But it is also a major factor contributing to deflation, since the new purchasing power thus created is 'frozen', in that it does not help stimulate the economy, which is its usually positive effect. OPEC's current account surpluses, which were dwindling prior to the second major oil price hike, will again climb sharply in 1980, and probably in 1981, as shown in Table 17.1. (Estimates made in 1980.)

Table 17.1 OPEC surplus (billions of dollars)

1978	1979	1980 (e)	1981 (a)	1981 (b)
11.6	66.1	131.6	132.9	105.5

(e), estimated.
(a) and (b), according to accepted assumptions.

The second oil price leap in 1979 will thus again profoundly modify the global economic situation. At the world level, this additional burden will release another wave of deflation, and a new recession, as soon as OPEC nations' supplementary revenues largely exceed their capacity to use them. Such a situation suggests little hope for a significant upswing in economic activity at the domestic level due to foreign demand. The available capital unemployed in domestic activities will, moreover, increase the risk of a global financial and monetary crisis because of the large sums involved.

17.2.2 Devaluation no longer an economic stimulant

Devaluation no longer has its former anti-deflationary impact and cannot stimulate economic activity. Because of the considerably greater international dependence and the existence of a floating exchange-rate system, no nation can restore economic equilibrium without exposing itself to a protective devaluation. Since the 1971 breakdown of the international monetary system set up at Bretton Woods, the floating exchange rates continuously devalue (or revalue) currencies, thus depriving nations of the time necessary to re-adapt their economies. In addition, the cost of imports is climbing, while exports are increasing only moderately due to the relative global recession: the downward re-valuations increase the balance of payments deficit and the maintenance of international currency parity thus becomes, for many nations, a factor contributing to the restoration of economic equilibrium.

17.2.3 Inflation no longer an economic stimulant

Stagflation is now rampant. Extensive under-employment coexists with high inflation, but inflation is the cause of, rather than the solution to, unemployment. The availability of a labour force no longer contributes to economic expansion. Up to now everything otherwise being equal, growth was sustained in proportion to the available work force, and full employment was considered normal. Unemployment could only be linked with immediate, short-duration economic situations, while expansion in foreign and domestic markets (hence growth) seemed destined to develop at a fast pace. Salary rises led to price hikes and clearly contributed to inflation, but also boosted employment. Now, large wage and price increases lead to a potential loss of competitiveness; thus a reduction in economic activity and higher unemployment.

Accordingly, the traditional economic mechanisms no longer seem to work, or rather now have a reverse effect, so that devaluation is no more effective to stimulate the economy than a degree of monetary *laissez-faire*, which fans inflation. Wage hikes do not stimulate the economy either, and the availability of an unemployed active work force is no longer a positive factor in stimulating expansion.

17.2.4 Upward outlook for oil prices

From a late 1980 standpoint, looking at the possibilities for balancing oil supply and demand in the last quarter of this century, the evolution following the 1973 crisis is significant. In 1973, the United States imported 30% of its oil and this figure is currently approaching 50%. The chances that it will not increase further due to the adoption and implementation of an effective, positive energy plan are difficult to assess. OECD reports, and most experts have noted, that few measures were actually taken to develop alternative energy sources, and only a world-wide recession has made it possible to slow down growing oil requirements to date. The OECD foresees a new oil crisis with much severer repercussions than before (early 1980–1990?), if governments fail to adopt more drastic measures and remain content with non-interventionist policies. In such a case, physical scarcity and possibly shortages leading to unbridled competition among energy importers is to be feared.

If OECD member countries do not successfully implement effective energy conservation policies and, for varying reasons, fail to make nuclear energy their main source of electricity, the consequences in terms of oil imports and reduction of economic activity will be very serious. The current slowdown in nuclear power plant construction programmes throughout the world, particularly in the United States, gives little cause for reassurance since, with time, the scope for turning to other energy sources than oil will become increasingly limited and the trend toward sharply rising oil prices (perhaps also for other energy sources) more entrenched.

From September 1977 to September 1980, nuclear power development in the Western World has progressed as follows: 34 reactors representing 28.5 GWe went critical and raised the total operational capacity to 122 GWe. In 1985, the tonnes of oil equivalent (toe) for power generation in OECD industrial countries will represent about one-fourth of oil imports, and over one-third in 1990, as shown in Table 17.2

Table 17.2 Nuclear power generation and oil imports in industrial nations (Mtoe)

	1979	1985	1990
Nuclear power generation	130	295–340	515–620
Oil imports	1250	1325	1400

Source: INFCE (International Fuel Cycle Evaluation).

However, only 36 nuclear plants (34.4 GWe) – ten in France – have been ordered since 1977, while contracts for 48 plants (50 GWe) – 32 for the United States alone – were cancelled. Global projections for nuclear power generation have to be scaled down nearly every year. Accordingly, the current installed nuclear capacity forecasts for 1990 are half of the 1973 estimates, made prior to the first massive oil price rise.

Yet nuclear energy is a necessary substitute source for industrial nations, and the lack of nuclear generation capacity increases the world oil deficit in a particularly sensitive manner for developing countries. It even seems plausible that, for want of an agreement between oil producing and consuming countries, a barrel of oil could cost as much as 45 to 60 dollars (unadjusted 1980 figure) at the end of the century, equalling the cost of synthetic fuels. The apparently inevitable increases may even include peaks which exceed these figures.

Coal may also be caught up in the spiral, rising from about $10 per barrel

Nuclear energy as an instrument of economic policy

of oil equivalent (boe) at present to $15 or $20, thus maintaining a clear-cut advantage over oil and continuing to provide the reference for evaluating the competitiveness of nuclear power. Consumer prices for natural gas could catch up with those for low-sulphur fuel oil before 1990 i.e. approximately 85% of the cost of oil.

17.3 NUCLEAR ENERGY AND INFLATION

The new oil challenge calls for greater attention to an aspect of nuclear policy hitherto inadequately emphasized: nuclear policy can help counteract inflation, and thus improve overall economic activity and stimulate employment. Economic analyses of inflation usually offer the following general breakdown:

1. Cost-related inflation, when prices rise independently.
2. Demand-related inflation, when demand exceeds supply.
3. Structural inflation related to organizational deficiencies in production, transportation and domestic distribution, or in international trade.

Table 17.3 gives economic data on power generation which require a

Table 17.3 Power generation economic data (estimates to January 1980)

	Type of plant		
	Nuclear	Oil	Coal
Investments in F/kWe	3853	2558	2968
Cost of kWh in centimes			
Capital investments	6.78	4.49	5.22
Operating costs	2.74[b]	2.26	2.55
Fuel costs	5.50[c]	21.67[a]	11.52
TOTAL	15.03	28.42	19.29
Fossil fuel desulphurization	—	2.60	2.60
Total, including desulphurization	15.03	31.02	21.89
Cost of imports per kWh	2.53	22.12	12.04
Added value in France	12.18	8.52	9.45
Cost per kWh if raw material prices double	17.79	52.69	33.41

(a) Annual mean increase of 1% delivered to site.
(b) Plus 20% as of 1980, due to Three Mile Island accident.
(c) Doubling of uranium prices between now and 2000.

number of comments:

First, the utilization of nuclear energy helps reduce cost-related inflation. Considering present oil and coal prices (only hydroelectricity, unfortunately limited, is generally much cheaper), the cost of 'nuclear electricity' is significantly lower than for fossil-fuel electricity. Even a steep increase in uranium prices would have only a relatively minor impact on generating costs for light water nuclear power plants, and would be virtually negligible for fast breeder reactors.

Confronted with the previously mentioned prospects for oil price rises, and considering the limited possibilities for competitive development of other energy sources, nuclear energy emerges as one of the few available means to channel energy costs to a moderate, ultimately stable level. It will also help maintain domestic industry's competitive edge (a particularly important factor in the export arena) and more generally sustain economic activity and employment in many countries.

Second, nuclear energy does not stimulate demand-related inflation. The utilization of nuclear energy may appear inflationary, if one looks at nuclear power plant capital costs above – about 30% more than for coal-fired plant. However, this additional investment is offset after less than three years of operation, by the lower running costs of nuclear energy.

A final consideration is that nuclear energy has a superior domestic added value (see Table 17.4). Nonetheless, the essential point is that nuclear energy helps improve the balance of payments and thus counteract structural inflation due to oil imports. The balance of payments, moreover, calls for a closer examination.

17.4 NUCLEAR ENERGY AND EXTERNAL CONSTRAINTS

The report on adjustments to France's Seventh Economic Plan contains the following remarks: 'Employment and growth are interrelated; moreover, growth is linked with the balance of payments, which is in turn related to adaptation of domestic industry.' The author feels that most industrial countries are prepared to accept such a message.

More specifically, given the present conditions and structures of the country concerned, there is a certain economic growth rate not high enough to avoid worsening unemployment, yet more or less corresponding to a trade balance situation. Beyond this rate, an additional 1% would create jobs (about 80 000 in France) but would cause a balance of payments deficit (about 6 billion francs for France).

If growth is stimulated to reduce unemployment, the balance of payments deficit would rapidly become intolerable. Yet the acceptance of restricted growth to ensure a trade balance would cause unemployment to rise unacceptably. Moreover, it will be increasingly difficult to escape this dilemma, as each new oil price hike will worsen the balance of payments deficit without providing new jobs in return.

The oil challenge and its impact on the balance of payments has made nuclear energy development a key to enhancing independence from external constraints, while promoting growth and employment.

Suppose, as a working hypothesis, that the objective is to provide consumers with the same service, namely to guarantee a specific level of electricity production according to a predetermined schedule. This can be achieved using different raw materials or primary energy sources: oil, coal or uranium. A power generation programme similar to the one implemented in France will be referred to as an example.

The general considerations developed above indicate the conditions in which growth, employment and trade balance are compatible. If a balance of payments deficit is allowed to accumulate, this will create a dynamic mechanism that will rapidly aggravate inflation, degrade currency, increase the trade deficit and, despite any corrective measures implemented, eventually cause a recession (instead of the expected durable growth) and further increases in unemployment (instead of the expected drop or eradication).

In the face of such a threat, the development of nuclear energy and, more generally, energy policy can promote higher growth and reduced unemployment, while maintaining the trade balance.

Differing energy policies clearly do not have the same effect on the balance of payments situation. It is possible to opt for an energy strategy which will permit higher growth and reduced unemployment, while satisfying trade balance requirements. Stated in other terms, a suitable energy policy – particularly one which calls for intensification of nuclear power

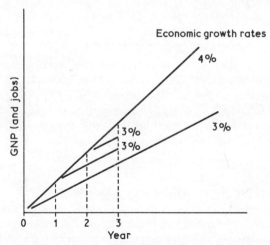

Fig. 17.1 Cumulative national wealth and employment creation process.

programmes – can provide countries experiencing balance of payments difficulties with a way to modify the structural background of growth and employment on which a trade balance is based. Such energy options can do much to reduce the risk of hazardous, unfamiliar dynamic mechanisms establishing themselves.

Figure 17.1 illustrates the process for cumulative creation of national wealth (GNP) and jobs. The 3% and 4% growth rates represented are only for illustrative purposes.

17.5 NUCLEAR ENERGY, FOREIGN CURRENCY AND EMPLOYMENT IN THE FRENCH CONTEXT

Annual foreign currency requirements for the nuclear, oil and coal strategies are plotted on Fig. 17.2 for economic conditions from 1 January 1980. It can be seen that the total foreign currency needs are higher as one goes from nuclear to coal, and from coal to oil.

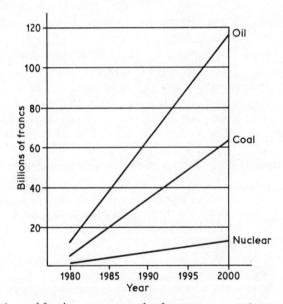

Fig. 17.2 Annual foreign currency outlay for power generation programmes.

To each possible energy programme stragegy, corresponds a given level of employment, which includes:

1. Operation of the power plants (direct employment).
2. Satellite activities (indirect employment) necessitated by plant construction and any fuel-related activities (e.g. in the nuclear branch, plant-related industry, construction and activities associated with the fuel cycle, including uranium).

No allowance is made here for the effects on unemployment of structural changes in the economy generated by the substitution of power generation strategies. Estimates are given in Table 17.4.

Table 17.4 Direct employment impact

Year	Oil or coal (1)	Nuclear (2)	(2) – (1)
1980	21 700	47 200	25 500
1985	23 850	50 400	26 550
1990	25 700	53 160	27 460
1995	27 300	55 560	28 260
2000	28 500	57 960	29 060

Yet the real factor determining the actual level of economic activity, hence employment, is the trade balance. As stated, Fig. 17.1 shows the cumulative process involved.

Table 17.5 shows the difference in foreign currency requirements for nuclear energy and coal, with reference to the programme considered. As indicated, if nuclear power had been abandoned and the same quantity of electricity produced with imported coal while maintaining a favourable balance of payments, it would have been necessary, starting in 1980, to envision less growth and additional unemployment.

Nuclear energy thus also helps stimulate economic activity through savings on imported oil or coal, provided a favourable balance of payments is maintained. In view of its consequences this growth, and consequent employment-sustaining role of nuclear energy, is essential. This explains how, restricting considerations to the Eighth Plan (1980–1984), the substitution of a coal-fired plant construction programme based on imported coal for the planned French nuclear power programme would produce 180 000 more job seekers in 1984 in comparison with present projections.

In conclusion, three points call for special attention:

1. The world is moving into an era of increasingly scarce and expensive energy sources, perhaps in the relatively near future. The second rise in oil prices showed that OPEC countries intend to control production to maintain or raise prices which inflation, or fluctuations of the dollar, will no longer undermine (as was the case between 1974 and 1979).
2. The industrial countries can no longer live indefinitely beyond their foreign currency means. From now on, revenue transfers resulting from the two major oil price hikes will be made in real terms, which implies sustained, persistent efforts to develop other energy sources to gradually mitigate their recessionist consequences.
3. Considering the global threat of greater oil supply difficulties, both with regard to quantities available and skyrocketing prices, a positive energy

policy in general and nuclear power development in particular are becoming an essential aspect of overall economic, monetary and social policy.

Table 17.5 Nuclear energy, foreign currency and employment

Year	Difference N−C in foreign currency needs (GF)*	Additional growth without trade deficit (%)	Growth induced employment	Additional direct and indirect employment
1980	3.87	0.65	52 000	about 27 350
1981	6.70	1.12	89 000	jobs for
1982	9.94	1.66	132 800	reference
1983	12.81	2.14	171 200	period
1984	13.60	2.27	181 600	

*Billions of francs.

BIBLIOGRAPHY

Le deuxième choc pétrolier (1980) Bulletin de conjoncture de la Banque de Paris et des Pays-Bas, July.
Thiriet, L. (1977) Réflexion sur le Programme Nucléaire Français et la Défense de la Monnaie, Note d'Information CEA, July–August.
Thiriet, L. (1978) Le Programme Nucléaire Français et la Croissance Economique, Note d'Information CEA, January.
Lattes, R. and Thiriet, L. (1979) Aspects économiques des politiques énergétiques. Revue Générale Nucléaire, July–August.
Lattes, R. and Thiriet, L. (1979 and 1980) Incidences macroéconomiques, monétaires et sociales de stratégies de production énergétique équivalente, Rapports CEA.

18

L.G. Brookes
LONG-TERM EQUILIBRIUM EFFECTS OF CONSTRAINTS IN ENERGY SUPPLY

18.1 INTRODUCTION

The conventional view of the energy situation since 1973 is that oil prices will continue to rise as demand and supply trends diverge, and that other fuel prices will follow suit. This, according to the conventional view is no bad thing; it will bring an overdue new respect for energy, leading to less wasteful practices and more energy-efficient processes. The high prices will also encourage the development of new energy resources which, up to now, have been considered too costly. These developments will bring supply and demand into balance at much lower levels of energy consumption per unit of economic output.

The author of this chapter takes a very different view. He argues that the main effect of high energy prices will be recurrent bouts of inflation that will do much economic damage. The only real solution is the development and exploitation of large new sources of low-cost energy – typically nuclear energy.

18.2 THE ECONOMIC ROLE OF ENERGY

The explosion in income per capita that started at the end of the 18th century had its origins, not so much in the widespread use of concentrated commercial fuels, as in the invention of the heat engine. This enabled industrial man to convert fuel into mechanical energy and emancipated him from the restrictions of human and animal power and dilute sources of mechanical energy, like wind and water mills. Ignoring price for the moment, we can thus see commercial energy as opening up entirely new dimensions by relieving purely physical constraints.

It is fundamental to man's use of commercial energy that, from the outset, he set out to use it inefficiently. The acme of thermodynamic efficiency is achieved when all processes are conducted infinitely slowly with infinitesimally small temperature differences – the slowest possible rate of progress towards maximum entropy.* This criterion of excellence does not suit industrial man at all. He adopted commercial energy and heat engines in order to do things more quickly – to travel more quickly, to produce many more goods in a shorter time, to liberate himself from the delay and uncertainties of sea transport under sail, etc. (These ideas have been expressed in papers and statements by Professor Malcolm Slesser [1] and Dr Alvin Weinberg [2].)

Moreover, man's concern has always been to optimize the use of other inputs – labour, materials and capital – in the sense of increasing output per unit cost of input. If we see output as depending upon a production function incorporating energy among these and other inputs, then it is highly probable that any improvement in the economic efficiency of energy use (in the sense of increasing output per unit cost of energy input) will be at the expense of reducing the economic efficiency of one or more of the other inputs. It is important to note here that energy is associated with two types of capital – one which is complementary to energy and another which is in substitution for it. The moral is clear: we should be wary of pressures to improve the efficiency of energy use almost as an end in itself. The result could be a reduction in economic efficiency of the other inputs to production resulting in a net loss of income per capita. We should be especially sceptical about the suggestion that improved efficiency of use can compensate for real price increases by factors of as much as 2 or 3. This would imply a very high return on the capital that is substituted for energy – and it may be incredibly high.

18.3 THE ECONOMICS OF ENERGY PRICE

Arguments about the role of energy in economic systems are incomplete without analysis of the effects of changes in the price of energy. It may be true (because energy can substitute for so many other things) that we can continue to increase output per capita as long as we are prepared to increase energy consumption per unit of output, but if the price of energy is higher than consumers are prepared to pay the process will come to a stop. It follows that the output at any point in time is the result of an equilibrium between a great many factors in the economy, one of which is the price of energy. The question that I now pose is whether the price of energy is an especially important factor.

*Entropy is a thermodynamic concept, often defined as a measure of disorder – with 'disorder' meaning 'lack of differentiation'. Entropy is increased when work is done by a heat engine because the input energy at high temperature becomes degraded into a low-temperature exhaust and low-temperature losses to the environment. The value of fuels lies in their ability to create 'islands' of low or negative entropy in a sea of high entropy. The ability to create such 'peaks' is an important quality to be taken into account in analyses of energy supply and demand. It is frequently overlooked by analysts who draw no distinction between electricity and low-temperature heat sources like the waste heat from power stations. Price is well correlated with thermodynamic quality.

Long-term equilibrium effects of constraints in energy supply 383

Conventional (rather superficial) studies suggest that it is not. Simple log linear regression analysis produces an income elasticity of energy consumption (more familiarly known as the 'energy coefficient') of about 1.0 and a short-term price elasticity* (the percentage change in demand for energy that goes with a 1% change in price) of about 0.3 or less. The trouble with this type of analysis (and indeed with most more sophisticated models of economic and energy systems) is that it takes GDP growth as autonomous and then sees energy demand as one of the things that flows from that growth, though modified by responses to changes in energy prices. Modellers tend not to consider the feedback upon the original assumption of GDP growth that might follow from a reduction in energy demand following a price increase. (An increase in price in the absence of an increase in demand constitutes a reduction in the availability of energy to the economic system, and it would be surprising if this did not adversely affect output.) At best they may simply deduct from the economy an amount equal to the value of the lost energy consumption. They make no allowance for the effect on the general level of economic activity of a smaller energy input into the economic machine.

George Kouris, when with the Economics Department at Surrey, noted that, for long-term trends, national income alone was sufficient to explain changes in the demand for energy [4]. It was only in the short term that the statistical explanation was improved by including a term for the price of energy. This phenomenon is partly explained by the familiar pattern of consumers responding immediately to a change in energy price, but fairly quickly back-sliding and resuming their old consumption patterns. The fact that the long-term trend can be wholly explained by national income changes, without the intervention of energy price, may lead some people to conclude that energy price is not an important factor in determining the level of energy consumption. It leads me to the opposite conclusion; or rather it leads me to the thought that it may be the level of energy consumption that is greatly influencing the level of output per capita (hence national income) with energy price influencing the long-term level of energy consumption. This means putting the causality the opposite way round from the familiar assumption, and it means that (at one remove) energy price may very well be a major influence upon the equilibrium level of economic activity. (Sam Schurr of Resources for the Future Inc. and the US Electric Power Research Institute has discussed this question in published papers and reached the tentative conclusion that it may be energy that drives modern economic systems rather than such systems creating a demand for energy.) It is important to think in terms of long term equilibria, because a substantial change in the price of energy in current money terms may very well lead to quite lengthy disequilibrium whilst the economy adapts to the new situation

*The Energy Modelling Forum based on Stanford University, California found long-term energy price elasticities that were close to 1.0 [3].

– with inflation in the prices of other goods and services bringing about a new real relationship (after inflation has been deducted) which substantially modifies the original price change. This is what has happened since the 1973 oil price hike: we are still in disequilibrium. There are three main reasons for this – not all of which are relevant to the main theme in this chapter. They are:

(a) The OPEC monetary surpluses

By the mid-1970s these were widely reported to be much lower than they were in the first year after the big price increases; and most observers have been surprised at the speed with which the imbalance between OPEC countries and the rest of the world was eroded. This erosion has been only partly due to OPEC countries increasing their imports from the oil-consuming countries. Much of the imbalance has been taken up by price inflation of goods imported by OPEC (which might be seen as part of the equilibrium-seeking process). The OPEC surpluses nevertheless remain important, because they are subject to economic multiplier effects that make their effect upon the world financial system more damaging than would at first sight appear. The cumulative surpluses, with the obligation to pay interest, will also hang over the oil-consuming countries for a long time, though the effect of the surpluses is a temporary one that should be solved in time. The problem has, however, re-established itself with the 1979 Iran crisis – a further major oil price rise occurring at a time when Saudi Arabia, at least, was showing some concern to keep inflation in check to avoid damage to Western economies, and was acting as a moderating influence within OPEC. 1982 has seen something of a collapse of the oil price. This was predictable according to the thesis given later in this chapter.

(b) Income effect

This is simply the effect of having purchasing power mopped up by essential spending on more expensive energy. Other things being equal, it will have a deflationary effect upon economies in the consuming countries. It is one of the ways in which the transfer of real income from consuming countries to the OPEC countries takes place. This effect too has been mitigated by inflation and by governmental measures to deal with unemployment (which also tend to be inflationary [5]).

(c) Price and substitution effects

These result in producers opting for somewhat less energy intensive methods than hitherto; for prices of goods in general to move somewhat in favour of the less energy intensive goods; and for final consumers of fuel to make marginal shifts in spending patterns between fuel and other things [6, 7].

Most commentators see the damage from the OPEC price rise in terms of (a) above; some also recognize (b); very few of them have paid much attention to the third effect. The reason is that the first effect swamps the other two and has been seen by governments as the big problem to be tackled. The second effect was seen more as an internal problem and has led to recognition of the phenomenon of inflation and deflation existing side by side – with steep price inflation mopping up spending power and having a deflationary effect upon demand and hence bringing about unemployment.

The third effect is the most difficult to detect because its influence tends to be long term. Producers cannot quickly change their technology and spending habits die hard. Nevertheless, the references given above suggest that a substantial, sustained, real increase in energy prices would have catastrophic effects upon national income in the long term. However, as the papers themselves indicate, such catastrophes are unlikely to occur because sustained, large, real price increases in energy are themselves very unlikely to occur. What happens in practice is that attempts to make substantial changes do a good deal of short term economic damage but the change is not sustained because of the very strong economic counterforces that are set up. The penalties of bad energy planning may therefore show themselves in halting and uncertain economic progress rather than economic catastrophe. The lesson to be drawn is that energy consuming countries should see to it that they do not allow situations to develop in which they are subject to arbitrary price changes – but this would be to defend themselves mainly against impermanent damage to their financial systems and bouts of inflation rather than against the permanent effect of a sustained, large, real price rise (because, as I have said, I do not believe that such price rises can be sustained – the point is developed in the next section).

It follows that the importance of studying the economics of energy prices lies in recognizing the limitations on the range of movement of energy prices in the long term, and planning in the light of that recognition. Any new energy system that looks as if its prices can never be brought down to the levels of today should, on this thesis, be abandoned – because real average energy prices are never likely to rise to the point that makes it economic. The new source should only be pursued if further development, large-scale production, and the learning process generally, seem likely to bring its costs down to where it can be offered on the market at prices not much different from what we pay today for established sources. This view of the situation needs to be qualified, of course, by some regard for energy quality and the possibility of a higher degree of specialization in energy use in future.

Series and graphs, published in the UK Digest of Energy Statistics, show that real average energy prices did not change very much between 1970 and

1976. On one measure the average price of industrial fuels fell over this period.

18.4 A FIRST ATTEMPT TO MODEL LONG TERM EFFECTS

A detailed discussion of the possible approaches and the derivation of the approach used here is given in reference [8]. The discussion and analysis in that paper will be summarized here.

The effects of changes in energy price upon productivity, and hence upon the level of economic activity and its rate of growth, were the subject of a special workshop held by the United States Electric Power Research Institute at Palo Alto, California, in January 1981. Widely differing views were expressed at that gathering, reflecting a wide range of opinions and judgements about how energy price changes influence the productivity of the economy. At one extreme, Professor Leonard Waverman argued that energy represents such a small part of total expenditure (less than 10%) that the relatively small changes in the real price of oil (and other fuels as they followed suit) could hardly account for the very large changes in the economic fortunes of the developed countries since 1973. In complete contrast, Professor Dale Jorgenson reported an econometric study that showed that the increase in energy prices more than accounted for the fall in productivity of the United States economy since 1973 (there were some offsetting factors that were overwhelmed by the effect of the energy price increases) [9].

More generally, some economists see no special role for energy in economic systems. (Christopher Allsopp of Oxford, speaking at a meeting of the British Institute of Energy Economics [10], maintained that an increase in the price of oil was no different from an increase in the price of copper) whilst others see an increase in energy price has having similar effects to an increase in the price of other commodities but, in addition, having the effect of damaging the economic productive process itself [11].

It follows that the analysis in this chapter is one among many that could have been made. It is up to readers to form their own conclusions about its soundness.

First, it is a highly simplified analysis. One of the speakers (Professor Zvi Grilliches of Harvard) at the EPRI Workshop mentioned earlier, suggested that the complexity of some models of the economic and energy systems was not justified. Models were, at best, very imperfect forecasting tools. There was a place for simpler models that helped to provide insights rather than actual detailed figures and forecasts. The model described here is a simple model of this type. It treats the world economy as a single entity and, to avoid complications of fuel substitution and differences in fuel quality, the assumption is adopted of a single homogeneous source of energy. In addition, a distinction is drawn between improvements in the efficiency of energy use that are part of general technical progress (they do not need the stimulus of raised energy prices or special governmental action) and those that take place as a direct response to an increase in energy price. The first category is subsumed in this analysis in a general concept of useful energy – in other words, an improvement in the efficiency of energy use that takes place as part of an underlying trend is treated as an addition to useful energy inputs. The second category of improvement is assumed to take place as a direct response to raised energy price and it is modelled by the use of a price elasticity of energy conservation response – with the elasticity being defined as the percentage improvement in the efficiency of energy use that takes place as a direct response to a 1% increase in the price of useful energy.

Next, it is assumed that the process of industrialization is one of energy- and capital-dependent activity penetrating a more primitive system with no energy inputs other than sunshine and human and animal muscle power. The arguments in support of such a model of industrial and economic development are given in reference [12]. They lead to the conclusion that the useful energy coefficient (percentage rate of growth of useful energy consumption divided by percentage rate of growth of economic output) tends to 1.0 from above. Strong statistical support for this hypothesis was found in work described by Brookes [12]. It leads to the model of energy consumption and economic activity shown in Fig. 18.1(a). Figure 18.1(b) is a linear approximation to this model.* It shares with the more complicated model the property that the energy coefficient tends to 1.0 from above, but is open to the objection that it effectively divides the economy into only two components – a relatively fixed non-energy-dependent component and a growing energy-dependent component with additional units of output requiring, on average, uniform increments of useful energy consumption. Despite its great simplicity, this more simplified model is a good approximation to actual experience in the real world [13].

If we add provision for the price elasticity of energy conservation response mentioned earlier this simple model may be expressed in the following form.

$$Y = AE\left(\frac{P}{P_0}\right)^c + B$$

where Y is economic output, E is useful energy consumption, c is the price elasticity of energy conservation response, and P is price, P_0 is price in year 0 and A and B are constants.

Few economists would object to the use of an inversion of this model to estimate the useful energy likely to be demanded when economic activity is at the level of Y. Most of them, however, would object to its use as a

---

*Figure 18.1(b) incorporates parameters that are a very rough approximation to real World values.

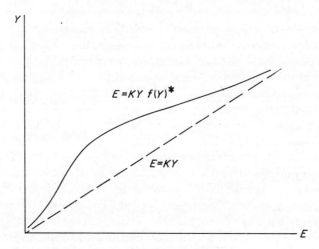

Fig. 18.1(a) A penetration model of energy consumption and economic activity [10]; $f(Y) = 0$, when Y is very small and $f(Y) \to 1$ as $Y \to \infty$.

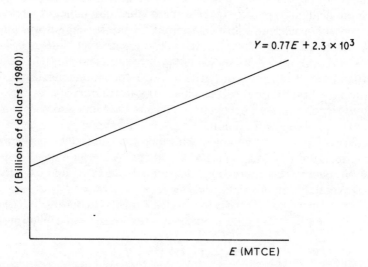

Fig. 18.1(b) A model of world energy consumption and economic activity.

'production function', with the level of economic output shown as depending upon a single variable – namely the level of energy consumption. They would point out that the conventions of macroeconomics call for a production function in which economic output is shown as a function of capital and labour inputs. Other inputs, like energy, can be included (subject to suitable safeguards). They would argue, however, that at the end of the day, aggregate incomes paid to labour plus aggregate incomes paid to capital must equate to the value of aggregate output, and that a production function that excludes labour and capital is basically unsound.

Those who argue, in addition, that there is nothing special about energy and that it is simply one of many commodities that are consumed in economic systems would have an additional reason for rejecting the model.

The author would argue in reply that his *a priori* hypothesis about industrial development given in [12] constituted a reasonable argument and that his own statistical validation and the satistical work of others – of which [13] was a good example – were impressive. He would further argue that capital and labour inputs are implicit in the model: the initial slope of the line reflects capital and labour inputs at year t_0; the changed position of the line (shown dotted in Fig. 18.2) reflects the capital and labour substitutions for energy that take place in response to raised energy price and whose magnitude is modelled by the value of the price elasticity of conservation response. In other words, the model is treated as bidirectional: an increase (or decrease) in the availability of energy will lead to an increase (or decrease) in the level of economic activity, just as an increase (or decrease) in the level of economic activity will lead to an increase (or decrease) in the demand for energy. The extent of the increase or decrease may be modified by the responses to changes in the price of energy.

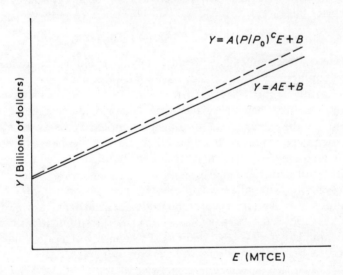

Fig. 18.2 A model of energy consumption and economic activity with provision for capital and labour substitution for energy.

The demand for energy is affected by its price as well as by the level of economic activity. We therefore need an additional model of the way in which energy consumption responds to price even when the level of economic activity remains constant. The model usually adopted is the following:

$$\frac{E}{Y} = K \left(\frac{P}{P_0}\right)^{-a}$$

where a is the price elasticity of demand for energy subject to income remaining constant and K is a constant.

This model should *not* be seen as modelling the improvement in the efficiency of energy use that takes place when energy price rises. Y is in the denominator simply to meet the stipulation that a is an indicator of response to price change only, with income remaining constant. The ratio E/Y can fall even when there is no increase in the efficiency of energy use – it might fall, for example, simply because the economy has shed some energy-dependent activity or gained some non-energy-dependent activity.

Finally, we need a model of energy supply. The one usually adopted is:

$$E = C \left(\frac{P}{P_0}\right)^{b}$$

where b is the price elasticity of energy supply and C is a constant.

These three equations, taken together, constitute a simple world model of the way in which world economic activity, world energy supply and the energy price interact.

18.5 WHAT IS A PRICE HIKE?

The large oil price hikes have been widely regarded as simply large step increases in the price of oil. (They have, of course, led to sharp increases – though not by so much – in the prices of other fuels.) Most studies of the effect of price hikes have taken the increase in price as exogenous (imposed from outside the system) and have then attempted to assess the effects of such exogenous increases. It follows that these studies do not distinguish between an increase in price that arises because of a change in demand and one that results from a change in supply.

In this chapter we shall take a different concept of a price hike – one which the author believes is much nearer the truth. We shall define a price hike as a shift in the supply curve as a whole (see Fig. 18.3) resulting in a higher price being demanded at all levels of output. This concept of a price hike reflects the author's belief that energy producers are not able to decree that 'from tomorrow the price of energy shall be twice what it is today'. They can only say 'if you want as much as you have been getting you will have to pay more for it. If you are not prepared to pay more you will have to make do with less'. It follows that the result of a price hike is for energy consumption to fall and for the real energy price to settle down eventually at a level somewhere between the original price and the new nominal price announced by energy suppliers.

If this concept of a price hike is allied to the model of energy consumption and economic activity derived in the previous section we have the result (observed in practice) that an increase in energy price results in a fall in the level of economic activity (unless there is a most improbable conservation response to raised price), a fall in the level of energy consumption and a real price for energy that is somewhere between the initial price and the one that the energy producers attempted to impose.

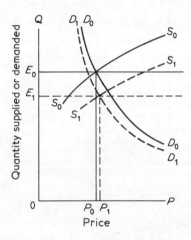

Fig. 18.3 Energy supply and demand functions: D_0/D_0 = demand at income Y_0, D_1/D_1 = demand at income Y_1, S_0/S_0 = supply before price hike and S_1/S_1 = supply after price hike.

This result is, of course, completely at variance with the projections of those forecasters who combine assumptions about economic growth with assumptions about future energy supply to produce forecasts that energy prices will rise in real terms by factors of between 2 and 3 in the next twenty years. These forecasts associate high levels of economic activity with high energy price. The explanation is that, in this case, the forecasters are making no allowance for the effect of restricted energy supply (as evidenced by high price) upon the level of economic activity and are assuming that the increase in energy price is due to increased demand, not an exogenous shift in the supply curve.

Events since 1973 support the hypothesis that the problem is one of a reduction in the availability of energy to the world economic system – in the form of a shift in the energy supply curve in a direction unfavourable to consumers. These events also support the hypothesis that a reduction in the availability of energy greatly damages the level of economic activity. (Begg, Cripps and Ward [14] support this view, but offer a solution – the maintenance of a high oil price – that the author would reject.)

18.6 MODELLING ENERGY PRICE HIKES

The demand model mentioned earlier can be shown diagrammatically as a series of curves logarithmically in parallel to each other (that is to say they would be parallel if plotted on logarithmic graph paper), with each curve representing a different level of economic activity. We might now consider how an energy price hike might work its way through the economic system:

1. Energy producers impose a 100% price hike – doubling the unit price at each level of supply. This change is shown in the shift of the supply curve in Fig. 18.4.

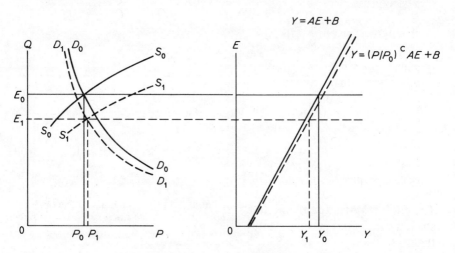

Fig. 18.4 An interactive model of energy consumption, economic activity and the conditions of energy supply.

2. This restriction upon the availability of energy – manifesting itself in the supply and demand curves intersecting at a lower level of output and consumption – leads to a reduction in the level of economic activity. This can be seen in the shift to a different point on the energy consumption/economic activity submodel.
3. The existence of a non-zero price elasticity of conservation response produces a change in the slope of the energy consumption/economic activity submodel that recovers some of the lost output.
4. The net reduction in economic activity (taking the highly probable case of the conservation response being insufficient to recover all of the initial loss) produces a shift to a lower member of the family of demand curves.

* This equilibrium can be one in which labour and capital are not fully employed if raised energy price makes some energy-using plant uneconomic and the clearing price of labour falls below a threshold at which the lowest paid workers offer themselves to the market.

5. Eventually, the system settles down to a new equilibrium* with lower levels of economic activity and energy consumption, the price settles at a level which is, perhaps, only 10% or 15% above its original level, and some improvement in the efficiency of energy use results from this somewhat higher price for energy.

The results depend very much upon the assumptions made about the initial position and slope of the energy consumption/economic activity submodel and the values chosen for the elasticities. The following values were taken representing a very rough approximation to the real world:

Y is expressed in $ billion (1980)
E is expressed in MTCE
$A = 0.77$
$B = 2.3 \times 10^3$
a lies between 0.6 and 1.0
b lies between 0.2 and 0.4
c lies between 0.2 and the value chosen for a. (Choosing a value for c greater than a produces nonsensical results: it implies that consumers over-compensate by improved energy efficiency for the reductions they make in energy purchases; the resulting increase in economic activity pushes up the energy price, producing yet another over-compensatory response and so on, with economic activity spiralling to an infinite level.)

The high long-run demand elasticities (0.6 to 1.0) are fully supported by studies conducted by the Energy Modelling Forum [3]. The fairly low (less than 0.5) values chosen for the supply elasticity reflect the fact that energy producers have limited scope for changing the level of output – at one end because they are limited by currently exploited resources and at the other end because they depend upon the revenue from their production. At the EPRI Workshop mentioned earlier, Professor Alan Manne suggested that oil producers might have backward sloping supply curves – implying that their concern for a given level of revenue was such that they would produce less when the price rose, and more when it fell. One might indeed expect such a pattern of behaviour from countries with small populations (like some of those in the Middle East) and dependent upon a depleting resource. In practice, however, the oil-producing countries have sold oil at prices that reflect movements in prices on the Rotterdam Spot Market – offering discounts on contract prices when spot prices fall and charging premia when spot prices rise. This implies low prices at low levels of output and high prices and high levels – a normal supply curve in other words.

It is very hard to identify a true price elasticity of conservation response. For a very long time – at least 30 years – the economy-wide efficiency of energy usage in the UK has been improving at the rate of about 1% per annum. This pattern continued during periods of falling energy prices. The reasons for it are complex: the onward march of technical progress and the substitution of inherently more efficient fuels for older less efficient ones undoubtedly each played a part. There has been some slowdown of technical progress since 1973 because investment in new plant has slowed down. The substitution of liquid for solid fuels (which no doubt accounted for much of the improvement in energy productivity in the last few decades) has gone into reverse. The substitution of natural gas for other home heating fuels has coincided with greater use of central heating – which usually results in higher energy use per capita.

Electricity usage has stayed level in the UK but has increased its share of total final energy consumption in some other industrial countries (e.g. the USA and France). Some apparent improvement in the economy-wide efficiency of energy usage takes place as energy intensive activity is shed in a recession that is associated with raised energy prices. With all these confusing factors it would be a bold man who claimed to have identified a response in the shape of improvements in the efficiency of energy usage that was due solely to the raised price. A success story reported by the brewing industry in the UK in the autumn of 1980 could be translated into a price elasticity of conservation response of 0.6. Reports from the same industry in 1982 record disappointment that the early success has not been maintained.

Government officials responsible for encouraging energy conservation tend, for the most part, to report disappointing responses to the various inducements offered.

For all these reasons a wide range for the price elasticity of energy conservation response was chosen. Remembering that this parameter applies strictly to the response to raised price (excluding the long-term trend of technical progress and the effect of reductions in the energy-intensive component of the economy) it seems likely that, in practice, the price elasticity of conservation response is less than 0.5 – perhaps significantly below this level.

Experiments with this model produce challenging results, (shown graphically in the Appendix) although it must be remembered that it has a purely hypothetical basis. Starting with an exogenous doubling of the energy price (shifting the demand curve so as to double the price at all levels of output) results (when equilibrium was regained) were typically as follows:

1. A fall in economic activity of between 6 and 16%.
2. A fall in energy consumption of between 11 and 22%.
3. A rise in the real equilibrium energy price of between 4 and 17%.

These results were obtained with a price elasticity of conservation response of less than 0.5. If this elasticity is assumed to be higher than this the loss of economic activity is mitigated (it is wholly recovered when the elasticity of conservation response equals the demand elasticity, but the economy must then find room for energy conservation investment that it did not previously need – an example of an increase in what welfare economists call 'regrettables').

Paradoxically, a higher energy conservation response *increases* energy consumption and produces a higher level for the real equilibrium energy price. This is not hard to understand if one recognizes that responding to high price by using energy more efficiently is a way of accommodating the raised price. For any given position of the supply curve this means striking a balance between supply and demand at a higher level of consumption and production. In other words using energy more efficiently in conditions of restricted availability of energy does not save energy or force its price down – it simply mitigates the loss of employment resulting from the imposed restriction upon energy supplies that a price hike constitutes.

By manipulation of terms it is easy to show that the proportion of national product expended upon energy is proportional to $\left(\dfrac{P}{P_0}\right)^{1-a}$. If the long term

Table 18.1 UK energy expenditure as a percentage of GDP, with and without taxation*

Year	Total energy expenditure without taxation (£ million)	Total energy expenditure with taxation (£ million)	GDP at factor cost (£ million)	Energy from GDP without taxation (%)	Energy from GDP with taxation (%)
1955	1429.2	1742.3	16894	8.46	10.31
1960	1893.3	2302.1	22615	8.37	10.18
1961	2064.7	2575.0	24198	8.53	10.64
1962	2126.2	2670.6	25252	8.42	10.58
1963	2241.9	2833.6	26863	8.35	10.55
1964	2252.6	2926.8	29182	7.72	10.03
1965	2476.6	3261.1	31212	7.93	10.45
1966	2648.6	3536.1	33083	8.01	10.69
1967	2793.3	3762.7	34877	8.01	10.79
1968	3162.2	4289.1	37390	8.46	11.47
1969	3359.9	4669.4	39338	8.54	11.87
1970	3459.2	4855.0	43368	7.98	11.19
1971	4210.4	5653.5	49151	8.57	11.50
1972	4548.8	6104.3	54958	8.33	11.11
1973	4722.0	6297.0	63492	7.44	9.92
1974	7502.2	9307.2	73652	10.19	12.64
1975	9226.9	11141.9	93078	9.91	11.97
1976	11239.5	13424.5	109080	10.30	12.31

*UK Department of Energy statistics for these parameters date only from 1967. A method was developed for figures going back to 1955 which, for consistency, was carried through to 1976. Discrepancies between figures for the overlapping period were not great. Means and standard deviations for 1955 to 1972 were respectively 10.8% and 0.54 with tax, and 8.26% and 0.27 without tax.

value of a is 0.6 or above, and the equilibrium real energy price increase following an attempt at a large price hike is less than 20%, this formula has the effect of keeping the proportion of national income spent upon energy within a very narrow band.

It would provide some support for the thesis advanced here if this proportion were in fact to stay within a fairly narrow band. Table 18.1 shows that it has in fact done so for a good many years in the UK. Oddly enough, the proportion fell in 1973 despite sharp increases in energy prices, but this was the year of the miners' strike and the three day week, which was introduced to meet restrictions on electricity supply caused by coal shortage. In the immediately following years there was an increase in the proportion by one or two percentage points, but these would be non-equilibrium values. There are signs now of a resumption of the old proportion. Dr Joy Dunkerly [15] reported similar stability in the USA.

18.7 IMPLICATIONS AND LESSONS FOR NUCLEAR ENERGY

Remembering once again that this is an exercise designed to producing insights rather than predictions, the implications are:

1. An energy conservation response to price rise can have only a mitigating effect at best.
2. The idea that raised oil price, in itself, brings a host of previously over-costly energy sources into the market is probably false. In practice, a price hike that takes the form of a shift in the supply curve works itself out in the form of inflation (blunting the energy price rise), a reduction in the levels of energy consumption and economic activity, and only a relatively small increase in the real equilibrium price of energy.

The further implications are:

1. A reduction in the rate of replacement of plant of all types (including electrical plant) because of the depressed state of national economies.
2. Very low or negative energy and electricity growth rates.

Paradoxically, the effect of an increase in the price of conventional fuels is to slow down the rate of introduction of new forms of energy like nuclear energy [16]. Table 18.2 shows what actually happened to nuclear power plans in the developed countries after 1973.

Thus, nuclear energy presents a dilemma as a source of relief to the restrictions upon energy availability that the world has suffered since 1973:

1. The only real solution to the problem is the exploitation of new additional energy sources at costs and prices that will have the effect of forcing the world all-energy supply curve back towards its original position. Of all the

Table 18.2 Forecast nuclear capacity in 1985 (GWe)

Country	1973 forecast	1975 forecast	1977 forecast
Belgium	5.5	9.5	4.9–5.6
France	32.5	56.0	31–35
Germany	38.0	44.6	32.38
Italy	18.0	26.4	6.4–7.4
UK	35.0	15.4	11–13
Spain	12.0	23.7	14–20
Switzerland	8.0	8.0	3.3–3.8
Sweden	16.0	11.3	7.8–9.0
Japan	60.0	49.0	35.1
Finland	4.6	3.9	?
USA	280.0	205.0	152

feasible new sources of energy, nuclear energy is outstandingly suitable for this role – it is relatively cheap and abundant, and offers the opportunity to continue the march of technical progress towards the more modern forms of energy that Mr Sam Schurr draws attention to [17].

2. But quite apart from the political problems caused by a vociferous antinuclear movement, combined with institutional arrangements that favour dissent in important countries, the economic conditions that favour rapid introduction of nuclear energy are lacking because of the very factor (restricted availability of world energy supply) that makes its introduction important.

The saving grace is that thermal nuclear power may be considered a mature technology and the fast reactor is at an advanced stage of development. All that is lacking in most countries is a readiness to take the remedial action that is called for – namely investment in the developed countries in the production of the commodity (fuel) that is giving cause for concern. In reference [18] Walt Rostow reported that on each of the previous four occasions in the last 200 years when the more developed countries found themselves faced with sharp rises in the prices of primary products they responded by investing in home production of those commodities. He noted that there were few signs of a similar response on this occasion. The lessons are clear in the experience of the one country, France, that has responded. As Table 18.3 shows France – with the largest nuclear power component in its electricity system and plans for further substantial increases in that component – has the highest rate of substitution of electricity for other forms of energy in Europe and the lowest electricity prices.

Table 18.3 (a) France: electricity production 1980, 1981 (TWh)

Year	Nuclear	Coal and oil fired	Hydro
1980	57.9	118.9	69.8
1981	99.5	92.1	72.4

(b) European Community: production in the first nine months – all plant

Country	1980	1981	% Change
West Germany	252.5	251.2	−0.5
France	176.6	189.7	+7.4
Italy	130.9	127.4	−2.7
Netherlands	45.4	44.6	−1.8
Belgium	37.4	34.8	−7.1
Luxembourg	0.8	0.89	+12.0
UK	194.4	187.3	−3.7
Ireland	7.6	7.5	−0.4
Denmark	17.6	11.3	−35.9

(c) Electricity prices to large industrial consumers (Source – NUS Survey of International Electricity Tariffs.)

Country	Cents/kWh
France	4.72
Netherlands	4.92
Italy	6.05
West Germany	6.16
Ireland	6.41
UK	7.09
Belgium	7.38

18.8 THE PRESENT REALITY

In early 1982 we have falling oil prices and some euphoria in consuming countries at what seems to be cracks in the OPEC cartel. According to the thesis in this paper, the softening of the energy market is only to be expected – it is part of the approach to the new equilibrium after the last price hike. No satisfaction can be drawn from the state of the world economy: growth rates are severely depressed and all countries are grappling with the problem of inflation – another manifestation of the approach to the new equilibrium, as new price relativities struggle to take shape.

There is no cause for complacency in the present state of the oil market. The price at time of writing (about $30 per barrel, with prices somewhat lower on the spot market) has to be associated with a level of OPEC output of less than 18 million barrels per day (mbd), against nearly 30 million barrels only two years ago. An output of 18 mbd linked to a price of $30 is a point on a distinctly less favourable supply curve than existed before the last price hike, and very much less favourable than the one that ruled before 1973. Possession of North Sea oil does not allow the UK to escape the sombre reality of the fall in the availability of energy to the World economic system. Our dependence on international trade, and hence our vulnerability to setbacks in the level of world economic activity, is too great for us to be able to draw more than modest comfort from the possession of that ephemeral piece of wealth.

The imperatives remain unchanged. No country can afford to neglect the exploitation of new modest cost energy sources. Unfortunately this important example of supply side economics is being subjected to demand side thinking in most countries – with France and the USSR as the outstanding exceptions. We are in great danger of chasing out tails downwards, as the rate of exploitation of new energy sources is tempered by regard for energy demand projections, that in turn follow from economic forecasts that are lower than they need be because of tardiness in exploiting new energy sources!

ACKNOWLEDGEMENTS

This paper was first presented at the annual conference of the International Association of Energy Economists at Cambridge UK, June 1982.

[1] Slesser, M. (1978) *Energy in the Economy*, Macmillan, London and New York.
[2] Weinburg, A. Energy Analysis Institute, Oak Ridge, Tennessee.
[3] E.M.F. (1980) *Aggregate Elasticities of Energy Demand*, Vol. 1, Energy Modelling Forum, Stanford University.
[4] Kouris, G. (1978) speaking at a seminar held by the Economics Department of the University of Surrey, January.
[5] Brookes, L.G. (1975) The Nuclear Power Implicatons of OPEC Prices, *Energy Policy*, June.
[6] Brookes, L.G. (1978) The Energy Price Fallacy and the Role of Nuclear Power in the U.K., *Energy Policy*, June.
[7] Brookes, L.G. (1979) Energy, Inflation and Economic Prospects, a paper to a conference Design, '79 held at the University of Aston-in-Birmingham, September.

[8] Brookes, L.G. (1981) *Energy Conservation Response to Price Increase – Is it Sufficient to Resolve a Problem of Energy Shortage?* published by A.P.G., 8 Ruvigny Mansions, Embankment, Putney, London, S.W.15.
[9] Jorgenson, D.W. (1981) Energy Prices and Productivity Growth, presented at the 2nd annual conference of the International Association of Energy Economists, Cambridge, U.K., June 1980 and distributed at the workshop on Energy, Productivity and Economic Growth, EPRI, Palo Alto, Ca, January 1981.
[10] Allsopp, C. (1981) speaking at a seminar held by the British Institute of Energy Economics in London, October.
[11] Pindyck, R.S. (1979) *The Structure of World Energy Demand*, MIT Press.
[12] Brookes, L.G. (1972) More on the Output of Energy Consumption, *Journal of Industrial Economics*, November.
[13] Smil, V. and Kuz, T. (1976) Energy and the Economy – a Global Analysis, *Long Range Planning*, Vol. 9, No. 3, June.
[14] Begg, I., Cripps, F. and Ward, T. (1982) Why Oil Prices Must Stay High, *Financial Times*, 6 January.
[15] Dunkerkey, J. (1979) speaking at the 1st annual conference of the International Association of Energy Economists, Washington, D.C., June.
[16] Brookes, L.G. (1974) *The Complementary Roles of Nuclear and Other Fuels*, a paper to an expert meeting on Alternative Strategies to Meet the Oil Crisis, held at the International Institute for the Management of Technology.
[17] Schurr, S.H. (1978) Energy, Economic Growth and Human Welfare, in the *EPRI Journal*, May. (Also in *Ethics and Energy*, the Edison Electric Institute 1979.)
[18] Rostow, W.W. (1978) *Getting from Here to There*, Macmillan.

APPENDIX

Illustration of the effects of a 100% energy price hike on the world economy

At zero conservation elasticity, the loss of output is dominated by the supply elasticity and is relatively insensitive to the demand elasticity. This is because the larger the supply elasticity the more nearly the supply curve approaches the vertical (on the axes used in the model) and the more nearly the equilibrium price increase approaches the imposed price increase. In an energy-driven world (such as is assumed here) the fall in output is due to the fall in energy consumption which in turn is due to the less favourable conditions of supply. The higher the demand elasticity the larger the fall in energy consumption. But the new equilibrium is at least as much affected by the shift in the demand curve (in response to a reduced level of economic activity) and, as explained in the text, this shift is insensitive to the value of the demand elasticity over the range given here.

When a constant conservation elasticity of 0.4 is adopted the picture is dramatically changed. This is because the definition of this elasticity is such that the potential loss of output is wholly recovered when it is equal to the demand elasticity. The curves therefore converge to zero at this point.

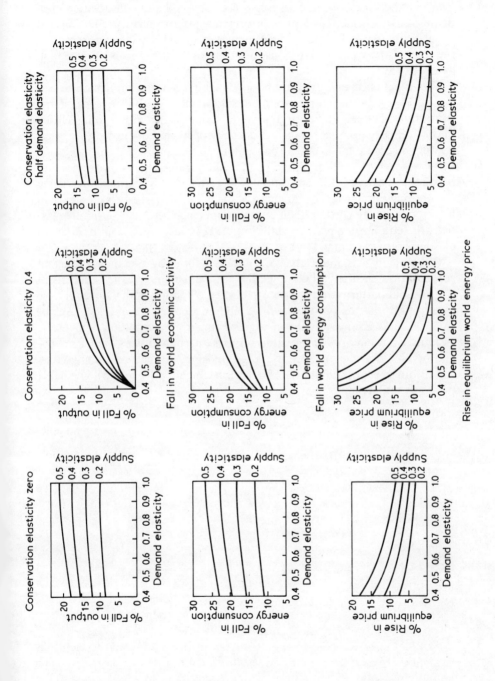

When the more realistic assumption of demand and conservation elasticities moving in step is adopted the original pattern is largely re-established – but with the falls in economic activity mitigated by the conservation response. But even with a response of as high as 0.5 the relief is no more than by about 25%.

The effect upon energy consumption is similar but the percentages are all larger. This is because the price hike acts directly upon the level of energy consumption. Percentagewise the fall in economic activity is mitigated by the existence of non-energy dependent activity in the economy – which is left untouched by the fall in energy consumption.

At zero conservation elasticity the equilibrium energy price rise lies between just over 3% and about 18%, the supply elasticity having a dominating influence. The influence of the demand elasticity is greater than it is upon energy consumption and economic activity. At 5% supply elasticity the equilibrium price rise falls from 18% to just over 8% as the demand elasticity moves from 0.4 to 1.0.

Once again a dramatic change is brought about by assuming a constant conservation elasticity of 0.4 regardless of the demand elasticity and again the original pattern is largely re-established by adopting the more realistic assumption that the two elasticities march in step. The changes in equilibrium price are, however, larger than when zero conservation response is assumed. An energy conservation response accommodates higher price – and results in a higher level of energy consumption because in any given conditions of supply equilibrium is struck at a higher price hence higher level of supply and demand.

CONCLUSIONS

L.G. Brookes and H. Motamen
CONCLUSIONS

When we planned this book we hoped that we would be able to communicate an objective understanding of the economics of nuclear energy as a fascinating and complex subject that has some unexpected twists and paradoxes. All too often, works on the economics of nuclear energy turn out to be over-simple comparisons between the costs of nuclear and fossil-fired power plants, with one or the other declared the 'winner' in a contest very largely determined by the assumptions fed into calculations, and with little or no account taken of the way in which nuclear and other plants play complementary roles in real world power systems.

In this volume we have shown where nuclear energy fits into the world energy system. We have gone one step further by highlighting the important policy considerations relevant for energy planning in a growth society. Important issues such as fuel substitution and price factors within the electricity industry are examined and the role of nuclear energy in this complex structure is discussed. Moreover, the economics of all the various aspects of nuclear power, such as uranium industry, nuclear fuel cycle services, the distinction between thermal and breeder reactors, the place of nuclear energy in less-developed countries, etc., are explained in detail.

In truth there is no simple answer to the question 'which type of plant (fossil or nuclear) produces the cheaper electricity?'. The answer depends on the circumstances of the region in which the plant is to be sited, including the level of development, degree of urbanization, natural resource endowments and even the nature of local institutions, including accounting practices and methods of fund raising for utility investment. While externalities on the social cost-benefit front are considered when deciding on an optimal energy system, the difficulties in quantifying and increasing them must be taken into account. On another level, decision making on the basis of perhaps more narrow, but measurable, economic calculations is explained.

Since the ideal degree of centralization and interconnection varies from region to region, a mismatch between power systems can have adverse effects on the costs and availability of electricity in different locations. The competitive factors among final-use forms of energy are significant for the way in which nuclear energy, transmitted through electricity systems, may meet such terms. Thus the characteristics of electricity systems employing nuclear energy, and the rates to which they lead, can be determined.

Some of our contributors have shown how sophisticated the methods of capital investment appraisal need to be to keep the evolution of electricity systems on an optimal path. Critics sometimes cry 'unfair' when they see nuclear and fossil-fuelled stations compared with a higher lifetime load factor assigned to the nuclear station than to the fossil station. However, it should be pointed out that this turns out to be one of the paradoxes of the subject. The fact that the nuclear stations have a higher average lifetime load factor than the fossil-fuelled stations works to their disadvantage in the calculations of investment appraisal. The reason is that a full appraisal has to take account of the effect of the new station on the economics of operating the stations already in the system. The nuclear station, entering the system at the top of the 'merit order', disturbs more existing plant than the coal-fired station entering somewhere down the 'merit order'. In short, there is a major distinction between short-term and long-term planning in the electricity supply sector which is relevant for the problems of optimizing the balance between nuclear and non-nuclear plants in electricity systems. Detailed planning is necessary to deal with uncertainties in demand and supply.

The subject bristles with those trade-offs that make economics such a fascinating subject. Large unit size for plant ultimately offers economies of scale but almost certainly at the expense of teething troubles with the first few of the larger plants and some requirement for more reserve plants because the breakdown of a single plant is more damaging if it is of larger size.

The range of activities of this industry are defined. It is brought to light how, for instance, the necessary components of nuclear power stations are produced by the heavy engineering industries. This remains true both for those countries that have adopted national architect/engineer companies and those that have left the design and construction of nuclear plants entirely to the private sector. A comparison is then drawn among different European and North American countries in their characteristics for maintaining a single national corporation responsible for nuclear power plant production or opting to leave it entirely in the hands of the private power plant companies. Although perceptions of the world's uranium supply capabilities are constantly changing, there are different methodologies that are used to assess it. Given the current 'state of the art' of uranium supply analysis one can arrive at a broad evaluation of the current global supply situation.

On the demand side, for uranium, firstly it is expected that nuclear power is likely to become an important component of all countries' electricity systems, particularly those capable of accepting single stations of around 1000 MW. Thus the rate of increase in demand will broadly depend upon the rate of world economic growth. In the analysis and projection of demand, a distinction must be drawn between thermal and breeder reactors, where the development of each type would depend upon the satisfactory security of international supply. The nuclear fuel cycle services are divided into front-end and back-end activities. The former is related to mining and producing 'yellowcake' and its conversion to uranium hexafluoride gas; enrichment of this gas, production of uranium dioxide and fuel fabrication. The latter is associated with spent fuel storage, reprocessing and waste solidification and disposal.

In the nuclear fuel cycle the trade-offs are even more esoteric. The sizes of plutonium inventories for fast reactors are kept down if the out-of-reactor time for fuel is kept to a minimum: that is to say, if fuel taken out of a reactor is reprocessed and returned as soon as possible. But the shorter the out-of-reactor time, the more highly active the material that has to be handled at the reprocessing plant, and the more sophisticated this plant has to be. In the last analysis the trade-off is between greater sophistication of plant and larger interest payments on nuclear materials inventories and perhaps, at the end of the line, some increase in demand for natural uranium. The right balance depends upon all the costs and prices involved. There is very rarely one single input whose use has to be optimized.

Whereas in thermal reactors it is not possible to reutilize the spent irradiated fuel to recover and recycle the uranium and plutonium, fast reactors are capable of reprocessing the spent fuel. Fast neutron reactors can be operated to produce or 'breed' more plutonium than they consume – hence the name fast breeder reactors. Since fast reactors require only a small supply of depleted uranium during operation, their use in place of thermal reactors reduces overall natural uranium needs. Thus, reliance on fast reactors over the world as a whole could possibly result in uranium requirements being supplied from low-cost high grade sources. This would in turn save the costs and efforts in searching for various renewable energy or the more expensive type of fossil fuels such as shale oil. Moreover, fast reactors will help to stabilize uranium prices by lowering demand, and thereby will assist in keeping nuclear electricity generating costs steady.

The question arises then, if uranium looks like becoming scarce does it automatically follow that nuclear system planners should opt for fast reactors (which can burn up to 60% of the heavy atoms in natural uranium) or CANDU reactors (which can burn up to 10%)? There is no simple answer. The most economically attractive system is the one that produces the cheapest electricity in the cost and price scenario in which it is likely to live out its life. Higher nuclear fuel costs may be acceptable if they are more than offset by lower capital and reprocessing costs. Thus, it should be emphasized that in an optimal system the combination of thermal and fast reactors would be desirable, for it is possible that some of the factors which help to reduce single station fast reactor generating costs, could increase total system expenditure.

Some interesting case studies are included in this book. The economics of electricity generation is analysed with special reference to the fast breeder reactors planned in France during the 1980s. These reactors are compared with the French pressurized water reactors, and the energy extracted from uranium by the two types of reactors are calculated. The cost benefit results of these analyses show that for the major industrial nations of the world the relatively small additional costs of producing electricity in the early years of fast breeder reactors could more than compensate the savings made when the first major uranium price hikes occur.

The economic comparison of nuclear and coal-fired electricity generation can best be demonstrated by examining the relative performance of power plants in the USA and in West Germany. The experience of the USA shows that nuclear power and coal are generally competitive. Indeed, in some parts of the country nuclear power should be the clear preference. In others (like Montana, South Dakota or Wyoming) coal might well be the choice on economic grounds. The uncertainties surrounding the regulatory conditions in the United States do not make nuclear power a viable option for new generating capacity ordered at present. Unforseen licensing delays can be extremely costly. Consequently, the atmosphere created by the absence of confidence is likely to hinder the development of nuclear units in that country in the future.

West Germany presents a slightly different case. Public utilities have a contract with the coal mining industry to increase the consumption of indigenous coal until 1995 by more than 50%. Import barriers are agreed by the government to be removed gradually to allow utilities to have access to additional coal imports. As a result, the electricity industry is in effect obliged to fulfil its commitments *vis à vis* coal-fired stations, even though nuclear power would be cheaper. Overall, the government energy policy is biased against the development of nuclear energy and the latter's role is only conceived as filling the gap that is left by conservation and use of coal.

If nuclear power on a large scale is developed it can affect electricity tariffs and have major commercial implications. Of course, some types of electricity usage may not be strongly influenced by pricing structures, for instance, the use of electricity for lighting. Similarly, for many miscellaneous domestic and commercial uses, energy costs play a smaller role because convenience is of more importance. There are many other cases, however, where electricity usage can be influenced by economic incentives based on differential pricing. Space and water heating are examples which are responsive to differential pricing. Large industrial loads can also be influenced through tariff incentives, e.g. metal melting furnaces, cold storage plants, induction heaters and pumping equipment. The use of advanced control systems allows modern industry to focus on cost optimization. Thus improvement in the price competitiveness of electricity will be a crucial factor in bringing about the substitution of electricity for other fuels. In this respect, the development of nuclear power will be important, which could potentially bring about lower tariffs for electricity users.

Why should Middle Eastern states, rich in oil and natural gas, opt for nuclear energy for their electricity supplies? The answer is not that they want the nuclear system to produce nuclear weapons clandestinely, as some would have us believe. It is much more likely that some of these countries are now getting short of gas and will only use their black gold for electricity production as a last resort. The opportunity cost of capital is often very low in these states – because there are so few alternative physical capital investment opportunities, so the true economic cost of the station is very low, and the only practical alternative to it may be a depreciating paper asset in an American bank.

The controversies that surround the link between energy growth and economic growth, and the latter's relationship with human welfare seem to rest on propositions that stand on shaky ground. Hence, the need for pursuing policies whose objectives are to surmount supply and environmental constraints rather than to succumb to their supposed inevitability, as well as the need to formulate strategies for environmental objectives.

Some energy economists believe that nuclear energy has a vital macroeconomic role to play in a world economic system which has evolved in an atmosphere of increasingly abundant inanimate energy and which now faces the prospect – without nuclear energy – of reversing engines and trying to adapt to reducing energy availability and rising energy prices. Will very low marginal cost electricity from nuclear energy provide the fourth great energy-based stimulus to the world economy in the last two hundred years – succeeding the alliance of coal with the steam engine, liquid fuels with the internal combustion engine and ubiquitous light industry and a consumer durables boom with the growth of centrally-generated electricity?

Given that the world is heading for a period of more scarce and costly energy resources, countries can no longer ignore the fact that traditional sources of fuel will not be as abundant as they were in the past. Hence, systematic energy policies at national level are required, to develop other energy sources that would mitigate the recessionary consequences of the major fossil fuel price rises in the 1970s.

Of course, no single form of energy source is indispensable, in the sense that it is necessary for the world social and economic system. In that respect, it should be emphasized that nuclear energy is not essential to the world future energy supply. However, it presents an opportunity among the future energy options that would essentially offer some relief from the hardship resulting from a tighter energy market.

Finally, it should be said that the various facets of nuclear energy economics are almost endless and we hope that we have succeeded in drawing at least some of them to the reader's attention.

INDEX

Added-value components, energy costs, 51, 53, 64, 375, 376
Advanced converters, defined, 8
Advanced gas-cooled reactors (AGRs)
   annual costs, 79
   burnup rates, 172
   characteristics, 110–11, 172–3
   decommissioning, 304–12
   equipment, 89–94, 99–100, 35–7
   see also Gas-cooled reactors
Aerodynamic techniques, uranium-enrichment, 171
Agricultural uses, nuclear power, 335, 348
Aid programmes, international, 21
Almelo Treaty, 169–70
American Society for Testing and Materials (ASTM), uranium resource evaluation standards, 124
America, *see* United States
Americium (Am), 173, 174, 253
Annuities, capital investment, 81, 82; *see also* Interest and depreciation charges
Approvals, mining, 151
ASEA–Atom, feasibility studies, 163, 349
Atelier de Vitrification à Marcoule (AVM), 180, 181
Atomic Energy Authority, development of reactor systems, 95
Atomic Industrial Forum (USA), 6
Australia
   foreign participation in nuclear programme, 152
   nuclear power programme, 345
   nuclear science capability, 343–5
   uranium
      production, 146, 147, 345
      resources, 22, 136, 137, 138, 151, 345–6
Availability rates, plant, 3, 275

Balance-of-payments deficit, 27, 373, 377–8
Bangladesh, nuclear power plants, 335–6, 348

Banking system, world, 26–7
Barge-mounted nuclear power plants, 355–6
Base-load plants, 1–2, 78
   comparisons of generating costs, 275–87
   generation costs, 81
   lifetime load factors, 81
   nuclear, 64, 219, 272, 320, 322
      cost advantage, 256, 327
Bending machines, boiler-tube, 100–2
Biological shields, constituents, 309, 310
Boilers, types, 90, 94
Boiling water reactors (BWRs), 109–10, 172, 351
Boron control rods, 168
Brazil
   hydroelectric power resources, 171
   nuclear power programme, 171, 185, 344
Break-even costs, coal versus nuclear fuel, 297, 299
Break-even distances, transmission costs, 37–8
Break-even load factors, 4–5, 13, 327
Breeding gains
   defined, 177–8, 214, 241–2
   quoted, 194–5, 228–9, 240–1
British Nuclear Forum, 6
British Nuclear Fuels Ltd (BNFL), 12, 96
   processing plants, 170, 172, 174, 175–6
Bulgaria, nuclear power capacity, 344, 352
Burnup rates
   advanced gas-cooled reactors, 172
   CANDU reactors, 172
   fast reactors, 180, 204, 227–30, 232, 240
   PWRs, 172, 241, 249, 281–2
   see also Utilization, fuel
Bus-bar costs, *see* Generating costs
Butt-welding machines, 100, 102, 104, 106

Canada
   Energy, Mines and Resources (EMR)
      mineral classification system, 122, 123
      mineral surveys, 124–5, 130, 134

409

exports
  nuclear plant, 167
  uranium, 154
foreign participation in uranium mining, 152–3
non-proliferation agreements, 154–5
nuclear power, 258–72
nuclear power capacity, 344
uranium
  production, 146–7, 345–6
  resources, 136, 138, 141, 146–7
CANDU reactors
  costs
    compared with coal-fired stations, 258–61, 264–5
    compared with light-water reactors, 267–72
    escalation, 261–4
  described, 17
  equipment, 89
  exported, 167
  fuel utilization, 17, 172, 196–8
Capacity, nuclear power
  developing countries, 344, 359
  fast reactors as proportion of, 217
  forecast, 397
  French, 59, 64–5, 344
  German, 302, 344
  UK, 105, 344
  world, 344
Capacity factors
  adjusted, 276–9
  coal-fired plants, 270, 275
  defined, 258
  nuclear plants, 262, 263, 264, 268–9, 271–2, 275
  system, 275
Capital
  availability, 39–40
  costs
    CANDU reactors, 259, 261, 263, 268–9, 270, 271
    defined, 257, 293
    developing countries, 353
    fast reactors, 200–2, 248
    fuels compared, 33–4, 44, 259, 279–80, 292, 375–6
    inflation of, 279–80
    per power unit, see Specific costs
    PWRs, 263, 268–9, 279–80
    transmission, 34, 44, 57
    uranium processing, 164–5, 175
    waste management, 184
  fuel costs ratio, 79–80
  intensity, 38, 238
  rates of return on, 79, 81
  substitution for energy, 25–6, 389

Carbide (UC/PuC) fuels, 195, 223, 226
Carbon dioxide coolants, 14, 168
Care-and-maintenance stage, decommissioning, 304, 309
Carrying charges, 276, 280–1, 283; see also Interest and depreciation charges
Cartels
  energy, 26, 398
  uranium, 166
Centralization
  benefits, 33, 35, 404
  defined, 31
  increase in, 42
Centrifuge processes, enrichment, 169–70
Ceramic fuels, 194
Chain reaction, described, 167
Cheap oil, 19, 24, 53
China, nuclear power programme, 349–50
Circuit-breakers, capital costs, 44
CIVEX process, 206
Classification, mineral resources, 117–21
  compared, 121–4
Clean-conditions manufacturing, 92, 99–100
Cleaning costs, 284
Climatic factors, 72
CNC bending machines, 100, 102
Co-processing, plutonium/uranium, 206
Coal
  energy–weight ratios, 10, 163
  environmentalists against, 106
  imports, 302
  as oil substitute, 106–7, 234
  prices, 282, 288, 295–6, 374–5
  railways expansion assisted by, 19, 367–8
  types, 279, 282, 285
  UK reserves, 178
  wood-fuel substitution, 367–8
Coal mines, distance from power stations, 259, 287, 293
Coal-fired plants
  capacity factors, 270, 275
  employment effects, 379
  generating costs, 212, 258–62, 263–4, 276–8, 283, 296–301
  pollution, 23–4, 163–4
Combined Development Agency (CDA), 146
Combined power–heat schemes, 1, 163
COMECON countries
  nuclear fuel leasing, 22, 339
  nuclear power capacity, 344, 350–2, 359
Commissariat à l'Énergie Atomique (CEA), 166, 167, 180
Commonwealth Edison Company (CECO), 274–87
Compagnie Générale des Matières Nucléaires (COGEMA), 166, 174
Compagnie Minière d'Akouta (COMINAK), 166

Index

Comparison methods, constant load-factor, 80–2
Comparisons, cost, factors affecting, 265–7, 290
Competition
  electricity market, 53, 54
  energy market, 32
  uranium markets, 9–10, 140–1
Computer analysis
  generation plant optimization, 84–5
  investment planning, 5
  merit-order loading, 78
  mineral production capability, 130
Concrete, AGR construction, 305, 307, 309
Conservation, energy, 28, 52–3, 104–6, 342, 366, 374, 396
  price-elasticities, 387, 389, 392, 393, 394, 400–2
  UK, 394
Consortia, nuclear power manufacturing, 97
Constant-currency accounting, 5, 48, 52, 62
Constraints
  economic calculation, 47
  energy supply, 367–9, 391–2, 397
  environmental, 85
Consumer behaviour, 325–6, 328
Consumption
  balanced by supply, 49, 326; *see also* Supply–demand balances
  energy
    measurements, 68
    industrial output influenced by, 383
    patterns, 325–6, 328–30
    related to economic growth, 364–5
Contamination, radioactive, 305, 307, 309
Contingency, transmission line, 44
Contracts
  coal supply, 289, 302, 407
  manufacturing, 97–8
  uranium supply, 146, 148, 151
Control rooms, reactors, 90–1
Control systems, electricity usage, 329, 407
Converter reactors, 8, 180
Cooling ponds, nuclear fuel rods, 173, 175
Cooling systems, 14, 168, 200, 202, 239
Cooling-down times, nuclear fuel, 211, 249
Core parameters, reactors, 194–5, 227–33, 239–40
Core sizes, reactor, 15, 168, 194–6, 200
Corrosion rates, decommissioned plant structures, 310
Cost structures, electricity supply, 318–26
Costs, enrichment, 249, 282, 296
Cranes, polar reactor, 89, 90
Credits, recovered uranium/plutonium, 212
Crisis, oil, economic effects, 52, 53, 63, 149, 369
Criteria, economic, itemized, 30–1, 47–8

Criticality
  control, 203, 206
  initiation, 194
  maintenance, 193
Cumulative demand curves, 79–80
Cumulative plant capacity curves, 77, 80
Cumulative requirements, uranium, 221, 224–5
Currency
  exchange rates, 27, 267, 292, 297, 373
  requirements, 378–80
Current-value accounting, 279, 294
Customers
  electricity, 39, 328–9
  nuclear power manufacturing, 93–6
Czechoslovakia
  IAEA membership, 358
  nuclear power capacity, 344, 352

Data
  presentation, 50
  requirements, 85, 265–7
Decanning, fuel elements, 175
Decay, radioactive, 173, 182–3, 205–6, 211, 307, 309
Decommissioning, 303–13
  costs, 207, 284, 292, 293–4, 312
  defined, 207, 303, 304
  monitoring decommissioned plant, 309–10
Deffeyes–MacGregor price elasticity model, 157, 158
Deflation, OPEC monetary surpluses contributing to, 372
Demand
  growth, 55, 146, 234, 289, 295, 342, 394, 396
  measurement, 68
  planning, 68
  variations, 1, 19, 318–19
Demand elasticities, long-run energy, 393, 400–2
Demolition plans, decommissioned plants, 311
Devaluation, monetary, 27, 372–3
Developing countries
  defined, 336–7
  nuclear power programmes, 20–2, 335–6, 346–59
  power system development, 83–4
  problems of, 337
Differential pricing, 20, 318, 327, 328–30, 407
Diffusion processes, enrichment, 169
Discounted generated costs, single-station, 208–9
Discounting rates, present-worth
  American calculations, 281
  Canadian calculations, 259, 263, 264
  French calculations, 237, 238

German calculations, 292
  mixed-system models, 48, 63, 78, 85
Discovery potential, uranium, 124–5
Distribution
  costs, 44
  as energy flow, 51
Diversification, generation plant, 106–7, 286
Domestic loss-of-power (outage) costs, 74
Doubling times
  electricity demand, 68, 76
  plutonium breeding, 16, 177, 214–15, 230, 243, 246–7
Dounreay fast reactors, 179, 180

East Germany, nuclear power capacity, 344, 352
Econometric studies, 367, 368, 386
Economic calculations
  conventions, 48
  examples, 58–61
  mixed systems, 46–51
  nuclear power, 107
Economic driving force, energy as, 383
Economic rent, old technology, 62
Efficiencies
  energy production, 8, 13, 17, 33–4, 68
  energy-use, 53, 365–6, 381–2, 387
  UK, 393–4
Egypt, nuclear power programme, 348–9
Electric Power Research Institute (EPRI), 386, 393
Elk River BWR, decommissioning, 303
Employment
  energy policies influencing, 376–9
  power plant industry, 112
Energy
  coefficients, defined, 383, 387
  defined, 68, 320
  economic role, 381–4
  interrelationship with economic activity, 24–5, 28, 363–5, 381–2
  types, comparison, 51–3, 61–5
  weight ratios, 10, 12, 13, 15–16, 163, 178
Energy Modelling Forum, 383, 393
Enriched-uranium fast reactors, 223–4, 242
Enrichment
  costs, 170, 218, 248–9, 282
  levels, 12, 167–8, 194, 241, 244–5
  plants, 10–11, 168–72, 245
Entropy, defined, 382
Environmental considerations
  nuclear power, 22–4, 30–1, 106–7
  uranium mining, 152, 158, 160
Environmental factors, 369
Environmentalists, objections, 106, 289
Equalities, cost, 61–2
Equipment
  generating, characteristics, 33–4, 44
  manufacture, 88–103
  transmission, capital costs, 34, 44
Equity participation, foreign, 152–3
Errors, load-forecasting, 69–71
Estimated additional resources (EARs), defined, 119, 164
Euratom Treaty, 154
Eurodif enrichment capacity, 169, 170
Ex-post evaluations, forecasting, 68, 71
Exploration, uranium, 147–8, 151
Exports
  gas and oil, 21, 22
  indigenous fuel resources, 346
  nuclear plant, 167, 175, 185
  uranium, 9, 154
Externalities, quantification, 31, 404

Fast breeder reactors (FBRs)
  defined, 8, 177, 196, 406
  French, 239–41, 406–7
Fast reactors
  capital costs, 200–2
  characteristics, 194–6, 200–4, 239–41
  development, 95, 139
  economics, 13–17, 156–60, 208–33, 247–52
  equipment, 100, 201
  fuel costs, 210–11, 248–51
  fuel utilization, 8, 13, 145–6, 157, 159, 178, 180, 197–8
  generation costs, 210
  linked to thermal reactors, 49, 146, 178, 212–13, 230
Fertile materials, defined, 193
Fire-fighting systems, sodium-cooled reactors, 201
Fissile materials, defined, 167, 193, 235
Fission
  cross-sections, 168, 193
  energy released by, 244
  processes described, 167
  products, 13, 173–4, 206–7
Fixed charge rates (capital equipment), defined, 43
Fixed costs, 3–4, 62
Floating nuclear power plants (FNPs), 355–6
Fluid fuels, nuclear energy manufacture, 55
Ford–Mitre report (on plutonium-recycling), 155, 176
Forecasting
  energy prices, 391
  load, 5–6, 68–71
Framatome, 167, 350
France
  energy policies, 178, 235, 376–80
  exports of nuclear plant, 185, 349–50
  fast reactors development, 178, 180, 239–41

Index

fuel-processing plants, 155, 169, 174, 180–1, 294–5
hydroelectric power generation, 59, 60, 64–5
nuclear power generation, 59, 64–5, 162, 178, 344, 397–8
Fuel costs
  analysed, 81, 171, 250
  CANDU reactors, 259–62, 271
  coal-fired plants
    German, 292, 295–6, 298, 301
    USA, 276, 282
  fast reactors, 14, 203, 210–11, 248–51
  heating costs constituent, 52–3
  inflation of, 65, 281–2, 294–7
  light-water reactors, 270, 271
  nuclear
    generally, 44, 81, 171
    German, 292, 294–7, 298, 300
    USA, 276, 281–2
  plant size affecting, 34
  proportion of generation costs, 8, 63–4, 145, 157
  PWRs, 248–51
  savings, 76–9
  thermal reactors, 248–51
Fuel cycle
  costs compared, 248–51
  defined, 10–13, 163, 405–6
  developing countries, 337–40
Fuel elements
  dimensions, 203, 226, 227–9
  fabrication, 12, 96, 171–2, 194, 202–4
  costs, 172, 249–50, 282, 296
Fuels
  compared
    capacity factors, 270–1, 275
    capital costs, 33, 44, 259, 279–80, 292–4, 375–6
    developing countries, 353–4
    foreign currency requirements, 378
    generating costs, 37, 41–2, 79–81, 258–61, 263–4, 274–87, 297–302, 343
    heating costs, 51–3
    profitabilities, 63–4
    types, listed, 33, 44, 275
Furnaces, electric heating, 329

Gag mechanisms, AGR boiler unit, 90, 93
Gas
  electricity generation by, 21, 105, 288, 321, 323
  exports, 21
  prices, 375
Gas-cooled reactors
  economics, 14
  start-up constraints, 2
  UK experience, 92, 95, 111
  see also Advanced gas-cooled reactors

Generating costs
  analysed, 256, 256–8, 259, 260, 261, 263, 298
  coal-fired plants, 212, 258–62, 263–4, 276–8, 283, 297–301
  fast reactor, single-station, 208–13
  LWRs, developing countries, 354
  mixed system, 214–26
  nuclear plants, 14, 53–4, 247–52, 267–72
  oil-fired plants, 354
  optimal unit size, 37
  system, 276–7
Germany, see East Germany, West Germany
Gold mining, uranium as co-product, 132, 146, 147
Graphite moderators, 168, 172, 180, 305, 310
Grid connection, break-even distances, 37–8
Gross domestic product (GDP), energy expenditure as proportion, 395–6
Gross national product (GNP)
  energy consumption related to, 364–5
  welfare indicators, 362
Growth, economic
  energy consumption linked to, 363–5, 381–2, 408
  energy supply constraints, 367–8
  human welfare linked to, 362–3, 408
Growth rates
  electricity demand, 68, 76, 289, 295, 394, 396
  developing countries, 342, 359
  nuclear power, 139–41, 343, 374
  developing countries, 354, 359

Handling equipment, boiler-tube, 100, 103
Hazards, radiological, 173, 304, 311
Heat transmission, 53–4
Heating
  costs of, 51–3, 328, 407
  district, 163
Heavy water
  costs, 258–60, 271–2
  as moderator, 168, 172
  reactors (HWRs), see CANDU reactors, Steam-generating HWRs
Helium coolants, 14, 180
High-sulphur (Illinois) coal, 282, 285
High-temperature reactors (HTRs), 159, 162, 172; see also Advanced gas-cooled reactors
Hinckley power station, 93, 94, 98, 99
Holdup times, fuel reprocessing, 213, 215, 222, 223, 231–2
Hong Kong, nuclear power programme, 350
House of Commons Select Committee on Energy, 107, 109
Hungary, nuclear power programme, 344, 350, 352

Hunterston power station, 91, 98, 99
Hydroelectric plants
  contribution to total capacity, 59, 60, 64, 105
  cost optimization affected by, 60, 236
  distance from users, 57, 171
  France, 59, 60, 64–5
  load-forecasting, 69
  merit order, 77, 78
  as peak-load plants, 2
Hydroelectric power resources
  Brazil, 171
  Canada, 255, 258

Imports
  coal, 288, 302
  dependence, 178, 277, 295
  oil, 27, 105–6, 178, 277, 373–4
Incremental approaches, 79, 82–3, 323–4
India
  IAEA membership, 358
  nuclear explosion, 153, 185
  nuclear power capacity, 344, 359
Individual generator devices, 31
Indivisibilities, plant-load, 75
Indonesia
  IAEA membership, 358
  nuclear power programme, 347–8
  uranium resources, 348
Industrial production
  losses due to outage, 74
  patterns of, 329
Industrialization
  defined, 387
  developing countries, 21, 341, 363
Inflation
  accelerated by oil price rises, 372
  effects on
    comparative generating costs, 375–6
    construction costs, 263, 279–80
    economic studies, 18
    fuel costs, 65, 281–3
    generating costs, 260–3, 277–8, 283–5
    investment charges, 280–1
  rates quoted, 280, 285–6, 294
  re-adjustments, 62, 149
  types of, 375
Inflation-proof characteristics, CANDU reactors, 260, 265, 277
Inspection requirements, 202, 310
Instantaneous start-up, 2
Insulation, thermal, 52, 53, 328
Interactive models, 392
Interconnections, system, 32, 55, 56–7
  break-even distances, 37–8
  degree of, 40, 404
  economics, 38
Interdependence, reactor types, 16–17

Interest and depreciation charges
  coal-fired plants, 259–60, 276, 280–1, 283
  defined 257, 276
  inflation of, 280–1
  mixed-system, 81, 82
  nuclear plants, 248, 259, 261–2, 276, 280–1, 283
Interest rates, effect on uranium supply, 149, 164
Intermediate product, electricity as, 73–4
International Atomic Energy Agency (IAEA)
  decommissioning studies, 303
  defined, 116
  inspectorate, 156, 187, 336, 341
  membership, developing countries, 358
  role of, 335–6
  uranium resource classification system, 118–23, 142
International Nuclear Fuel Cycle Evaluation (INFCE), 117, 141–2, 155–6, 184, 337, 342–3
International trade
  electricity, 57–8
  nuclear fuel, 185
  oil, 58
International Uranium Resources Evaluation Project (IUREP), 125, 133, 137
Interruptions, power, see Outage
Intervention, market, 10
Inventories, plutonium, 179, 212–13, 227–9
Inventory, total radioactivity, 304–5
Investment
  appraisal, 3, 5, 323, 405
    AGR/PWR, 110
  planning, dynamic, 5, 290
  public sector, 73
Italy
  joint research projects, 348
  nuclear power capacity, 344

Japan
  exports of nuclear plant, 347
  floating nuclear power plants, 356
  nuclear power capacity, 344
Jet nozzle uranium-enrichment plants, 338
Judgemental decisions, 31

Korea
  industrialization, 21
  nuclear power capacity, 344, 350, 355, 359
Kraftwerk Union (KWU), 167, 293, 347

Labour
  productivity, 33, 266
  substitution for energy, 25, 389
Land requirements
  mining and milling, 163–4
  waste disposal, 183

Index

Laser techniques, uranium-enrichment, 171
Leaching, nuclear wastes, 182–3
Lead-times, generating plant construction, 75, 108–9, 318
Learning curves, utility, 68, 265
Leasing, nuclear fuel, 22, 339
Leisure activities, energy requirements, 366–7
Leisure time, loss, 73, 74
Levelizing calculations, 208, 236–8, 252–4, 297–9
Licensing, nuclear plant manufacture, 95, 167, 355
Life-index concepts, mineral reserves, 125–6
Lifetime trends, unit energy costs, 262–3
Lifetimes, operating, 210, 249, 259, 261, 292, 294
Light-water reactors (LWRs)
  cost comparison with heavy-water reactors, 267–72
  defined, 267
  fuel cycle, 339
  see also Pressurized-water reactors; Boiling-water reactors
Linkages, user, 53–5
Liquid metal fast breeder reactors (LMFBRs), 14, 177–9, 199–233, 239–41; see also Fast reactors
Load
  balances, 74–5, 323
  characteristics, electricity costs affected by, 35
  densities, 40
  duration curves, 60–1, 217, 319–24, 330
  manipulation, 328–30
  shedding
    marginal costs, 73
    see also Outage
Load factors
  classified, 40
  constant, 81
  defined, 4–5, 32, 68
  effects on, 39
  generating costs affected by, 297, 299–300
  lifetime, 81–2, 213, 405
  mean annual, 218–19, 292
  system, 255, 323
Location, consumers, 54, 171
London Club (nuclear fuel suppliers), 185–6
Long-distance transmission, electric power, 57, 171
Long-run marginal costs, 19, 322–6
Long-term forecasting
  load, 69
  uranium production capability, 137–40, 142
Loop designs, LMFBR heat-exchangers, 14, 200

McKelvey system, resource classification, 118
Macro-economics, 24–8, 361–402
Magnox reactors
  development, 98, 162
  plant availability, 3
  reprocessing of fuel, 176, 204
  start-up constraints, 2
  withdrawal from service, 303–4
Management efficiency, effect on costs, 266
Manufacturers, nuclear power equipment, 7, 88–93, 110–111
Marginal analysis, 80, 317, 321
  developing countries, 83
Marginal costs
  nuclear power influencing, 327
  ratios between various fuels, 66
  seasonal variation, 65–6, 321
Marginal-cost pricing, 19, 325–7
Margins, spare capacity, 71–2
Marine propulsion, 1, 355
Markets
  energy, 51–5
  electric-power share, 66
  macroeconomic analysis, 26
  power equipment, 103–4
  uranium, 9–10, 138, 140–1, 165–6, 222
Massachusetts Institute of Technology (MIT) studies, 284, 345–6
Merit order concepts, 3, 76–8, 213, 217, 219, 321, 405
Mexico
  IAEA membership, 358
  nuclear power programme, 344, 349
Micro-economic studies, 17, 208–13
Mid-merit plants, 2–4
  fuel-cost savings, 78
  lifetime load factors, 81
Military uses, 23, 145, 146, 184, 186, 341
Mine-by-mine analysis, uranium resources, 130, 132, 134–5
Mineral Area Planning Study (MAPS), 130
Mining, uranium, 7, 146–52, 163–6, 338
Mix, generation plant, 75–6, 321
  optimization, 79–80, 323
Mixed (nuclear/non-nuclear) systems
  economic calculation, 46–51, 216–20
  energy market, 51–5
  generating costs, 214–26
  optimal planning, 67–85
  supply characteristics, 56–62, 404
Mobile power sources, 31–2, 354, 368
Models
  consumption-price, 389–90
  dynamic, 49–50, 84–5
  econometric energy, 368–9
  electricity-generation, 199, 216
  electricity-supply, 80
  energy supply, 390

energy-economic, 383, 386–90
large-scale, 49–51
merit-order loading, 78
price elasticity, uranium, 157
resources, uranium, 133–4
supply reliability, 73–4
Moderators, 167–8, 172, 180, 305, 310
Monasinghe reliability/outage model, 73–4
Monopolies Commission report, nuclear power plant industry, 107
Monopolistic pricing, 317–18
Multi-criteria analysis, 47

National income, 377, 385
National Nuclear Consortium (NNC), 96–8
National product, energy expenditure as proportion, 364–5, 395
National Uranium Resource Evaluation (NURE), 124
Net-effective-cost methods, 82–3, 198
Neutron
  capture cross-sections, 168, 193
  economy, 159, 192
  flux, 305
    monitoring equipment, 90, 92
  irradiated materials, 305, 307
Niger, uranium production, 136, 138, 141, 166, 346
Non-Proliferation Treaty (NPT), 12, 153–6, 186, 336
Nuclear capacity, forecasts, 397
Nuclear Energy Agency (NEA), 116
  uranium resource classification system, 118–23, 142
Nuclear energy industry, defined, 6–7, 88–93, 405
Nuclear Installations Inspectorate, 202
Nuclear Non-Proliferation Act (NNPA), 155, 156, 186

Objectors, nuclear power, 23, 106, 111–12, 146, 289
Obsolescence, plant, 5, 325
Off-peak electricity, prices, 20, 318, 321
Offshore Power Systems (OPS), 355–6
Oil
  'banana-price', 20
  cheap, 19, 24
  crisis, 52, 53, 63, 149, 369
  electricity generation using, 105, 288, 353–4
  energy industry, 7
  exports, 22
  imports, 27, 105–6, 178, 277, 373–4
  price-rises, 108, 234, 295–6
  prices, 26–7, 28, 52–3, 234, 295–6, 374–5, 396, 399

producers
  nuclear plant purchases, 21, 408
  supply characteristics, 393
  see also OPEC
refining industry, 10
requirements, 219–20
shortages, 106, 107, 108
On-power fuelling, 270
Once-through fuel cycles, 163, 165, 173, 188, 205, 221, 244–5
  uranium utilization, 192
Ontario Hydro, heavy-water reactors, 255–72
OPEC
  collapse of cartel, 384, 398
  monetary surpluses, 22, 372, 384
  price-rises, 369
  reduction of production, 21, 379, 399
Operation and maintenance costs
  coal-fired plants, 259–60, 276
  defined, 257, 283
  nuclear plants, 259, 261, 262, 271, 276, 292
  variation with plant size, 34, 265
Opportunity costs, 22, 74, 408
Optimization
  dynamic, 49–50
  electric power system, 6, 42, 322–3
  production structure, 60–2, 323–4, 405
Orders, nuclear power plants, 96–7, 344, 374
  cancellations, 150, 170, 374
  uncertainty, 98–9, 103, 111
Ore-grades, uranium, 157–8, 163–4
Outage
  costs, 41, 73, 74
  probabilities, 72

Panama Canal, floating nuclear power plant, 355
Parametric analyses, 40–2
Paris Agreement, fast reactor technology, 179–80
Part-load operations, 2–3
Peak load
  curves, 85
  marginal costs, 19–20, 321
  plants, 2, 78, 272, 288, 321
    lifetime load factors, 81
    running costs, 3, 78
  of system, 36–7, 319
Penetration models, 388
Petrochemicals industry, growth, 19
Phenix fast reactors, 180, 251
Philippines
  IAEA membership, 358
  nuclear power programme, 344, 355
Planning
  energy price-changes, 385
  power system, 50, 67–71
Planning plant margins, 71–3

Plant extension programmes, constraints, 76
Plant/load balances, 74–5
Plutonium
  as fuel, 177, 193–5, 235
  mass equivalent in fuel cycle, 240, 242–3
  prices, 212, 235, 249, 296–7
  production, 11, 16, 146, 173–4, 235
  stockpiling, 13, 187, 220–1
  toxicity, 203
Poland, nuclear power programme, 344, 352
Policies, national energy, 47, 103–9, 216, 318, 369–71, 408
  French, 178, 235, 376–80
  German, 289, 302
Political factors, 22–3, 64, 145, 153–5, 160
  effect on nuclear power programme, 22–3, 287, 289, 302
  effect on oil supplies, 64
  effect on power plant industry, 103–9, 111–12
  uranium supplies, 145, 152–5, 160, 222
Pollution, 23–4, 163–4
Pool-type fast reactors, 200–1, 239
Power, electricity, definition, 68
Power requirements, enrichment processes, 169, 170, 171
Pressure equipment, manufacture, 111
Pressurized water reactors (PWRs)
  characteristics, 109–11, 172–3
  compared with CANDU reactors, 269–72
  cranes, 90
  economics, 63–4, 157, 159, 196–8
  French, 236, 241
  fuel costs, 248–51
  recycled plutonium fuel, 159, 205, 245–6
  Russian designs, 350–2
Price elasticities
  energy, 383
  energy conservation, 387, 389, 392, 393, 394, 400–2
  uranium, 157
Price mechanisms, 317–18, 382–6
Price-hikes, effects of, 28, 377, 390–6, 400–2
Prices
  coal, 282, 288, 295–6, 374–5
  demand response, 366
  electricity, 20, 52–3, 65, 316–17, 328–30, 397–8
  energy, 25, 65
  gas, 375
  oil, 26–7, 28, 52–3, 234, 295–6, 374–5, 396, 399
    political pressures on, 64
  plutonium, 212, 235, 249, 296–7
  uranium, 15, 147–9, 234, 296–7
    fast reactors influencing, 218, 222, 225
  yellowcake, 249, 282

Primary power
  classification,
    France, 59, 64, 156, 162, 178, 397–8
    Hungary, 350, 352
    UK, 105, 162
    USA, 162
  nuclear proportion, 59, 64–5, 105, 328, 359
Probability analyses, 59–60, 71, 72, 85
Process heating
  direct from nuclear plant, 54–5, 162–3, 342
  electric, 329, 407
Production capacity, uranium, 58–60, 130–4, 296
Productivity, industrial, energy price effects, 386
Profitability techniques, 62–3, 324–5
Project delays, specific capital costs affected by, 265
Proliferation, nuclear weapons, 12, 23, 153–6, 184–7
Public debate, nuclear power, 111–12, 252, 289, 357
Public sector, investment, 73
Pumped storage plants, 2, 57
Purchasing power, erosion by energy price-rises, 384

Radioactivity levels, 173, 182–3, 187, 205, 211
  decommissioning, 304–5
Railways
  electrification, 51
  interrelationship with coal supplies, 19, 367–8
Random errors, load-forecasting, 69–70
Randomness, optimization calculation, 50
Rating, linear, fuel pin, 195, 211, 230, 231
Reactors, types, 159, 166, 172–3, 196–8
Reasonably assured resources (RARs), 119, 164, 338
Recession
  causes, 22, 372
  oil requirements affected by, 373
  uranium market affected by, 9
Recycling, nuclear fuel, 155, 176–7, 196, 198, 221
Red Books (OECD estimates of U/Th resources), 116, 120, 134
Regional characteristics, 39–41
  parametric analysis, 41–2
Regional nuclear fuel cycle centres (RFCCs), 187
Regulatory authorities, 340–1
  effect on costs, 267, 277, 289, 407
Reliability
  assessment of, 85
  electricity supply, 5, 32, 40–1, 58–9, 71–5, 85; see also Security of supply

equipment, 18, 92
  models for, 73–4
Renewable energy sources, 24, 275
Reprocessing, nuclear fuel, 10–13, 163, 173–6, 204–7
  cessation by USA, 155, 186
  costs, 175–6, 249, 295
  by developing countries, 340
Required rates of return (RRRs), 78, 81, 207, 238
Research and development
  co-ordination of, 335
  efficiency, 266
Reserve capacity, 35–6, 56–7
  margin, 72
Reserves
  depleted uranium, 235
  mineral, defined, 125, 128
Reserves And Production Projection (RAPP) model, 133
Resource case, fast reactors, 15, 159, 220–2
Resources
  availability, 35, 39–40
  mineral, classification, 117–24
  mineral fuels, shortages, 369, 408
  uranium, 22, 151, 222, 345–6, 348
    assessment, 8–9, 116–42
Restbedarfsphilosophie (German energy policy), 289
Retrofitting, costs of, 264, 283–4, 289
Revenue requirement methods, 290–1
Risk calculations, forecasting, 69, 287
Robotic welders, 102, 107
Robustness, background plan, 80
Rock, storage of nuclear wastes, 181–3
Rolls-Royce barge-mounted nuclear power plants, 355–6
Romania
  IAEA membership, 358
  nuclear power programme, 344, 352
Running hours, 3, 209, 249
  reduction, 78
Rural electrification schemes, 30
Russia (USSR)
  nuclear power capacity, 344, 351
  nuclear-powered water-purification plant, 163, 351
  uranium enrichment plants, 169
  uranium exports, 170
Rutherford, (Lord) Ernest, quoted, 18

Safety requirements, 92, 202, 340–1
Salt mines, storage of nuclear wastes, 174, 183–4, 294
Scale economies
  capital costs, 33–4, 38
  fuel cycle, 170, 174, 187, 236
  generating costs, 32, 33–4

Schwantag-Formel cost formula, 296
Scrap values, decommissioned plant, 207, 312
Sea
  dumping of nuclear wastes, 181, 183
  uranium extraction, 158
Seasonal variations, 1, 255, 318–19
  marginal costs, 65–6, 321–2
Secrecy, nuclear work, 111, 169
Security of supply
  energy, 371
  fuel, 9, 15, 165, 187
Self-generation recycles, 196, 198
Sensitivity analyses
  generating cost components, 209–10, 285–6, 299–301
  generating-system, 199
  planning, 71, 80
Separative work units (SWUs), defined, 169
Sequences, generating plants, 83–4
Service industries, energy requirements, 366
Service industry, nuclear fuel cycle, 10–13, 163–88
Sharing, energy production system, 33
Short-term outlooks, 112, 140–1
Shortages, gas and oil, 106, 107, 108, 369, 408
Shutdown, nuclear plant, 202
Sizes
  fuel-processing plants, 170, 175
  nuclear reactors, 54, 359
  plant extension programme, 75
  power plants, 18, 40, 76
    costs affected by, 33–7, 265
    optimization, 33–7, 38
    reliability affected by, 18
Small nuclear power plants, 346–7, 348, 355–6.
Social factors, 22–3
Societal decision-making, 30, 289, 302, 357
Societal endorsement, nuclear power projects, 267
Sodium coolants, 14, 177, 200, 239
Sofratome, 349
Solar-powered generators, 31, 32, 275
South Africa
  mineral resource classification, 122, 132
  nuclear power programme, 344
  uranium enrichment plant, 171
  uranium resources, 136, 138, 141
Specific (per unit of power) capital costs
  base-load plants, 3, 259
  compared, 33, 44, 259, 375
  nuclear, 201, 210, 261, 263, 268–9, 375
  transmission equipment, 34, 44
Speculative resources, uranium, 120–1, 125, 128–9, 133, 137–9
Spot calculations
  fuel-cost savings, 78
  load-forecasting, 69

Index

Spot prices, uranium, 149
Stagflation, 373
Standby units, generation, 56–7
Start-up times, power plant, 2
Statistical studies, 18, 387, 389
Steam generating heavy water reactors (SGHWRs), 95
Steels, AGR construction, 305, 307
Stockpiling
  electricity, inability to, 56, 318
  plutonium, 13, 187, 220–1
  uranium, 9, 10, 141, 147, 150
Strategies, alternative, assessment, 217–25, 244–6
Structural integrity, nuclear plant, 310
Substitutability, energy, 31–2
Substitution
  energy, 25, 28, 382, 384
  factors of production, 25–6, 28
  oil, 108, 109, 234, 342
Superphenix (SPX) power plants, 178, 180, 240–1, 248–51
Suppliers
  experience-record, 7, 111, 266
  nuclear power manufacturing, 96–103
Supply
  electricity, economic characteristics, 56–62
  energy, effect on economic growth, 367–8
Supply elasticities
  energy, 393
  uranium, 159, 165
Supply–demand balances
  electricity, 49, 61–6, 74–5
  energy, 390–2
  uranium, 145–60
Surge impedance loading (SIL), defined, 44
Surges, demand, 2
Sweden
  nuclear industry, 163, 181, 349
  nuclear power capacity, 344
System generating costs, 276–7, 406
System development
  economic and technical determinants, 32–9
  regional socio-economic characteristics, 39–41

Tails assay, uranium enrichment, 169, 245
Taiwan, 21, 344
Tariffs, electricity, 19–20, 66, 316, 325–6, 328–30, 407
Taxation, effect on consumer, 48
Teamwork, development, 98
Technologies, effect on
  generation costs, 20, 49, 75–84, 109–11
  lifetime uranium requirements, 159, 196–8
  supply curves, 61
Tenders, competitive, 97
Tennessee Valley Authority, 30

Terminology, mineral resource classification, 119–20, 122, 125, 128
Territorial rights, 151, 152
The Nuclear Power Group (TNPG), 97, 98
Thefts, plutonium, 187, 206
Thermal efficiencies, nuclear reactors, 172, 180, 200, 204, 240–1
Thermal oxide reprocessing plant (THORP), 174, 176, 184
Thermal reactors
  compared with fast reactors, 13–14, 16–17, 159, 247–52
  defined, 7–8, 168, 193
  generation costs, 210, 247–52, 258–72, 297–301
  see also Boiling-water reactors; CANDU reactors; Pressurized-water reactors; Steam-generating HWRs
Thermodynamic efficiency, energy-use, 381–2
Thorium
  as fuel, 8, 17, 180, 342
  resources, 116, 180
Three Mile Island accident, effect on nuclear programme, 149, 277, 279, 375
Time-scales
  decision-making, 7
  decommissioning, 312
  load-forecasting, 68–9
  manufacturing, 99, 102–3, 108–9, 112
  mineral supply, 130–1
  plant construction, 75
  radioactivity, 182–3
  uranium supply, 135–6, 165
Total unit energy cost (TUEC)
  calculation methods, 256–8
  quoted, 259, 261, 271–2
Total-cost criterion, 47
Toxicity, plutonium, 203
Trace-heating, pipework, 201
Training, 336, 337
Transformers, capital costs, 44
Transmission
  capital equipment for, 44
  as energy flow, 51
  heat, 53–4
  long-distance, 57, 171
  specific capital costs, 34
Transport
  costs, 258
  nuclear wastes, 175, 205, 211

Uncertainties
  nuclear power plant manufacturing, 98–9, 103, 111
  planning, 59, 69–71, 405
  reactor costs, 209
Unit-cost concepts, 62–3, 251–2

420   *Index*

United Kingdom (UK)
  electricity usage, 394
  Magnox nuclear plants, 2–3, 98, 102
  nuclear fuel processing, 170, 172, 174, 176, 179–82
  power generation, 105, 162, 344
United Nations
  Conference on Trade and Development (UNCTAD), 363
  mineral resource classification systems, 123–4
United States
  coal–nuclear comparisons, 274–87, 407
  Department of Energy (USDOE)
    NURE programme, 124
    resource classification system, 122
    uranium production capability studies, 130
  energy consumption per unit GNP, 364–5
  energy supply, 367–8
  General Accounting Office (USGAD), 129
  Geological Survey (USGS) resource classification scheme, 118, 124
  nuclear fuel processing, 169, 174, 180
  nuclear power generation, 275–87, 344
  Nuclear Regulatory Commission (NRC), 355, 356
  oil imports, 277, 373
  shortages of natural gas, 369
  uranium production, 138, 151, 346
  uranium resources, 136, 139
Uranium
  distribution, 9–10
  enrichment, fast reactors, 223–4, 242
  enrichment sales organization (URENCO), 170
  hexafluoride, 12, 168–9
  market
    affected by fast-reactor programme, 222
    affected by recession, 9
  marketing, 9–10, 138, 140–1
  natural, 8, 167, 168
    requirement, 196–7, 230–1
  prices, 15, 147–9, 234, 296–7
  processing, 10–13, 164, 168–72, 338–9
    losses, 123, 129, 148
  production capability, 7–9, 58–60, 130–4, 296, 345
  requirements, 214, 220–5
    minimization, 226
  resources, 8–9, 22, 116–42, 151, 222, 345–6, 348
    forward-cost evaluation, 123
    forward-reserve concepts, 126–8
  stockpiling, 9, 10, 141, 147, 150
  supply
    estimates, 136–42, 338, 405
    glut, 150, 165

Uranium Institute, 9, 156
Uranium Marketing Research Organization (UMRO), 166
Utilization, fuel
  CANDU reactors, 17, 172, 192, 197
  fast reactors, 8, 13, 145–6, 157, 159, 178, 180, 197–8, 246
  PWRs, 186–8, 244–6

Vertical integration, fuel-processing and power-generation, 166
Visual impact
  decommissioned nuclear plant, 309
  wind-power generators, 24
Vitrification, nuclear wastes, 181–2

Wastes
  disposal
    decommissioned plants, 312
    costs, 184, 207, 282, 285–6, 295–6
  management, 13, 180–4, 206–7
  storage, 90, 95, 181–4, 286, 294
Water, availability, 77, 85
Wave-power generators, 24
Welders, types, 102, 106–7
Welfare, economic growth linked to, 362–3
Welfare economics, 73, 395
West Germany
  comparative generating costs, 292–302, 407
  exports of nuclear plant, 185, 347
  fuel-processing plants, 170, 174, 183–4, 294
  nuclear power capacity, 302, 344
  PWR development, 167
Westinghouse reactor designs, 166–7, 241, 350
  fuel-element fabrication, 172, 282
Wind generators, 24, 275
Windscale Inquiry Report, 175, 179, 184
Wood, as fuel, 367
World, defined, 136, 146
World Bank, load-forecasting study, 69, 70, 71
World economy
  effect of energy price rises, 26–8
  growth, 363
World Energy Conference (WEC) studies, 117, 134–5

Xenon, build-up, 2

Yellowcake, 11–12, 164
  demand for, 13
  prices, 249, 282
  production, 11–12, 164